Physical Process in the Interstellar Medium

Physical Processes in the Interstellar Medium

LYMAN SPITZER, Jr.

Princeton University Observatory

WILEY CLASSICS LIBRARY EDITION PUBLISHED 1998

A Wiley-Interscience Publication

JOHN WILEY & SONS, INC.

New York • Chichester • Weinheim • Brisbane • Singapore • Toronto

This text is printed on acid-free paper. ♾

Wiley Classics Library Edition Published 1998.

Published simultaneously in Canada.

Library of Congress Cataloging in Publication Data:

Spitzer, Lyman, Jr., 1914–
Physical processes in the interstellar medium.
"A Wiley-Interscience publication."
Based in part on an earlier book, Diffuse Matter in Space, published in 1968.
Includes index.
1. Interstellar matter. I. Title.
QB790.S67 523.1'12 77-14273
ISBN 0-471-02232-2
ISBN 0-471-29335-0 Wiley Classics Paperback Edition

Printed in the United States of America

10 9 8 7 6 5 4 3 2

To the younger generation,

from whom I have learned so much.

Preface

During the eight years since *Diffuse Matter in Space* was published, research on interstellar matter has expanded almost explosively. Entirely new fields have opened up, including radio measurements of pulsars, X-ray studies of supernova remnants, observations of infrared emission from interstellar clouds, and ultraviolet measures both of absorption lines and of selective extinction. Important new results have also been obtained in established fields of research, including 21-cm H observations, radio analysis of continuous and recombination line emission from H II regions, and detection of faint Hα galactic emission. Theoretical studies during the last few years have transformed many aspects of our understanding of these various fields.

Although it is clear that a new graduate text on interstellar matter problems is appropriate at this time, it is also clear that any treatment must be very incomplete if the book is to be of reasonable size. Hence the emphasis of the present book has been somewhat altered from the previous one, with additional physical processes described, while the coverage of observations is even less full than before. Especially in a field which is developing rapidly, a sound knowledge of relatively constant physical principles is more useful than a wide familiarity with observational facts which may soon be outdated. As part of this emphasis, the observational results have to some extent been presented here as examples of the ways in which physical theory can be used to provide information about the interstellar gas, although in each area these results are supposedly described in sufficient detail to provide at least preliminary observational guidance to a scientist interested in interstellar research.

Because of both space limitations and inadequate knowledge on the author's part, the fascinating field of high-energy astrophysics has been omitted entirely. The discussion of the various processes involving relativistic particles would involve entirely different areas of physics from those relevant to most interstellar matter research. Hence while some basic observational information on cosmic rays and synchrotron emission is summarized in the book, especially in Chapter 1, there is very little discussion of these topics throughout most of the book. Inevitably the author's background and interests have influenced the relative emphasis given to various topics. Thus measures of interstellar absorption lines, especially in the ultraviolet, have received much more attention than either molecular line emission or infrared studies. The level of condensation also reflects the author's individual preference; as in the earlier book, a fuller, more detailed presentation would make easier reading, but might not increase the educational value of the work.

The organization of this book has been altered in various respects from that of the earlier text. In particular, observational results on all phases of the interstellar medium have been summarized in Chapter 1, to indicate the general nature of the physical system which is subject to the various processes discussed in the following chapters. Chapters 2 through 6 discuss various local phenomena occurring in the gas, whereas Chapters 7, 8, and 9 do the same for the dust grains. While elastic collisions are not often the subject of much interstellar research, Chapter 2 is entirely devoted to this subject, since the existence of kinetic equilibrium as a result of elastic collisions is absolutely basic in all subsequent theoretical developments. Chapter 3 discusses those interactions of the gas with radiation which can be understood without any detailed analysis of excitation conditions, which are treated in Chapter 4, based on the steady-state equation. Essentially the same equation provides the basis for analyses of ionization and dissociation in Chapter 5 and of gas heating in Chapter 6. The optical properties of grains and the corresponding results for grain radii, composition, and mean density are presented in Chapter 7; results of optical polarization studies are grouped together with the alignment analysis in Chapter 8. The theory of such grain properties as electric charge, temperature, radiative acceleration, and likely, if somewhat speculative, evolution is treated in Chapter 9. Chapters 10 to 13 discuss the overall equilibrium and dynamics of the interstellar gas, including explosions of H II regions and of supernova shells and the processes of star formation.

As a result of the many changes in scope, emphasis, and organization, most of this book has been newly written. However, in each chapter certain paragraphs presenting basic theoretical concepts follow very closely the corresponding text in *Diffuse Matter in Space*, especially in later chapters where there are relatively few discussions of current observational results. The known errors of the previous text, for example, in the discussion of shockwaves, have been corrected, while other theoretical discussions, including the Oort limit in Section 1.6 and the expansion of an H II region in Section 12.1b, have been much simplified and shortened. Among the new theoretical topics treated are interstellar scintillation, optical pumping of H_2 rotational levels, ionization of dusty H II regions, spin-up of grains, Rayleigh-Taylor instability, magnetostatic equilibrium in the galactic plane, accretion flow, and spiral density waves.

The notation used in *Diffuse Matter in Space* has been followed in most cases. Among the few important changes which have been made are the use of a subscript "d" in place of "g" to denote quantities pertaining to the interstellar dust grains to avoid confusion between grains and gas. The subscript "G" is used to denote the gas in dynamical discussions where the gas and dust grains move together. Also, to conform to the now almost general usage, the fractional polarization P is used instead of the polarization p in magnitudes. The numerical values in all equations are based on cgs units except where noted otherwise.

References to books and published papers, which are grouped at the end of each chapter, are selected primarily with a view to generality and availability to the English-speaking student. Thus a review paper will often be referred to instead of the relevant individual papers. Only a very small fraction of all the significant papers on any subject are listed, of course. In Chapter 1 the works referred to are mostly those not listed in subsequent chapters. Frequently references are made to other sections in the text. Square brackets are used both for external and internal references, thus [7] denotes the seventh work listed in the references, whereas [S4.2b] refers to item b of Section 4.2 of Chapter 4 in the text. The relatively complete index and the list of symbols should help in finding, respectively, discussions of specific subjects and equations defining various symbols.

It is a pleasure to acknowledge the cooperation of the many scientists who have assisted during the preparation of this manuscript, including J. E. Baldwin, R. A. Chevalier, L. L. Cowie, D. L. Crawford, B. T. Draine, G. B. Field, T. Gehrels, J. J. Gelfand, A. Hewish, R. Hunt, M. Koren, C. M. Leung, F. J. Lovas, R. McCray, D. E. Osterbrock, J. P. Ostriker, N. Panagia, E. E. Salpeter, P. R. Shapiro, C. H. Townes, W. D. Watson, P. R. Woodward, H. F. Weaver, and C. G. Wynn-Williams. H. Cohn, M. Hausman, E. Martin, and A. Stark made many corrections and helpful suggestions during a term course based on this material. Particularly important suggestions concerning Chapters 3 to 5 were made by A. Dalgarno and concerning Chapters 7 and 8 by E. M. Purcell. A critical reading of much of the manuscript by M. Jura was especially helpful. The photographic prints used for Figures 1.3, 1.4, and 1.5, provided by G. Wallerstein, H. C. Arp, and P. van der Kruit, respectively, are gratefully acknowledged.

This book was written in draft form during the period August 1975 through January 1976, when I was a visitor first during two months at the

Institute of Astronomy, Cambridge, England, and next during four months at the Institut d'Astrophysique, Paris, France. The author is greatly appreciative of the hospitality extended by these two institutions and particularly by Professors D. Lynden-Bell and M. Rees at Cambridge and Professor J. C. Pecker at Paris.

LYMAN SPITZER, JR.

Princeton, New Jersey
June 1977

Figure Credits

Fig. 1.2 Reproduced by permission of the Hale Observatories.

Fig. 1.4 Photograph by Halton Arp, Hale Observatories.

Fig. 1.5 Adapted from ref. 13, Chap. 1, by P. van der Kruit.

Fig. 1.6 Adapted from ref. 14, Chap. 1, by courtesy of the International Astronomical Union.

Fig. 1.7 Reproduced from *Stars and Stellar Systems*, published by the University of Chicago. See ref. 18, Chap. 1. Copyright © 1965.

Fig. 3.1 Reproduced from *Ap. J. Supp.*, published by the University of Chicago. See ref. 22, Chap. 3. Copyright © 1972.

Fig. 3.2 Reproduced from ref. 34, Chap. 3, by courtesy of *Ann. Rev. Astron. Astrophys.*

Fig. 4.3 Adapted from ref. 25, Chap. 4, by courtesy of the Royal Astronomical Society (London).

Fig. 4.4 Adapted from ref. 33, Chap. 4, by courtesy of *Ann. Rev. Astron. Astrophys.*

Fig. 6.1 Adapted from *Astrophysics of Gaseous Nebulae* by Donald E. Osterbrock, W. H. Freeman & Co. Copyright © 1974.

Fig. 6.2 Reproduced from ref. 6, Chap. 6, by courtesy of *Ann. Rev. Astron. Astrophys.*

Fig. 7.3 Adapted from ref. 18, Chap. 7, by courtesy of the *Publications* of the Astronomical Society of the Pacific.

Fig. 8.1 Adapted from *Stars and Stellar Systems*, published by the University of Chicago. See ref. 3, Chap. 8. Copyright © 1968.

Fig. 8.2 Adapted from ref. 5, Chap. 8, by courtesy of the *Astronomical Journal*.

Fig. 8.3 Reproduced from *Stars and Stellar Systems*, published by the University of Chicago. See ref. 11, Chap. 8. Copyright © 1963.

Fig. 9.1 Adapted from *Ap. J. (Lett.)*, published by the University of Chicago. See ref. 17, Chap. 9. Copyright © 1974 by the American Astronomical Society.

Fig. 11.1 Reproduced from *Ap. J.*, published by the University of Chicago. See ref. 23, Chap. 11. Copyright © 1974 by the American Astronomical Society.

Fig. 12.1 Reproduced from *Ap. J.*, published by the University of Chicago. See ref. 20, Chap. 12. Copyright © 1976 by the American Astronomical Society.

Fig. 13.1 Adapted from ref. 1, Chap. 13, by courtesy of the Royal Astronomical Society (London).

Fig. 13.2 Adapted from ref. 4, Chap. 13, by courtesy of the Royal Astronomical Society (London).

Fig. 13.3 Reproduced from *Ap. J.*, published by the University of Chicago. See ref. 8, Chap. 13. Copyright © 1975 by the American Astronomical Society.

Contents

1. Interstellar Matter—An Overview

In the enormous volume between the stars in our Galaxy there occur many different physical processes. Energy generated in the stars is absorbed and reemitted by the interstellar medium in ways that can be used to indicate the physical conditions within this medium. Material enriched in heavy elements is expelled from the stars, is mixed with the existing gas, and condenses to form new stars, determining the evolution of our Galaxy over many billions of years. The subsequent chapters in this book analyze some of these processes, especially those which seem relatively well understood, and discuss how the theory can be used to interpret the observed data and to draw conclusions about the nature and evolution of interstellar matter.

To view these theoretical discussions from a proper perspective it is helpful to have a general knowledge of the physical system which is being analyzed. The present chapter is designed to provide such an understanding in a rather broad and nonquantitative way. Hence the following sections describe what is known about the interstellar gas in each of three different phases, classified according to the ionization state of hydrogen, the dominant constituent. The discussion includes observations of the different constituents of the gas, including in addition to the gas of hydrogen and helium, atoms of the lighter trace elements, the small solid particles or grains, the interstellar magnetic field, the energetic charged particles confined by this field, and the gravitational field produced by the interstellar medium.

1.1 NEUTRAL GAS

Measurement of the radiation emitted at 21.11 cm by neutral H atoms has given extensive information on the distribution of this phase of the interstellar gas, often referred to as "H I regions." The overall mean value of n(H I), the particle density of neutral hydrogen, is about 1 atom per cm^3 in the solar neighborhood [S11.1a]. The "effective thickness," $2H$, of this gas increases somewhat with increasing distance from the galactic center, and it is about 250 pc near the Sun; this effective thickness is defined as N(H I), the "column density" of neutral hydrogen in a cylinder 1 cm^2 in cross section extending through the Galaxy normal to the galactic plane, divided by n(H I) in the midplane of the Galaxy. Most of the mass of the interstellar medium, probably more than 95 percent, is believed to be in the hydrogen and helium of this neutral phase.

More detailed information on the distribution and velocity of gas in H I regions within about a kiloparsec from the Sun is provided by measures of atomic absorption lines both from ground and from space. These measures show that much of the neutral gas tends to occur in separate regions or "clouds," characterized by a range of values of the radial velocity. The dispersion of radial velocities of the more conspicuous H I clouds is about 6 km s^{-1} [S11.1a]. Some clouds have substantially higher values of the radial velocity v_r (in the local standard of rest), exceeding 50 km s^{-1} in a few cases. Consistent with these velocity data, some clouds are observed more than 750 pc above the galactic plane.

The kinetic temperature characterizing the random velocity of the atoms within these clouds ranges between about 50 and 150°K, averaging about 80°K; similar values are obtained both from the absorption of 21-cm radiation in the spectrum of extragalactic radio sources, and from the ratio of orthohydrogen to parahydrogen [S4.3b] in interstellar H_2, whose ultraviolet absorption lines are observed with space telescopes. For denser, more opaque clouds, measures of CO emission at 0.26 cm show a gas temperature usually less than 40°K, with mean values about 30°K for clouds surrounding H II regions and about 10°K for dark clouds, presumably with no strong internal source of energy. The cores of these clouds are likely to be still colder. These temperatures begin to approach the 2.7°K measured for the universal blackbody radiation [1]. There is also evidence both from the 21-cm data and from ultraviolet absorption lines that an appreciable fraction of the neutral hydrogen may be distributed in a more nearly uniform medium with n(H I) in the range from 0.05 to 0.2 cm^{-3} and with T as great as 6000°K [S11.1a]. Evidently an enormous range of both densities and temperatures is possible in the H I gas.

Of the other constituents besides atomic hydrogen in interstellar clouds, one of the more evident is dust, a collection of small solid particles or grains with radii less than about 10^{-4} cm. These grains absorb and scatter light of all wavelengths. Since the effective cross sections for these optical processes generally increase with decreasing wavelength, the colors of stars seen through the clouds are altered, with light of shorter wavelengths selectively weakened; the mass of the grains present in a 1-cm^2 column extending from the Earth to the star may be measured by the color excess E_{B-V}, defined as the difference in magnitude between the color index $B-V$ of a star and the color index of a close unreddened star with the same spectral type. As shown in Fig. 1.1, the correlation between N_H, the column density of hydrogen atoms along the line of sight to a star, and E_{B-V} is reasonably good, much better than would be expected from the rather poor correlation of either quantity with distance. Some of the scatter in this plot may be due to the fact that some of the grains along the line of sight may be in ionized regions, which make no contribution to N(H I). Interpretation of these data with the use of theoretical cross sections for absorption and scattering gives a total ratio of gas mass (including He) to dust mass equal to about 160 in interstellar space, consistent with the point of view that most heavier atoms have condensed on grains together with about one-third of the C, N, and O atoms [S7.3b]. The H_2 molecule is also an important constituent of the conspicuous H I absorption clouds. Interstellar absorption lines of H_2 indicate that the fraction of hydrogen in molecular form is as great as two-thirds in the more opaque clouds examined. In the clouds which are too opaque to permit analysis in this way most of the hydrogen must be molecular.

The chemical composition of the gas, as determined from extensive absorption line measures, especially in the ultraviolet, differs somewhat

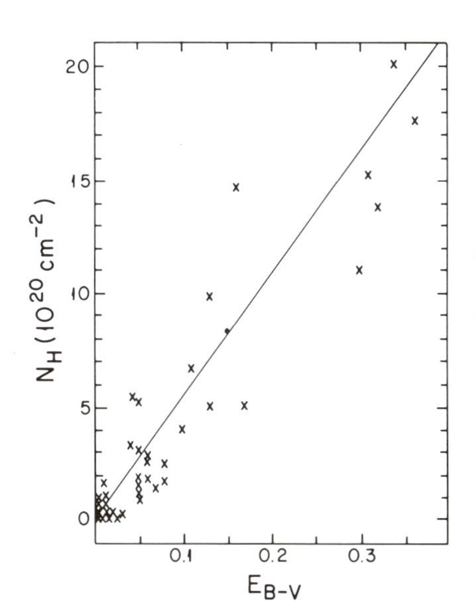

Figure 1.1 Correlation of N_H with E_{B-V}. The column density of H atoms per cm^2 in the line of sight to various stars is plotted against the color excess E_{B-V}, which measures the total amount of dust in each line of sight. Each plotted point, taken from ref. [32] of Chapter 3, represents a sum of N(H I) found from $L\alpha$ absorption and $2N(H_2)$ from the H_2 molecular lines; N(H II), the column density of ionized H, is ignored.

from the "cosmic composition" shown in Table 1.1 [2, 3], based on measures of the Sun, stars, and meteorites; the "cosmic abundance" values give $\log_{10}N_X$ for each element X, normalized to 12.00 for H. For comparison the measured values of the depletion are shown for the atoms along the line of sight to the reddened star ζ Oph [3] (Type O9.5V, $E_{B-V} = 0.32$ mag), which evidently shines through an H I cloud.

Table 1.1. Cosmic Composition Compared with That of Interstellar Gas

Element X	He	Li	C	N	O	Ne	Na
Cosmic Abundances*	11.0	3.2	8.6	8.0	8.8	7.6	6.3
Depletion,† ζ Oph		−1.5	−0.7	−0.7	−0.6		−0.9
Element X	**Mg**	**Al**	**Si**	**P**	**S**	**Ca**	**Fe**
Cosmic Abundances*	7.5	6.4	7.5	5.4	7.2	6.4	7.4
Depletion,† ζ Oph	−1.5	−3.3	−1.6	−1.1	−0.3	−3.7	−2.0

*Cosmic abundance values: $12+(\log N_X/N_H)_{\text{cosmic}}$.
†Depletion: $(\log N_X/N_H)_{\zeta\ \text{Oph}} - (\log N_X/N_H)_{\text{cosmic}}$.

The depletion is defined as the logarithmic ratio of column density for element X to that for H, minus the corresponding ratio for cosmic composition. In the cloud toward ζ Oph, the gases which appear to have condensed out include about three-fourths of the C, N, and O atoms together with all but a few percent of the more refractory elements, such as Fe, Ca, and Al. Toward the unreddened stars the depletion seems less, especially for the lighter elements [S3.4c], and is roughly consistent with the observed dust-to-gas ratio noted above, if the "missing atoms" are in dust grains.

Statistical information on the distribution of clouds is best obtained from measurements of the dust, since values of E_{B-V} are available for several thousand stars of known distances. The data can be represented by two types of clouds, relatively transparent "standard clouds" and more opaque "large clouds," with photographic extinctions of about 0.2 and 1 stellar magnitude per cloud, respectively. The large clouds may be identified with visible obscuring regions in the Milky Way, some 70 pc in diameter. A typical standard cloud might have a diameter of some 10 pc. Hydrogen densities of about 20 cm^{-3} seem indicated for clouds of both types.

Examination of Milky Way photographs (see Fig. 1.2) or detailed studies of individual clouds indicate that in fact the physical properties of H I regions extend over a very large range of values, and a model with only two types of clouds is a very crude approximation. Density concentrations are clearly present with all dimensions from a few kiloparsecs down to the limit set by the photographic plate. Study of individual clouds which produce many measurable atomic and H_2 lines (often called "diffuse clouds") indicates that the densities range from 10 to 1000 cm^{-3} or more. In the "dark clouds" which produce the observed molecular lines, the opacity due to dust grains is usually too large to measure in visible light, since any source shining through the cloud is largely obscured. The values of n_H in these dark clouds, obtained from analysis of molecular excitation conditions, are generally in the range from 10^3 to 10^6 cm^{-3}. In some of these dense clouds, especially those associated with some energy source, whose presence is indicated by continuous emission at radio or infrared wavelengths, intense emission from OH and H_2O molecules is observed from very small sources. This emission is characterized both by variability and by high polarization, and is generally attributed to maser amplification.

1.2 PHOTON-IONIZED GAS

Near stars of high surface temperature, which are sources of strong ultraviolet radiation, the hydrogen is ionized by photons shortward of 912 Å; that is, with energies exceeding 13.60 eV, the ionization energy of H. The emission from such regions has been extensively observed [4] at many different wavelengths, including the Balmer lines and the Balmer continuum, forbidden lines of N II, O II, and other atoms, continuum emission at infrared and radio frequencies, and H recombination lines in the radio region.

Such regions tend to be very conspicuous around the early O stars, since these are the most luminous in the far ultraviolet. A photograph of one such region around the O-star association, I Mon, is shown in Fig. 1.2. Since the O stars have been formed recently in clusters or associations, the gas and dust left over from the star formation process are often still in the vicinity, and obscuration by dense, dark, and presumably neutral clouds is evident in the figure. Regions of ionized gas around O and B stars are often called H II regions. Planetary nebulae, which are not discussed much in this book, are similar in many ways [4] to H II regions, except that the ionization level tends to be somewhat higher; the planetary nuclei which are thought to have expelled the gas constituting these nebulae are hotter than the O stars, and their photons are consequently more energetic.

The intensity of the radiation emitted by an H II region can be used to give the mean value of n_e^2 in the ionized gas. Ratios of some of the forbidden lines give the local value of n_e. The most detailed analysis has

Figure 1.2 Section of H II Region NGC 2237 (Rosette nebula), photographed in red light with the Mt. Palomar Schmidt telescope at the Hale Observatories. The gas is ionized by the O-star association I Mon, containing nine O stars from type O5 (HD 46223) to O9, centered on the star cluster NGC 2244 visible at the upper right. The system is some 1500 pc from the Sun.

been carried out for the Orion nebula (NGC1976), where n_e determined from line ratios decreases from 1.6×10^4 cm^{-3} near the center of the Trapezium to 2.6×10^2 cm^{-3} at a distance of some 24 arcminutes (3 pc). The root mean square n_e determined from the radiation intensity is about one-sixth of the value computed from the line ratios, suggesting that in each spherical shell the emission comes from only a small fraction (about 1/30 of the volume), a very clumpy radiation source. In other H II regions the evidence for clumpiness is much less marked.

In regions far from any O stars there is a diffuse but weak galactic Hα emission whose intensity corresponds to a value between 0.005 and 0.015 cm^{-6} for $\langle n_e^2\rangle$. Similar mean values of n_e^2 are deduced from the galactic thermal absorption of radio waves from extragalactic sources. This ionized gas may be attributed to H II regions around B stars and possibly other ultraviolet sources such as the nuclei of planetary nebulae and other dying stars. Such regions may account for much of the mean electron density of 0.03 cm^{-3} deduced from the dispersion of pulsar signals. For example, if such regions occupy 10 percent of the galactic disk and have each a mean density of 0.3 cm^{-3}, the mean n_e deduced from pulsars is accounted for [S3.6a].

The kinetic temperature of the photon-ionized gas has been determined in many ways—from ratios of emission lines, both in the visible and the radio regions of the spectrum, and from measures of the Balmer and radio continuum. The resultant values are mostly between 7000 and 10,000° in H II regions, with 8000° K as a reasonable mean value [S11.1a]. In addition to the spread of values between different objects, there are differences between different determinations for the same object, perhaps a result in part of temperature gradients in the ionized gas.

The presence of dust within H II regions is indicated both by the presence of scattered starlight, giving an observable optical continuum, and by the presence of observable infrared emission. Analysis of the scattered starlight gives a dust-gas ratio of about 1/100 by mass, roughly the value obtained for H I clouds. The large scatter in the observed ratios, from 1/20 to 1/700, may reflect uncertainties in the interpretation; alternatively, the very low value observed in the central part of the Orion nebula may be real, especially since extinction measures indicate that the smaller grains normally present are lacking in this region. The densest, brightest photon-ionized clouds, called "compact H II regions," are so heavily obscured by dust that they often cannot be seen at visual wavelengths, and are detected primarily by their free–free radio emission. Most of the energy from such regions appears in the infrared, probably in large part from H I clouds surrounding the ionized gas. The intensity of the radio emission observed from such regions and from visible H II regions as well is consistent with the view that much of the stellar ultraviolet radiation is absorbed by hydrogen rather than by dust. A dust-gas ratio of about 1/100 may be consistent with these measures as well as with the observed scattered starlight.

The chemical composition of the Orion nebula and of two other H II regions appears close to the cosmic composition of Table 1.1 for the observed elements: H, He, N, O, Ne, and S. The refractory elements, whose relative abundance seems low in H I clouds, have not been measured in H II regions, but must presumably be depleted to provide the atoms which constitute the dust grains observed in these regions.

The total fraction of the interstellar medium, by mass, in the form of photon-ionized gas has been estimated [4] as 0.01 for the Galaxy as a whole. Along the line of sight to early-type stars, this fraction is, of course, much greater. The fraction of the volume of the galactic disk occupied by low-density H II gas may be as great as 0.1 (see above).

1.3 COLLISION-IONIZED GAS

When a star expels gas with a velocity exceeding 1000 km s^{-1}, a shockwave results, which heats the interstellar gas to kinetic temperatures exceeding 10^6 degrees. Collisions between atoms and electrons at these temperatures would be expected to ionize H and He, generate X-rays, and produce highly ionized atoms of elements heavier than He. The X-rays anticipated from such collisionally ionized gas have been directly observed from supernova remnants, whereas the highly stripped ions have been observed in absorption along the line of sight to most O and B stars.

Supernova remnants may be identified [5] by their nonthermal radio emission and by the clouds or filaments of gas which are moving outward with velocities ranging from 300 km s^{-1} for the Cygnus Loop [6], an old remnant, up to 6000 km s^{-1} for Cas A, believed to be only a few hundred years old. Soft X-rays (0.15 to 2 keV) have been observed from many of these. From the Crab Nebula (produced by a supernova observed in 1054 A.D.) the X-rays are polarized, have a power-law spectrum, and must be nonthermal, but the X-rays from remnants older than a few thousand years can be fitted by thermal emission from a hot gas [S3.5a]. For three of the best known older remnants, the computed values of the kinetic temperature T behind the shockwave, together with other parameters obtained [7] by fitting the data with theoretical nonradiating models [S12.2b], are listed in Table 1.2. A photograph [8] of the Vela X supernova remnant is reproduced in Fig. 1.3. For more recent supernovae, with higher expansion velocities, the temperatures must exceed 10^7°K, but fitting the spectrum is more complicated.

Table 1.2. Parameters Obtained from X-Rays Emitted by Old Supernova Remnants

Object	Temperature (°K)	Initial n_H (cm^{-3})	Energy Release (ergs)	Age (years)
Pup A	7×10^6	0.6	3×10^{50}	4,000
Cyg Loop	3×10^6	0.15	4×10^{50}	17,000
Vela X	4×10^6	0.08	4×10^{50}	13,000

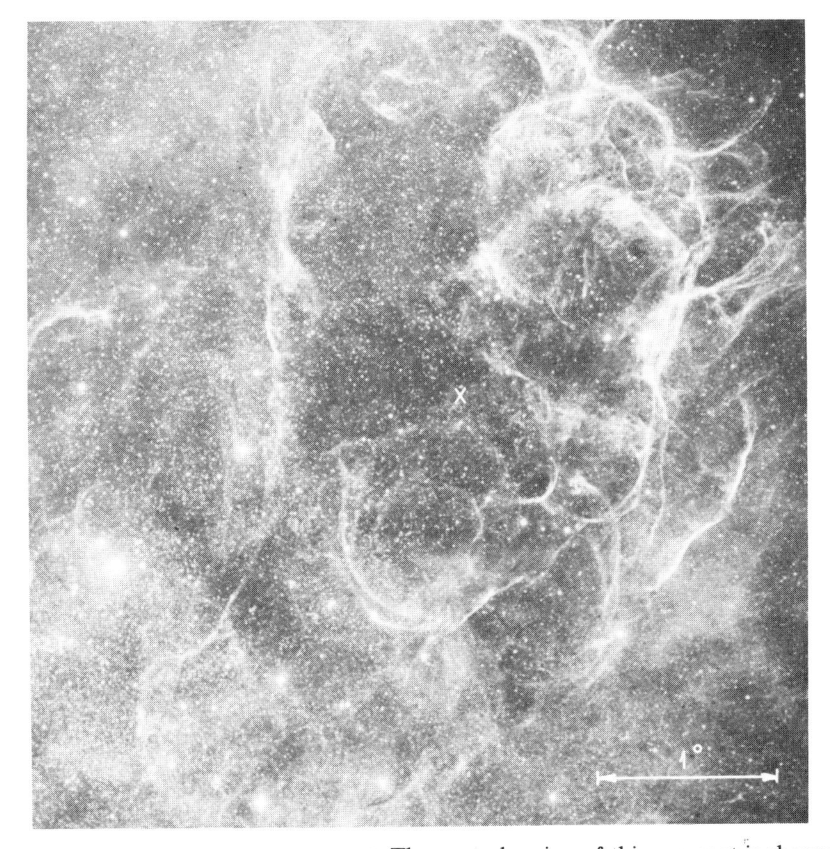

Figure 1.3 Vela X supernova remnant. The central region of this remnant is shown, photographed [8] with the University of Michigan Curtis-Schmidt telescope at the Cerro Tololo Interamerican Observatory. The $\times$ near the center indicates the position of the pulsar PSR 0833-45, whose apparent age, $dt/d(\ln \text{period})$, is about 20,000 years.

The presence of collisionally ionized gas at somewhat lower temperatures is indicated by the ultraviolet interstellar absorption lines of O VI, which have been measured in most of the O and B stars observed. The absence of the similar lines of S IV, Si IV, and N V gives a lower limit of 2×10^{5}°K for the kinetic temperature [S5.2b], whereas the observed absorption line widths give upper limits of about 10^{6} degrees. The mean density of such ions in the galactic plane is about 10^{-8} cm^{-3}, and the corresponding value of n_{H}, if the abundance of O is cosmic, is at least 10^{-4} cm^{-3}. Evidently this "coronal" gas seen in absorption is somewhat less hot than the gas which produces the soft X-rays. Some of this coronal gas must occupy the central zones within H II regions surrounding stars with high mass loss rates. In addition, coronal gas produced by supernovae fills some, perhaps most, of the volume between H I clouds and could extend several kiloparsecs from the galactic plane. In any case, this phase probably does not contribute more than 10^{-3} of the mass of the interstellar medium within 100 pc of the galactic plane and within 1000 pc of the Sun.

1.4 MAGNETIC FIELDS AND COSMIC RAYS

The mean value of the magnetic field in the Galaxy is determined from measures of pulsars. The Faraday rotation gives $\langle n_e B_{\parallel}\rangle$ along the line of sight to the pulsar, while the dispersion gives $\langle n_e\rangle$ along the same line; n_e is the electron density, while $B_{\parallel}$ is the component of **B** parallel to the line of sight. The values of $B_{\parallel}$ show much scatter, but the indicated mean field is 2.2×10^{-6} G parallel to the galactic plane in the direction $l=94\pm11°$. In several H I clouds with n(H I) exceeding 100 cm^{-3}, values of $B_{\parallel}$ ranging up to 7×10^{-5} G are found from Zeeman splitting of the 21-cm line in absorption [S3.4a].

One of the many effects produced by this interstellar magnetic field is the alignment of interstellar grains. Detailed information on the direction of **B** is given by the orientation of the polarization produced by these grains in the light of distant stars. The data indicate that the direction of the apparent mean field is $l=50°$, significantly different from that found from Faraday rotation. For comparison, the direction of the Orion spiral arm is generally taken to be $l=70°$ (see the following section).

Another result of the magnetic field is the production of synchrotron radiation by relativistic electrons within the galactic disk. Such electrons, with energies between 10^{9} and 10^{12} eV, are observed reaching the Earth [9]. If the particle densities of these relativistic particles throughout the galactic disk are assumed the same as that in the solar system, a uniform field of at least 10^{-5} G would be required [10] to explain the observed synchrotron radiation. A highly nonuniform magnetic field, perhaps with the energetic electron density n_{eR}, correlated with **B**, could explain the observations; the synchrotron emission, which varies about as B^2, would then originate in regions of relatively high n_{eR} and B. The polarization of the synchrotron radiation [11] is consistent with a magnetic field whose mean direction is in the galactic plane along the line $l = 70°$, intermediate between the other two determinations.

All three of these magnetic field indicators show that the magnetic field has a fluctuating component comparable with the mean field. Fluctuations can be found at almost any scale. Thus the measures of polarization direction in the double cluster h and χ Per suggest [S8.2c] fluctuations in field direction by about 30° over less than a parsec. Larger fluctuations occur with scales of 10 to 50 pc. The difference in direction of the mean field observed in different ways is presumably the result of such spatial nonuniformities, perhaps over an even larger scale. As a result of all these nonuniformities, the mean scalar or total B averages about 3×10^{-6} G [S11.1a], significantly greater than the mean systematic or vector field.

Most of the energetic particles or "cosmic rays" reaching the Earth are protons [9], which outnumber the electrons by about a factor 100 at 10^9 eV. The total energy density U_R of the cosmic radiation is estimated as 1.3×10^{-12} ergs cm^{-3}. Nuclei of all elements are apparently present, with relative abundances somewhat different from those in stars, a result in part of spallation reactions resulting from collisions with the interstellar gas during travel times of 10^6 to 10^7 years in the galactic disk. The cosmic radiation appears isotropic up to the maximum observed proton energy of 10^{20} eV.

Collisions of these energetic particles with the interstellar gas produce γ rays. The observed flux of γ rays from the galactic disk (after an isotropic extragalactic component has been subtracted out) gives information on the galactic distribution of cosmic rays [S1.5].

1.5 GALACTIC DISTRIBUTION

The large-scale distribution of the various components of the interstellar medium is best studied in external galaxies, which can readily be seen as a whole. Photographs of M31, the closest spiral galaxy, show that bright young O stars, the emitting H II regions that surround them, and the dust clouds which obscure starlight generally are all concentrated within the spiral arms. Figure 1.4, photographed with an Hα filter [12], shows conspicuous H II regions along a spiral arm in M31. Between the arms in spiral galaxies there appears to be no conspicuous obscuration nor emission nebulosity. Probably the neutral hydrogen and helium are also concentrated to the spiral arms, although the resolution of most 21-cm H-line

Figure 1.4 H II emission regions in M31. This photograph was obtained [12] with the 48-in. Schmidt telescope at the Hale Observatories; a filter with a band-pass of 100 Å, centered at Hα, was used. The H II regions along a spiral arm, together with some obscuration by dust particles, are clearly evident.

surveys is too low to verify this. Since the O stars responsible for the brighter H II regions cannot shine for more than about 10^6 years, photographs such as Fig. 1.4 are generally taken as evidence that formation of these massive stars occurs in the relatively dense interstellar gas found in spiral arms. It should be noted that the pattern of spiral arms in M31 is somewhat irregular, with more than two arms present, and with some elongated features not clearly related to the overall structure.

The spiral galaxy M51 has a somewhat more regular two-arm pattern than does M31. The continuous emission at 1415 MHz (21.2 cm) from this object, believed to be largely synchrotron radiation, has been measured [13] with a beam width of about 28 arcseconds, corresponding to about 500 pc at the distance of M51. The resultant contours, superposed on an optical photograph, are shown in Fig. 1.5. There is a clear apparent concentration of the synchrotron radiation to the inner edges of the spiral arms.

Figure 1.5 Regions of peak radio emission from M51. Superposed on the photograph of M51, obtained with the Hale Telescope on Mt. Palomar, are white lines indicating apparent ridges of peak emission observed [13] at 1415 MHz with the Westerbork, Netherlands, Synthesis Radio Telescope. The beam shape at half peak power is indicated in the lower left of the figure.

Within our own Galaxy uncertainty as to the distance of any emitting feature makes it difficult to determine the overall structure of the gas. The difficulty is compounded at some wavelengths by the presence of obscuration which is sometimes difficult to estimate. For H II regions the distances can be obtained with reasonable accuracy from observations of the central stars, whose spectra can be measured; the resultant spectral type yields an absolute magnitude, whereas the measured color index gives a color excess E_{B-V}, which is proportional to the amount of obscuration by dust grains. The locations determined for H II regions and for other spiral arm "tracers" are assembled together in Fig. 1.6 [14], projected on the galactic plane. Included are data on O-B0 star associations and clusters, dark clouds, and a few individual young stars, mostly within about 3000 pc from the Sun. Three rough concentrations seem indicated: the Orion (or Ori-Cyg) local arm, just beyond the Sun, the Perseus (or Per-Cas) arm, some 2000 pc further out, and the Sagittarius arm, some 2000 pc further in.

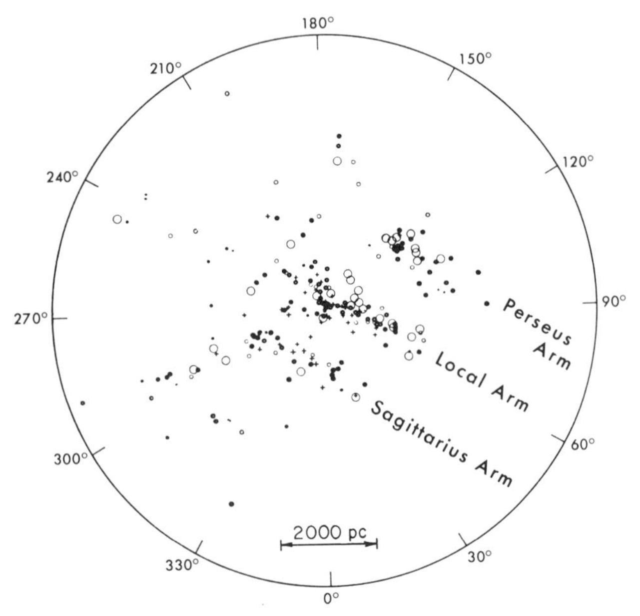

Figure 1.6 Spiral arm tracers near the Sun [14]. The locations of young stars, gas and dust, which are mostly confined to spiral arms, are shown projected on the galactic plane for distances within about 5000 pc from the Sun. The large open and large filled circles represent O-B0 associations and clusters, while Bpe and bright Cepheids are indicated by small open and small filled circles, respectively. H II regions and dark clouds are represented by large filled circles with central holes and by plus signs. The galactic center is by definition in the direction $l = 0$.

However, an interconnection seems suggested between the Orion and Sagittarius arms, and as in M31, the distribution evidently cannot be described in terms of entirely separate arms that spiral about the galactic center.

Similar plots can be obtained for more distant H II regions [4] and for H I clouds [15], using the observed radial velocities of these regions to determine the distances. In these determinations the gas is assumed to have an equilibrium circular velocity v_θ around the galactic center at a distance R, with the centrifugal force v_θ^2/R just balanced by the gravitational acceleration g_r; from an assumed model for the Galaxy, g_r can be computed. This method suffers from two serious uncertainties. The mass distribution assumed for the Galaxy may be incorrect, and the gas velocities may differ appreciably from the circular velocities. Hence these results, which have the advantage of extending to much greater distances, indeed across the entire Galaxy, are of lower weight than the ones presented in Fig. 1.6. In fact, there is general qualitative agreement between these methods for regions within 2500 pc.

Evidence on the distribution of cosmic rays within the galactic disk is provided by the intensity of the γ rays observed from different regions of the galactic disk. These data can be fitted with a cosmic ray density which increases with decreasing R, and at $R=5$ kpc is between two and four times the measured value near the Sun. The larger increase is obtained [16] if n_H, the total particle density of H nuclei in all forms, is assumed proportional to n(H I) obtained from 21-cm observations, while the smaller increase results [17] if an increasing fraction of H is assumed to be molecular in the denser regions near the galactic center, and n_H in these regions is assumed proportional to the observed particle density of CO molecules.

1.6 GRAVITATIONAL MASS

Many possible forms of interstellar matter, such as highly ionized atoms, many types of molecules, and any solid objects much bigger than a few microns in size, would be difficult to detect directly from radiation reaching the Earth, even if the total mass in these forms were much greater than in the form of neutral hydrogen atoms. In discussing physical processes in interstellar space it is very helpful, therefore, to have a limit on the total mass of material present between the stars. Such a limit can be obtained from the total gravitational attraction of the matter within a few hundred parsecs of the Sun. Since our Galaxy is a flattened disk, measurement of the gravitational acceleration g_z perpendicular to the disk gives directly the total mass of the material in the solar neighborhood, including both interstellar matter and stars. In particular, g_z is determined by measuring the density gradient in the z direction of some group of stars whose velocities in the z direction can also be measured.

The analysis becomes particularly simple if we approximate the Galaxy in the solar neighborhood as a one-dimensional system, where g_z, the number density n of stars per unit volume, and the velocity w_z of a star in the z direction are all functions of z only. If we assume a statistically steady state, then the equation of hydrostatic equilibrium for the group of stars under consideration gives

$$\frac{1}{n}\frac{d}{dz}\left(n\langle w_z^2\rangle\right)=g_z(z), \tag{1-1}$$

where $\langle w_z^2\rangle$ denotes the mean value of w_z^2 for all the stars in the group at the distance z from the midplane of the Galaxy. The total mass density $\rho(z)$ can then be obtained from Poisson's equation, which for a one-dimensional system becomes

$$\nabla^2\varphi=-\frac{dg_z(z)}{dz}=4\pi G\rho(z), \tag{1-2}$$

where φ is the usual gravitational potential, here a function of z only.

Equation (1-1) has been applied to K giants [18], which form a reasonably homogeneous group and are sufficiently bright and numerous to provide good statistics. The resultant values of g_z are plotted in Fig. 1.7.

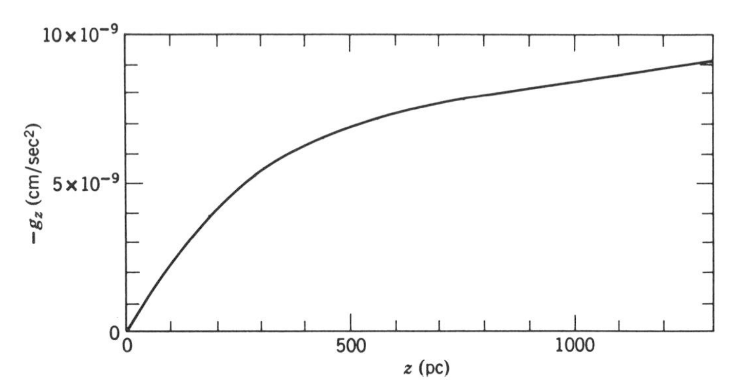

Figure 1.7 Gravitational acceleration perpendicular to the galactic plane [18]. The curve shows values of $-g_z$ in cm s^{-2} as a function of z, the height above the galactic plane in pc, deduced from the measured distribution of K giant stars with z.

From equation (1-2) the total mass density at the midplane is found to be 10.0×10^{-24} g/cm^3. The mass density of known stars is about 4×10^{-24} g/cm^3. Since the compilation of stars may be incomplete at the faint end, we have for ρ_{Int} the total density of interstellar matter in the solar neighborhood,

$$\rho_{\mathrm{Int}} \leqslant 6.0 \times 10^{-24} \text{ g/cm}^3, \tag{1-3}$$

corresponding to 2.6 H atoms per cm^3, if a He/H ratio of 10 percent by number is assumed. This upper limit on ρ_{Int} is sometimes known as the "Oort limit." When account is taken of the small thickness of the interstellar material—about 200 pc perpendicular to the galactic plane—and the uncertainty of the observations which have been differenced to obtain the smooth curve plotted in Fig. 1.7, a value of ρ_{Int} as great as 10×10^{-24} g/cm^3 would not be entirely excluded [18, 19]. The gravitational limit on ρ_{Int} is of fundamental importance in the study of the interstellar medium. The value of ρ_{Int} obtained from direct measurements of the gas in the galactic disk within 1000 pc from the Sun is about half the Oort limit, indicating that the undetected mass is comparable with the masses of the known stars and of the measured interstellar matter.

REFERENCES

1. P. J. E. Peebles, *Physical Cosmology*, Princeton University Press (Princeton, N.J.), 1971, Chapt. 5.
2. V. Trimble, *Rev. Mod. Phys.*, **47**, 877, 1975.
3. D. C. Morton, *Ap. J. (Lett.)*, **193**, L35, 1974.
4. D. Osterbrock, *Astrophysics of Gaseous Nebulae*, W. H. Freeman (San Francisco), 1974.
5. L. Woltjer, *Ann. Rev. Astron. Astroph.*, **10**, 129, 1972.
6. R. P. Kirshner, *Publ. Astron. Soc. Pac.*, **88**, 585, 1976.
7. P. Gorenstein and W. H. Tucker, *Ann. Rev. Astron. Astroph.*, **14**, 373, 1976.
8. E. W. Miller and J. C. Muzzio, *Sky Telesc.*, **49**, 94, 1975.
9. P. Meyer, *Ann. Rev. Astron. Astroph.*, **7**, 1, 1969.
10. G. Setti and L. Woltjer, *Ap. Lett.*, **8**, 125, 1971.
11. F. F. Gardner and J. B. Whiteoak, *Ann. Rev. Astron. Astroph.*, **4**, 245, 1966.
12. H. Arp and F. Brueckel, *Ap. J.*, **179**, 445, 1973.
13. D. S. Mathewson, P. C. van der Kruit, and W. N. Brouw, *Astron. Astroph.*, **17**, 468, 1972.
14. Th. Schmidt-Kaler, *Trans. IAU* **XIIB**, D. Reidel (Dordrecht), p. 416, 1964.
15. F. J. Kerr and G. Westerhout, *Stars and Stellar Systems*, Vol. 5, University of Chicago Press (Chicago), 1965, p. 167.
16. G. F. Bignami, C. E. Fichtel, D. A. Kniffen, and D. J. Thompson, *Ap. J.*, **199**, 54, 1975.
17. F. W. Stecker, P. M. Solomon, N. Z. Scoville, and C. E. Ryter, *Ap. J.*, **201**, 90, 1975.
18. J. H. Oort, in *Stars and Stellar Systems*, Vol. 5, University of Chicago Press (Chicago), 1965, p. 455.
19. R. J. Gould, T. Gold, and E. E. Salpeter, *Ap. J.*, **138**, 408, 1963.

2. Elastic Collisions and Kinetic Equilibrium

The interstellar medium is in many ways very far from thermodynamic equilibrium. In particular, while the mean energy density of starlight corresponds to thermodynamic equilibrium at a temperature of about 3°K, the mean energies of these photons are several electron volts, corresponding to the temperatures of thousands of degrees found in stellar atmospheres. Under these conditions the relative populations of atomic and molecular energy levels do not correspond to their values in thermodynamic equilibrium. The detailed understanding of interstellar processes generally requires analysis of the many different transitions that are possible between these various levels.

There is fortunately one basic simplification possible in this relatively complex topic. The velocity distributions of atoms, electrons, and molecules are closely Maxwellian, with a single kinetic temperature applicable to all these components. This circumstance results mainly from the enormous predominance of H and He relative to the other elements. Collisions at energies of 10 eV or less among these atoms or between their ions and the free electrons are almost perfectly elastic, and thus translational kinetic energy is exchanged back and forth many times before an inelastic collision can occur with a heavier atom, a molecule, or a grain. This is just the condition for the establishment of a Maxwellian distribution and for equipartition of translational kinetic energy between different particles.

Since the existence of a local kinetic temperature at each point (or more properly in each small region, several mean free paths in extent) of the interstellar gas is of such fundamental importance, this present chapter is devoted to elastic collisions and their effects. Successive sections in this chapter treat elastic collisions between charged particles, elastic collisions involving neutral atoms, and the extent of possible departures from the Maxwellian velocity distribution. The use of thermodynamic equilibrium at the kinetic temperature as a reference state for the actual excitation and ionization is discussed in a final section.

2.1 INVERSE-SQUARE FORCES

As a particle moves through a gas, its velocity $\mathbf{w}$ changes frequently as a result of encounters with other particles. To investigate the statistical effect of such encounters it is customary [1] to consider a small group of identical particles, called "test particles," all with the same initial velocity $\mathbf{w}_t$, and to compute the mean change of $\mathbf{w}$ and its moments per unit time for this group. The particles in the gas with which these test particles interact are often called "field particles," with particle density denoted by n_f; the velocity of such a field particle is denoted by $\mathbf{w}_f$.

The mean velocity change of the test particles per second, as a result of successive two-body encounters with the field particles, is denoted by $\langle \Delta w_{\|} \rangle$, where the subscript $\|$ denotes the component of $\mathbf{\Delta w}$ parallel to $\mathbf{w}_t$; it is clear from symmetry that the mean component of $\mathbf{\Delta w}$ perpendicular to $\mathbf{w}_t$ must vanish, provided that the distribution of $\mathbf{w}_f$ is isotropic. We define a "slowing-down" time t_s by the equation

$$-\langle \Delta w_{\|} \rangle t_s = w_t. \qquad (2\text{-}1)$$

The minus sign is required because $\langle \Delta w_{\|} \rangle$ is negative. The slowing down of a group of particles moving through a gas is sometimes called "dynamical friction." If the test particles are much lighter than the field particles (as for electrons interacting with protons), the binary encounters scatter the test particles and reduce their directed momentum with relatively little change in kinetic energy; t_s is then approximately the time required for an rms deflection of about 90°. In the opposite case, where the test particles being considered are much heavier than the field particles with which they are colliding, the slowing down represents a loss of kinetic energy, and t_s equals $-w_t/(dw_t/dt)$.

We now apply these concepts to a gas of charged particles, for which the electrostatic attraction or repulsion produces appreciable velocity changes in a mutual encounter. The mass and charge of the test particles will be denoted by m_t and $Z_t e$, with m_f and $Z_f e$ denoting corresponding quantities for the field particles; e denotes the proton charge in esu. To bring out the physical ideas involved, we derive an approximate formula [2] for t_s in the case of light test particles (such as electrons) interacting with much heavier field particles (such as heavy positive ions) with comparable kinetic energies. In this case w_f is relatively small and may be ignored compared to w_t. The orbit of the test particle is a hyperbola, and the angle between the asymptotes is 90° if the "impact parameter" p, defined as the distance of closest approach in the absence of mutual attraction, equals p_0, where

$$p_0 = \frac{Z_t Z_f e^2}{m_t w_t^2}. \tag{2-2}$$

Thus the test particle is deflected through a right angle if the potential energy at the distance p equals twice the initial kinetic energy. Evidently $\Delta w_{\parallel}$ equals $-w_t$ for one such encounter; if we assume that t_s is the mean interval of time per test particle between encounters with p less than p_0, we find

$$t_s = \frac{1}{n_f w_t \pi p_0^2} = \frac{m_t^2 w_t^3}{\pi n_f Z_t^2 Z_f^2 e^4}. \tag{2-3}$$

A more detailed analysis [1,2] of such encounters shows that the cumulative effect of many encounters more distant than p_0 outweighs the closer encounters by one or two orders of magnitude. For electrons as test particles colliding with protons as field particles, the slowing-down time, which we denote by $t_s(e,p)$, is given by

$$t_s(e,p) = \frac{m_e^2 w_e^3}{4\pi n_p e^4 \ln(\Lambda m_e w_e^2/3kT)} = \frac{1.241\times 10^{-18} w_e^3}{n_p \ln(\Lambda m_e w_e^2/3kT)}\,\mathrm{s}. \tag{2-4}$$

The quantity Λ, given in general by

$$\Lambda = \frac{3}{2Z_f e^3}\left[\frac{k^3 T_e^3}{\pi n_e}\right]^{1/2}, \tag{2-5}$$

is a "cutoff factor," resulting from shielding by electrons of the mutual electrostatic forces beyond some critical distance (the "Debye shielding distance" [2]) from positive-ion field particles. Under typical interstellar conditions, with n_e and T equal to 1 cm^{-3} and about 10^4°K, respectively, $\ln\Lambda = 23$. Evidently the exact value of $t_s(e,p)$ is less than found from equation (2-3) by about a factor $1/(4\ln\Lambda)$, or roughly $1/100$. For particle energies exceeding about 40 eV, the cutoff factor given in equation (2-5) must be modified [2].

For electrons more energetic than the average, equation (2-4) is approximately valid also for collisions of electrons as test particles with electron field particles, provided that n_e is substituted for n_p, and we have

$$t_s(e,e) = \frac{n_p}{n_e} t_s(e,p), \qquad \text{if } m_e w_e^2 > 3kT. \tag{2-6}$$

The slowing-down time $t_s(e,e)$ approximately equals the energy loss time $-E_e/(\Delta E_e)$ for the velocity range over which equation (2-6) is valid, the difference amcunting to about 10 percent at most if $m_e w_e^2$ is more than twice its mean value. In contrast, as pointed out earlier, the interactions between electrons and heavy ions produce relatively little change in the mean electron energy, and $t_s(e,p)$ corresponds primarily to a deflection time. Equation (2-6) for $t_s(e,e)$ may be used to give $t_s(p,p)$, provided that subscripts e are changed to p in equations (2-4) and (2-6). The ratio of $t_s(e,e)$ given in equation (2-6) to its exact value [2] equals 0.61 when $m_e w_e^2$ equals $3kT$, increasing to 0.89 and 1.00, respectively, for kinetic energies two and four times this great.

The quantities $t_s(e,e)$ and $t_s(p,p)$ provide a measure of the time required to establish a Maxwellian velocity distribution for electrons or protons of energies equal to or greater than the mean kinetic energy. Use is made of these properties in Section 2.3 below.

For heavy ions of charge $Z_i e$ moving through electron field particles, $t_s(i,e)$ is independent of the test particle velocity, and we have

$$t_s(i,e) = \frac{3m_i(2\pi)^{1/2}(kT_e)^{3/2}}{8\pi m_e^{1/2} n_e Z_i^2 e^4 \ln \Lambda} = \frac{503 A_i T_e^{3/2}}{n_e Z_i^2 \ln \Lambda} \text{ s}, \tag{2-7}$$

where A_i is the ion mass in atomic units. Equation (2-7) may also be used for the motion of heavy ions i through any lighter, more rapidly moving ions j, if subscripts e are changed to j and Z_i^2 is multiplied by an additional factor Z_j^2.

For some purposes one wishes to know the rate at which two groups of particles, each with a Maxwellian velocity distribution, but at different temperatures, approach equipartition of kinetic energy with each other; that is, the rate at which the two temperatures approach each other. To compute this rate for proton-electron interactions, for example, we consider the change of energy of protons interacting with electron field particles. In each separate two-body encounter the change of proton energy is given by

$$\Delta E_p = m_p \mathbf{w}_p \cdot \mathbf{\Delta w}_p + \tfrac{1}{2} m_p (\Delta w_p)^2. \tag{2-8}$$

If we now sum over collisions with the electron field particles per unit time, keeping $\mathbf{w}_p$ fixed, the first term on the right-hand side gives $-2E_p/t_s(p,e)$ from equation (2-1). According to equation (2-7), $t_s(p,e)$ is independent of w_p. Hence averaging this negative term over a Maxwellian distribution of test particle velocities is trivial, giving $-2\langle E_p\rangle/t_s(p,e)$. The second contribution gives a positive term on this double integration. When T_p and T_e are equal, the net flow of kinetic energy from electrons to protons must vanish, and these two terms must then be equal and opposite. Since it may be shown that this second term is independent of T_p, and since $\langle E_p\rangle$ is proportional to T_p, we must therefore have

$$\frac{dT_p}{dt} = -\frac{2(T_p - T_e)}{t_s(p,e)}. \tag{2-9}$$

The time $t_s(p,e)/2$ may be called the "equipartition time" for protons interacting with electrons. Since electron-proton collisions cannot directly alter the total kinetic energy density, $3k(n_pT_p + n_eT_e)/2$, the corresponding equipartition time for electrons is $t_s(p,e)n_e/2n_p$. In general, t_s computed from equation (2-7) for heavy test particles interacting with lighter, much more rapidly moving field particles equals twice the corresponding equipartition time.

2.2 SHORT-RANGE FORCES

For elastic collisions of neutral particles with other particles, either charged or neutral, quantum mechanical effects become important. As a result, the values of t_s given for such collisions are necessarily less exact than those obtained for encounters governed by inverse-square forces. We again make use of $\langle\Delta w_\parallel\rangle$, but express this in terms of the usual "momentum transfer" cross section, which we denote by $\sigma_s(u)$, a function of the relative velocity $\mathbf{u}$ between a test particle and a field particle. Evidently

$$\mathbf{u} = \mathbf{w}_t - \mathbf{w}_f. \tag{2-10}$$

For any particular value of a collision cross section, $\sigma(u)$, the corresponding collision rate per test particle can be readily computed. We assume first that $\mathbf{u}$ is the same for all field particles. Each test particle may be regarded as located at the center of a circle of area $\sigma(u)$, whose plane is perpendicular to $\mathbf{u}$. The cylindrical volume swept over during unit time equals $u\sigma(u)$; multiplication by the particle density n_f gives the number of field particles in this volume which will be encountered per unit time, and yields a collision rate of $n_f u\sigma(u)$ per test particle per unit time.

In the present case $\sigma_s(u)$ is defined so that the collision rate, multiplied by the "available" momentum $m_r u$, equals the mean loss of momentum of a test particle per second. Here m_r is the familiar reduced mass, given by

$$m_r = \frac{m_f m_t}{m_f + m_t}, \tag{2-11}$$

and $m_r u$ is the momentum of either the field or test particle relative to the center of mass of the two colliding particles. With this definition of $\sigma_s(u)$, we can write, if $\mathbf{u}$ is again taken to be the same for all collisions,

$$m_t \langle \Delta \mathbf{w}_t \rangle = -n_f u \sigma_s(u) m_r \mathbf{u}. \tag{2-12}$$

It is evident physically that $m_t \langle \Delta \mathbf{w}_t \rangle$, the total momentum change of the test particle per second, must be parallel to $\mathbf{u}$ under these assumptions.

It may be noted that the value of $\sigma_s(u)$ depends both on the range of the forces involved and on the distribution of deflections experienced. Thus if the colliding test and field particles are impenetrable elastic spheres, of radii a_t and a_f, respectively, the field particles are scattered isotropically with respect to the center of mass and carry off no momentum on the average. As a result, the mean momentum transfer per geometrical collision in this case actually equals the available initial momentum, $m_r \mathbf{u}$, and $\sigma_s(u)$ equals the geometrical cross section $\pi(a_t + a_f)^2$.

We now average equation (2-12) over all directions and magnitudes of w_f, assuming a Maxwellian distribution [see equation (2-17)]. If the mean w_f is either much greater or much less than w_t, we may write approximately

$$\langle \mathbf{u} u \sigma_s(u) \rangle = \mathbf{w}_t \langle u \sigma_s(u) \rangle. \tag{2-13}$$

Here the mean value of $u\sigma_s(u)$ will be called a "slowing-down coefficient." In the particular case where $u\sigma_s(u)$ is a constant, equation (2-13) is exact, as is evident from equation (2-10) and the assumed isotropic distribution of $\mathbf{w}_f$. With these same assumptions $\langle \Delta \mathbf{w}_t \rangle$ in equation (2-12) becomes $\langle \Delta w_{\parallel} \rangle$ times a unit vector in the direction of $\mathbf{w}_t$; i.e., the components transverse to $\mathbf{w}_t$ cancel out by symmetry. Use of equations (2-1) and (2-13) as well as (2-12) now gives

$$t_s = \frac{m_t + m_f}{n_f m_f \langle u \sigma_s \rangle}, \tag{2-14}$$

where the argument u of σ_s has been omitted. In subsequent discussion the argument of σ_s, where given, refers to the interacting particles, with H, H^+, e, and i referring to neutral H atoms, protons, electrons, and heavy ions, respectively. If w_f much exceeds w_t, u may be set equal to w_f, and the average of $w_f\sigma_s$ in equation (2-14) may be taken over a Maxwellian distribution. If w_f is much less than w_t, u may be set equal to w_t, with no further average taken.

In the special case of impenetrable elastic spheres, σ_s is independent of the velocity, as pointed out above. The mean value of $\mathbf{u}u$, averaged over all $\mathbf{w}_f$, may now be computed exactly [3]. If we use equation (2-10), and if the square root obtained for u is expanded for w_t much less or much greater than w_f, we obtain

$$\langle \mathbf{u}u \rangle = \begin{cases} \frac{4}{3}\mathbf{w}_t \langle w_f \rangle & \text{if } \langle w_f \rangle \gg w_t, \\ \mathbf{w}_t w_t & \text{if } \langle w_f \rangle \ll w_t. \end{cases} \tag{2-15}$$

Thus if w_t is small, equation (2-13) gives only three-quarters the correct result in this case. For a heavy test particle such as a dust grain, colliding with more rapidly moving field atoms of much lower mass and radius, equation (2-14) is replaced by

$$t_s = \frac{3m_t}{4n_f m_f \sigma_s \langle w_f \rangle}. \tag{2-16}$$

Equations (2-15) and (2-16) are also valid if the atoms are reflected diffusely from the grains, provided that the mean kinetic energy of the escaping atoms corresponds to a grain temperature which is much less than that of the approaching atoms.

The slowing-down time given in equation (2-14) or (2-16) may be divided by 2 to give the equipartition time between the components involved, since equation (2-9) is valid in this case as well as for particles interacting with inverse-square forces.

While values of $\langle u\sigma_s \rangle$ for the reactions of interest, as a function both of test particle velocity and of the mean square velocity for the field particles, are not available, some mean values of $u\sigma_s$, which have been computed for elastic collisions involving H atoms, are given in Table 2.1. For H–H encounters, the momentum exchange cross sections have been computed quantum mechanically and averaged over $u^3 f^{(0)}(u)$ to give $\bar{\sigma}_s$[4]; values of $\langle u \rangle \bar{\sigma}_s$ are given in the table at the indicated temperatures; values of $\langle u \rangle$ may be obtained from equation (2-19). The values of $\sigma_s(C^+–H)$, which are needed in discussions of ambipolar diffusion [S13.3e], were determined [5] from classical orbits of H atoms relative to C^+ ions, using the known potential function $U(r)$, which varies as r^{-4}. For this potential the collision radius $(\sigma_s/\pi)^{1/2}$ is roughly the distance at which the potential energy

Table 2.1. Calculated Slowing-Down Coefficients for H Atoms and C^+ Atoms. Units of 10^{-10} cm^3 s^{-1}

Temperature T (°K)	10°	30°	100°	300°	1000°
$\langle u\rangle\bar{\sigma}_s$ for H–H	3.3	5.1	7.4	10.2	13.6
$\langle u\sigma_s\rangle$ for C^+–H	22.	22.	22.	22.	22.

equals the kinetic energy, and it varies as $u^{-1/2}$. Hence $u\sigma_s(C^+\text{–H})$ is independent of u, and equation (2-13) is exact. The values of $u\sigma_s(C^+\text{–H})$ in Table 2.1 may be used for other heavy positive ions, and at temperatures less than about 100° they give roughly the values obtained for H–H^+ impact [6]. For collisions between He atoms and positive ions, the lower polarizability of helium gives values of $\sigma_s(C^+\text{–He})$, about half those for $\sigma_s(C^+\text{–H})$.

2.3 VELOCITY DISTRIBUTION FUNCTION

The distribution of translational velocities for any group of particles in a gas may be expressed in terms of the velocity distribution function $f(\mathbf{w})$ for that group, where $f(\mathbf{w})\mathbf{dw}$ is the fractional number of particles whose velocity lies within the three-dimensional volume element $\mathbf{dw} = dw_x dw_y dw_z$ centered at the velocity $\mathbf{w}$. In thermodynamic equilibrium, $f(\mathbf{w})$ is equal to the Maxwellian function $f^{(0)}(w)$, defined by

$$f^{(0)}(w) = \frac{l^3}{\pi^{3/2}} e^{-l^2 w^2}, \tag{2-17}$$

where

$$l^2 = \frac{m}{2kT} = \frac{3}{2\langle w^2\rangle}. \tag{2-18}$$

Since $f^{(0)}(\mathbf{w})$ is isotropic, w replaces $\mathbf{w}$ as the argument. Equation (2-17) also gives the distribution of the relative velocity u between two groups of particles, provided that the reduced mass in equation (2-11) is used in equation (2-18). For H atoms colliding with particles of mass Am_H, the mean value of u is given numerically by

$$\langle u\rangle = \left[\frac{8kT}{\pi m_r}\right]^{1/2} = 1.46\times 10^4 T^{1/2}\left(1+\frac{1}{A}\right)^{1/2} \text{ cm s}^{-1}. \tag{2-19}$$

Thermodynamic equilibrium is set up within a region if energy is exchanged back and forth between all different forms in the region, and a steady state is reached with no energy flux. If translational kinetic energy is exchanged back and forth between all the particles within a gas, and no transfer of energy occurs to and from other forms, the velocity distribution approaches the Maxwellian function, which is the most probable state for constant total kinetic energy. Under interstellar conditions there is some transfer of energy from translation to radiation, as a result of collisional excitation of atomic or molecular energy levels, followed by spontaneous deexcitation and emission of a photon. However, this process is relatively infrequent compared to elastic collisions, and one would expect the deviations from the Maxwellian distribution to be correspondingly small.

We now estimate the magnitude of such deviations, computing approximately the departure of $f(\mathbf{w})$ from the Maxwellian function $f^{(0)}(w)$ for electrons (and for neutral H atoms in H I regions). In a steady state, $(\partial f/\partial t)_{el}$, the rate of change of f resulting from elastic encounters, must be equal and opposite to $(\partial f/\partial t)_{in}$, the corresponding change from inelastic collisions. The first of these two quantities may be determined from a simplified kinetic model [7], which assumes that, on the average, elastic collisions among electrons during the time interval $t_s(e,e)$ [or during $t_s(\mathrm{H,H})$ for H atoms] tend to establish a Maxwellian distribution for electrons (or H atoms). Thus we may write

$$\left(\frac{\partial f}{\partial t}\right)_{el} = \frac{f^{(0)}(w) - f(\mathbf{w})}{t_s}. \tag{2-20}$$

Since it is not clear just which collision time should be used, equation (2-20) is uncertain by a factor of order unity.

The $(\partial f/\partial t)_{in}$ term may be evaluated simply if we assume that only one excited level need be considered. Collisional excitation of this level, with an excitation energy E_k above the ground level ($E_j = 0$), reduces the number of particles with an energy greater than E_k. Hence we have

$$\left(\frac{\partial f}{\partial t}\right)_{in} = -\frac{f(\mathbf{w})}{t_{ex}}, \qquad \text{for } \tfrac{1}{2}mw^2 > E_k, \tag{2-21}$$

where we have [S4.1a]

$$t_{ex} = \frac{1}{nw\sigma_{jk}}. \tag{2-22}$$

The particle density n refers to the neutral or ionized atoms which are being excited by the electrons (or H atoms). Since the mass of the excited atoms much exceeds the electron mass (or the mass of an H atom), we equate here the relative velocity u and the electron (or H atom) velocity w.

Equation (2-21) neglects inelastic collisions by particles of energy greater than $2E_k$, which would yield a positive contribution to $(\partial f/\partial t)_{\rm in}$. We also neglect deexcitation collisions, a valid approximation when the fraction of atoms in the excited state is very small; in thermodynamic equilibrium such collisions provide detailed balancing for $(\partial f/\partial t)_{\rm in}$. As a result of these approximations the present analysis gives a rough upper limit on departures from $f^{(0)}(w)$; use of the full equations for $\partial f/\partial t$ in place of equations (2-20) and (2-21) gives [8] about the same results as those obtained here.

If we now combine equations (2-20) and (2-21), together with the condition that the net $(\partial f/\partial t)$ must vanish, we obtain, again for $mw^2/2 > E_k$,

$$\frac{f^{(0)}(w)-f(\mathbf{w})}{f(\mathbf{w})}=\frac{t_s}{t_{\rm ex}}. \tag{2-23}$$

Both $t_{\rm ex}$ and t_s in this equation are functions of $\mathbf{w}$. For low velocities $f(\mathbf{w})$ is increased above $f^{(0)}(w)$, an effect which we do not consider here; for the realistic case that E_k substantially exceeds the mean kinetic energy, the relative departure of $f(\mathbf{w})$ from $f^{(0)}(w)$ is appreciable only in the high-velocity tail of the velocity distribution, at kinetic energies exceeding E_k.

We now compute numerical values from equation (2-23), first in H II regions, and second in H I regions. We use the values of $t_s(e,e)$ and $t_s(\mathrm{H},\mathrm{H})$ obtained above, together with the values of σ_{jk} given in Chapter 4. In H II gas, O II can be one of the more effective cooling agents. For excitation of the two ^2D levels in O II, with an energy of 3.32 eV, Table 4.1 together with equation (4-10) gives a value of 1.33×10^{-16} cm^2 for σ_{jk} at a value of $1.08\times10^{8}\,\mathrm{cm\,s^{-1}}$ for w, corresponding to a kinetic energy of 3.32 eV. For electrons of this velocity, the time $t_s(e,e)$ given by equations (2-4) and (2-6) equals $6.6\times10^{4}/n_e$. Hence equations (2-22) and (2-23) give

$$\frac{f^{(0)}(w)-f(\mathbf{w})}{f(\mathbf{w})}=1.0\times10^{-3}\frac{n_i}{n_e}. \tag{2-24}$$

Evidently in H II regions, where n_e exceeds n_i, the density of O II ions, by about 10^3, relative deviations from a Maxwellian distribution of velocities are about 10^{-6}.

In H I regions n_i is about equal to n_e, increasing the relative deviations from a Maxwellian velocity distribution. However, $t_s(e,e)$ is reduced by a counterbalancing factor of 10^{-3} when T is reduced from 8000°K down to 80°K. The important inelastic process in H I regions is the excitation of

C II ions by electrons. While the collision strength for this reaction is comparable to that for O II excitation (see Table 4.1), σ_{jk} is increased at the lower electron velocity, and detailed calculations show that $t_s(e,e)/t_{\mathrm{ex}}$ is still less than about 10^{-5}. For H atoms the conclusion is similar though less strong. Use of the slowing-down times in Table 2.1, together with the excitation rates obtained from Table 4.2, shows that the excitation of C^+ or of H_2 molecules can yield a value of $t_s(e,e)/t_{\mathrm{ex}}$ in equation (2-23) as great as 10^{-2}. Even this much larger deviation from the Maxwellian velocity distribution is generally unimportant.

We conclude that atoms and electrons in the interstellar gas usually have velocity distributions close to the Maxwellian, a result of fundamental importance for all interstellar matter studies. In addition, one kinetic temperature is generally applicable for different types of particles [9].

2.4 THERMODYNAMIC EQUILIBRIUM

As a result of the Maxwellian velocity distribution among the components of the "thermal" gas (distinguishing these from the high-energy "suprathermal" or cosmic-ray particles), the relative populations of various atomic and molecular energy levels will have some tendency to approach the values they would have in thermodynamic equilibrium. This tendency is particularly marked where transitions resulting from emission or absorption of photons are relatively unimportant, and collisional excitation and deexcitation by the thermal gas are dominant. For example, excited states of the H atom at a very large total quantum number n are populated and depopulated primarily by collisions with electrons. Similarly, low-lying excited levels in complex atoms, such as the ^1D and ^1S states of the ground term of C I, O I, N II, O III, and the excited fine-structure states of the ^3P level in these same atoms (see Fig. 4.1) are metastable, and they tend to be populated and depopulated by collisions with electrons or neutral H atoms. In these cases the fact that the velocity distribution is Maxwellian [S2.3], just as in thermodynamic equilibrium at the temperature T, tends to establish relative atomic populations which also agree with the values in thermodynamic equilibrium at the same T.

As a result of this tendency it is frequently useful to relate $n_j(X^{(r)})$, the particle density of atoms of element X ionized r times and in a particular state j, to its value in thermodynamic equilibrium. We define an "equivalent thermodynamic equilibrium" (ETE) as a state of thermodynamic equilibrium in which the temperature equals the actual kinetic temperature, and n_e equals the actual electron density. Also, the particle density of atoms of element X ionized $r+1$ times is set equal to its actual interstellar value; for excitation of neutral H, this requirement gives the proton density n_p in ETE equal to the actual value. The value of $n_j(X^{(r)})$ in this reference system will be denoted by $n_j^*(X^{(r)})$ or, more simply, by n_j^*. The overall mass density in the ETE system will generally be different from the actual interstellar value, since the relative distribution of atoms in different stages of ionization is different. We now define

$$b_j \equiv \frac{n_j}{n_j^*}. \tag{2-25}$$

Since the reference ETE system is chosen with respect to the next higher stage of ionization, b_j approaches 1 only when collisions dominate ionization and recombination, as for neutral H of very high total quantum number n. For excitation of a metastable state k from a ground state j, b_k/b_j tends to approach unity for the higher interstellar densities.

For reference, we give here the equations determining the relative values of n_j^* in thermodynamic equilibrium. Again, we consider an element X which is ionized r times. The relative populations of two levels in the same atom or ion are given by the Boltzmann equation

$$\frac{n_j^*(X^{(r)})}{n_k^*(X^{(r)})} = \frac{g_{rj}}{g_{rk}} e^{-(E_{rj}-E_{rk})/kT}, \tag{2-26}$$

where E_{rj} and g_{rj} are the energy and statistical weight, respectively, of level j in ionization stage r. Hence from equation (2-25) we obtain, omitting here the r designation,

$$\frac{n_k}{n_j} = \frac{b_k}{b_j} \frac{g_k}{g_j} e^{-h\nu_{jk}/kT}, \tag{2-27}$$

where ν_{jk} is the frequency of the photons emitted or absorbed in radiative transitions between levels j and k. As in most of this text, we adopt the convention that E_j is less than E_k.

If we take the sum of $n_k^*(X^{(r)})$ over all k to obtain $n^*(X^{(r)})$, then the fraction of $X^{(r)}$ atoms which are excited to level j becomes

$$\frac{n_j^*(X^{(r)})}{n^*(X^{(r)})} = \frac{g_{rj}}{f_r} e^{-E_{rj}/kT}, \tag{2-28}$$

where f_r, the partition function for ion $X^{(r)}$, is defined by

$$f_r = \sum_k g_{rk} e^{-E_{rk}/kT}. \tag{2-29}$$

The distribution of atoms of element X over different stages of ionization in thermodynamic equilibrium is given by the Saha equation

$$\frac{n^*(X^{(r+1)})n_e}{n^*(X^{(r)})} = \frac{f_{r+1}f_e}{f_r}, \tag{2-30}$$

where f_r and f_{r+1} are defined in equation (2-29), and where f_e, the partition function of the free electrons per unit volume, may be shown to be

$$f_e = 2\left(\frac{2\pi m_e kT}{h^2}\right)^{3/2} = 4.829 \times 10^{15} T^{3/2}. \tag{2-31}$$

The various symbols have their usual meaning (see the list of symbols at the end of this book).

If the partition functions f_r and f_{r+1} are approximated by their first terms, with the ground level denoted by the subscript 1, the Saha equation takes the more familiar form

$$\frac{n^*(X^{(r+1)})n_e}{n^*(X^{(r)})} = 2\frac{g_{r+1,1}}{g_{r,1}}\left(\frac{2\pi m_e kT}{h^2}\right)^{3/2} e^{-\Phi_r/kT}, \tag{2-32}$$

where Φ_r, the energy required to ionize the ion $X^{(r)}$ from its ground level, is given by

$$\Phi_r = E_{r+1,1} - E_{r,1}. \tag{2-33}$$

One property of systems in thermodynamic equilibrium which we shall use frequently in later chapters is that of detailed balancing; that is, the number of transitions upward from state j to state k per second is precisely equal to the corresponding number downward in the reverse direction, from state k to state j. This principle makes it possible to express the transition probability for a downward transition in terms of that for the corresponding upward transition.

The term "local thermodynamic equilibrium" (LTE) is frequently used to indicate the situation where $b_j/b_k = 1$ for bound levels. Such a state may differ from true thermodynamic equilibrium in that the radiation intensity may differ by an arbitrarily large amount from the Planck value [S3.1]. As noted above, LTE is a good approximation for certain problems, where as a result of frequent collisions $b_j = 1$ or $b_j/b_k = 1$ for certain selected levels. However, in interstellar space, where the radiation is generally much weaker than in thermodynamic equilibrium, there are very few physical situations where all bound levels within a given atom or molecule are populated according to the Boltzmann formula. While the assumption of LTE has frequently the advantage that it makes a problem tractable, by ignoring detailed processes of excitation and deexcitation, it must be used with caution. Since radiative processes, especially radiative recombination, usually dominate the ratio of ions to neutrals, the assumption of LTE does not imply that the Saha equation is applicable.

REFERENCES

1. S. Chandrasekhar, *Principles of Stellar Dynamics,* University of Chicago Press (Chicago), 1942, Chapter 2 and Section 5.6.
2. L. Spitzer, *Physics of Fully Ionized Gases,* Wiley (New York), 2nd edition, 1962, Chapt. 5.
3. M. J. Baines, I. P. Williams, and A. S. Asebiomo, *M. N.,* **130**, 63, 1965.
4. A. C. Allison and F. J. Smith, *Atomic Data,* **3**, 317, 1971.
5. D. E. Osterbrock, *Ap. J.,* **134**, 270, 1961.
6. A. Dalgarno, *Phil. Trans. Roy. Soc.* (London), Ser. A, **250**, 426, 1958.
7. D. L. Bhatnagar, E. P. Gross, and M. Krook, *Phys. Rev.,* **94**, 511, 1954.
8. R. J. Gould and R. K. Thakur, *Phys. Fluids,* **14**, 1701, 1971.
9. L. Spitzer and M. G. Tomasko, *Ap. J.,* **152**, 971, 1968.

3. Radiative Processes

Analysis of the interaction between radiation and matter serves two purposes in the study of the interstellar medium. Radiation reaching the Earth provides most of our direct knowledge about the universe outside our solar system; a detailed understanding of how this radiation was produced gives information on the emitting and absorbing materials along the line of sight. In addition, analysis of the various physical processes in interstellar space which determine the density, temperature, composition, velocity, and so on must take into account the absorption and reemission of photons by atoms, molecules, and dust grains.

The basic concepts of radiative transfer are reviewed in the first brief section of this chapter. The subsequent three sections review the absorption and emission of photons in transitions between bound levels, and show how these concepts are used in interpreting observations first of emission and then of absorption lines. After a brief review of continuous emission and absorption processes, the chapter ends with a discussion of the refraction of electromagnetic waves in a gas containing free electrons, and use of the theory to interpret data on pulsars and other radio sources. The effect of radiative processes on the physical state of the interstellar gas is included in subsequent chapters, using many of the concepts and principles developed in the first two sections below.

3.1 RADIATIVE TRANSFER

The photons traveling by a point $\mathbf{r}$ at a time t will each have a different direction, denoted by the unit vector $\boldsymbol{\kappa}$, and a different frequency ν. To characterize the radiation field we must specify the energy passing by as a function of all four of these physical variables. We define the specific intensity $I_\nu(\boldsymbol{\kappa}, \mathbf{r}, t)$ so that $I_\nu \, d\nu \, d\omega \, dA \, dt$ is the energy of those photons which during a time interval dt pass through the area $\mathbf{dA}$, whose frequency lies within the element $d\nu$ about ν, and whose direction is within the solid angle $d\omega$ about $\boldsymbol{\kappa}$; the area $\mathbf{dA}$ is located at the position $\mathbf{r}$ and is perpendicular to the photon direction $\boldsymbol{\kappa}$.

The change of I_ν resulting from interaction with matter is governed by the equation of transfer. This equation is derived by considering the flow of energy in and out the ends of a cylinder of length ds, with the use of the absorption and emission coefficients κ_ν and j_ν. The emission coefficient, or emissivity, $j_\nu(\boldsymbol{\kappa}, \mathbf{r}, t)$ is defined so that $j_\nu \, dV \, d\nu \, d\omega \, dt$ is the energy emitted by the volume element dV (equal to $ds \, dA$) in the intervals $d\nu$, $d\omega$, and dt, whereas $\kappa_\nu I_\nu \, dV \, d\nu \, d\omega \, dt$ is the corresponding energy absorbed from a beam of specific intensity I_ν. If we assume that the photons travel in straight lines, the change of I_ν along a distance ds, taken along a light ray, then equals [1, 2]

$$\frac{dI_\nu}{ds} = -\kappa_\nu I_\nu + j_\nu. \tag{3-1}$$

We define the "optical depth" τ_ν backward along the ray path by the expression

$$d\tau_\nu = -\kappa_\nu \, ds; \tag{3-2}$$

at the observer $\tau_\nu = 0$, and, if κ_ν is positive, τ_ν increases toward the source. If we consider the radiation received from a region or cloud of total optical thickness $\tau_{\nu r}$, equation (3-1) may be integrated to yield

$$I_\nu = I_\nu(0) e^{-\tau_{\nu r}} + \int_0^{\tau_{\nu r}} \frac{j_\nu}{\kappa_\nu} e^{-\tau_\nu} \, d\tau_\nu. \tag{3-3}$$

In this equation $I_\nu(0)$ denotes the value of I_ν on the far side of the emitting region from the observer, where $\tau_\nu = \tau_{\nu r}$. The ratio j_ν / κ_ν may vary with τ_ν.

Under some conditions j_ν and κ_ν have the same relative values as they would in strict thermodynamic equilibrium, where I_ν equals $B_\nu(T)$, given by the familiar Planck function

$$B_\nu(T) = \frac{2h\nu^3}{c^2} \frac{1}{e^{h\nu/kT} - 1}, \tag{3-4}$$

and T is the kinetic temperature of the gas [S2.3]. Since dI_ν / ds must vanish if I_ν is constant, we see from equation (3-1) that the ratio j_ν^* / κ_ν^* in the ETE system [S2.4] (and in thermodynamic equilibrium generally) is given by

$$j_\nu^* = \kappa_\nu^* B_\nu(T), \tag{3-5}$$

a relationship known as Kirchhoff's law.

If equation (3-5) is assumed to hold for j_ν/κ_ν in equation (3-1), and if the temperature T is constant over the pathlength, equation (3-3) becomes

$$I_\nu = I_\nu(0)e^{-\tau_{\nu r}} + B_\nu(T)\{1 - e^{-\tau_{\nu r}}\}. \tag{3-6}$$

For measurements in the radio spectrum, I_ν is frequently replaced by the "brightness temperature" T_b, usually defined as the temperature at which $B_\nu(T_b)$ equals the observed I_ν. If $h\nu/kT$ is much less than unity, as it often is for radio frequencies, we may expand the exponent in equation (3-4), neglecting terms of second and higher order in $h\nu/kT$, obtaining the familiar Rayleigh-Jeans law

$$B_\nu(T) = \frac{2\nu^2 kT}{c^2} \qquad \text{for } h\nu \ll kT. \tag{3-7}$$

If this relation is applicable, then in terms of T_b equation (3-6) takes the simple form

$$T_b = T_{b0}e^{-\tau_{\nu r}} + T(1 - e^{-\tau_{\nu r}}), \tag{3-8}$$

where T_{b0} is the brightness temperature of the incident radiation on the far side of the emitting and absorbing region.

If j_ν/κ_ν obeys Kirchhoff's law, equation (3-8) is a good approximation even if kT_b is much less than $h\nu$, provided that $h\nu/kT$ is small compared to unity and that equation (3-7) rather than (3-4) is taken for defining T_b. Since

$$\frac{h\nu}{kT} = \frac{1.439}{\lambda(\text{cm})T(°\text{K})}, \tag{3-9}$$

equation (3-7) becomes inapplicable for lines of millimeter wavelength and H I temperatures as low as 5 to 10°K.

3.2 EMISSION AND ABSORPTION COEFFICIENTS

To determine I_ν in any physical system we must evaluate j_ν/κ_ν in equation (3-3). In this section we consider how j_ν and κ_ν depend on the following more basic physical quantities: the particle densities n_j and n_k in the lower and upper levels, respectively, of the transitions involved; the frequency of the emitted photon, equal to $1/h$ times the energy difference involved; and three radiative transition probabilities, whose values are interrelated.

We consider first the emission of radiation. The Einstein radiation probability for a spontaneous downward transition per second per particle in the upper level k is denoted by A_{kj}. If ν_{jk} denotes the frequency of the photons emitted by an atom or molecule of type X, ionized r times, then the total energy radiated spontaneously per cm^3 per unit solid angle per second is given by

$$\int j_\nu \, d\nu = \frac{h\nu_{jk} n_k(\mathrm{X}^{(r)}) A_{kj}}{4\pi}, \tag{3-10}$$

where ν_{jk} is the frequency at the line center; isotropic emission has been assumed and the integral extends over the line. The quantity $n_k(\mathrm{X}^{(r)})$ is again [S2.4] the particle density for atoms of element X, ionized r times, and in the energy level k.

When most atoms of element X are ionized one stage higher than the emitting particles, it is convenient to express $n(\mathrm{X}^{(r)})$ in equation (3-10) in terms of $n_e n(\mathrm{X}^{(r+1)})$, using the correction factors b_k which refer to the ETE system with these values of n_e and $n(\mathrm{X}^{(r+1)})$ [S2.4]. If equations (2-25), (2-28), and (2-30) are used, we find

$$n_k(\mathrm{X}^{(r)}) = \frac{b_k g_{rk} e^{-E_{rk}/kT} n_e n(\mathrm{X}^{(r+1)})}{f_{r+1} f_e}. \tag{3-11}$$

As in the case of emission, the total energy absorbed in a line, per unit solid angle per cm^3 per second, may be obtained from the integral of $\kappa_\nu I_\nu \, d\nu$ over the line. The relevant quantity here is not the stimulated absorption by itself, but rather the difference between the stimulated absorption and the stimulated emission, since the photons produced in this latter process are coherent with the stimulating photons; that is, they have the same direction and frequency and essentially replace the photons absorbed. We assume that I_ν is relatively constant over a frequency band equal to the intrinsic line width, which is normally several orders of magnitude smaller than the Doppler width of the line profile. Then I_ν can be taken out of the integral over $d\nu$, and set equal to $I(\nu_{jk})$, its value at the central line frequency ν_{jk}. Thus we obtain, for atoms of each velocity,

$$I(\nu_{jk}) \int \kappa_\nu \, d\nu = \frac{h\nu_{jk}(n_j B_{jk} - n_k B_{kj}) I(\nu_{jk})}{c}, \tag{3-12}$$

where we have introduced the two Einstein probability coefficients for induced upward and downward transitions, respectively. Evidently in equation (3-12) stimulated emission appears as negative absorption. Replacement of I_ν by $I(\nu_{jk})$ in this equation becomes invalid when very strong absorption lines reduce $I(\nu_{jk})$ almost to zero [S4.3b].

To find a simple relationship between these three Einstein probability coefficients we use the condition that in thermodynamic equilibrium the rates at which energy is emitted and absorbed must be equal. If we set the right-hand sides of equations (3-10) and (3-12) equal to each other, with I_ν and n_k/n_j in thermodynamic equilibrium found from equations (3-4) and (2-27), respectively, we find

$$g_j B_{jk} = g_k B_{kj} = \frac{c^3}{8\pi h\nu_{jk}^3} g_k A_{kj}. \tag{3-13}$$

It may be noted that with the definitions adopted here, B_{jk} (or B_{kj}) must be multiplied by the total energy density U_ν per cm^3 per frequency interval to give the probability of an upward (or downward) transition per second per particle in the appropriate state.

a. Absorption Coefficient κ_ν

These results may now be combined to give a relatively simple formula for the absorption coefficient. We express κ_ν in the form

$$\kappa_\nu = n_j s_\nu = n_j s\phi(\Delta\nu), \tag{3-14}$$

where s is related to s_ν, the absorption cross section per particle, by

$$s = \int s_\nu \, d\nu, \tag{3-15}$$

integrated over the line. The quantity $\phi(\Delta\nu)$, whose integral over $d\nu$ equals unity, is a function of $\Delta\nu \equiv \nu - \nu_{jk}$, where ν_{jk} is again the frequency at the line center. This function, which describes the frequency distribution of s_ν, depends both on the intrinsic line width and on the distribution of Doppler shifts resulting from particle velocities. If we integrate equation (3-2) along the line of sight, using equation (3-15), we obtain

$$\tau_{\nu r} = N_j s\langle\phi(\Delta\nu)\rangle \equiv N_j s\phi_a(\Delta\nu), \tag{3-16}$$

where the "column density" N_j is defined by

$$N_j = \int n_j \, ds, \tag{3-17}$$

integrated over the pathlength ds along the line of sight. The line profile function $\phi_a(\Delta\nu)$ is an average of $\phi(\Delta\nu)$ along the line of sight; these functions will differ if, for example, large-scale macroscopic motions of the gas vary along the line of sight. If we integrate equation (3-16) over the line, making use of the normalization of $\phi_a(\Delta\nu)$, we obtain

$$N_j = \frac{1}{s}\int \tau_{\nu r}\, d\nu. \tag{3-18}$$

If the line profile function is entirely determined by Doppler broadening, then $\phi_a(\Delta\nu)\,d\nu$ equals $P(w)\,dw$, the fraction of atoms whose radial velocity lies within the velocity range dw; evidently w and $\Delta\nu$ are related by the usual Doppler formula

$$\frac{w}{c} = \frac{\Delta\nu}{\nu_{jk}}. \tag{3-19}$$

In the special case where the velocity distribution is Maxwellian, we have

$$\phi_a(\Delta\nu) = \lambda_{jk} P(w) = \frac{\lambda_{jk}}{\pi^{1/2} b} e^{-(w/b)^2}, \tag{3-20}$$

where λ_{jk}, equal to c/ν_{jk}, is the central wavelength. For thermal broadening, the velocity-spread parameter b is related to the atomic weight A by the relation

$$b = \left(\frac{2kT}{m}\right)^{1/2} = 1.290 \times 10^4 \left(\frac{T}{A}\right)^{1/2} \text{cm s}^{-1}. \tag{3-21}$$

If σ denotes the dispersion of radial velocities, and $\Delta\nu_h$ denotes the full width of $\phi_a(\Delta\nu)$ at half maximum intensity [with the corresponding Δw_h obtained from equation (3-19)], we have the relations

$$\sigma = \frac{b}{2^{1/2}}, \qquad \Delta\nu_h = \frac{2(\ln 2)^{1/2} b}{\lambda_{jk}}. \tag{3-22}$$

To express s, the integrated absorption cross section, in terms of atomic constants, we may substitute equation (3-14) in the left-hand side of equation (3-12) with equations (2-25), (2-26), and (3-13) on the right-hand side, obtaining

$$s = s_u \left\{ 1 - \frac{b_k}{b_j} e^{-h\nu/kT} \right\}, \tag{3-23}$$

where by definition

$$s_u \equiv \frac{h\nu B_{jk}}{c}\,; \tag{3-24}$$

we omit the subscripts jk from ν in the rest of this section. The quantity s_u is an integrated absorption cross section uncorrected for stimulated emission; in terms of the usual upward oscillator strength f_{jk},

$$s_u = \frac{\pi e^2}{m_e c} f_{jk} = 2.654 \times 10^{-2} f_{jk}. \tag{3-25}$$

If the downward oscillator strength is defined by the relation

$$f_{kj} = -\frac{g_j}{g_k} f_{jk}, \tag{3-26}$$

then from any level j, the sum of all the oscillator strengths obeys the usual sum rule

$$\sum_k f_{jk} = \text{Number of jumping electrons in level } j. \tag{3-27}$$

Oscillator strengths have been computed for H [3, 4, S4.2c], H_2 [5], and either computed or measured for various other interstellar atomic lines [6].

For radio molecular lines it is customary to express s_u in equation (3-23) in terms of the dipole matrix element $\boldsymbol{\mu}_{jm} = e\mathbf{r}_{jm}$, where e is the electron charge (in esu) and $\mathbf{r}_{jm}$ is the matrix element of the radius vector between states j and m. If we denote by $|\boldsymbol{\mu}_{jk}|^2$ the sum of $|\boldsymbol{\mu}_{jm}|^2$ over the degenerate states of an upper level k, we have

$$s_u = \frac{8\pi^3 \nu_{jk}}{3hc} |\boldsymbol{\mu}_{jk}|^2. \tag{3-28}$$

Values of $|\boldsymbol{\mu}_{jk}|^2$ have been tabulated for various molecular lines of astrophysical interest [7, S4.1d].

Equation (3-23) may be simplified further if $h\nu$ is much greater than kT [equation (3-9)], a generally valid assumption for the optical region of the spectrum. In this case, $\exp(-h\nu/kT)$ is very small and can be ignored in equation (3-23); stimulated emission is then negligible even in thermodynamic equilibrium.

b. Effect of Stimulated Emission on κ_ν

When $h\nu/kT$ is much smaller than unity, the exponent in equation (3-23) can be expanded, with only the linear term retained, giving

$$s = s_u \frac{h\nu}{kT}\chi, \tag{3-29}$$

where by definition

$$\chi \equiv \frac{b_k}{b_j} + \frac{kT}{h\nu}\left(1 - \frac{b_k}{b_j}\right). \tag{3-30}$$

Equation (3-29) contains two correction factors for stimulated emission. The first is the normal $h\nu/kT$ factor, which is valid in thermodynamic equilibrium; the second is an additional factor required by departures from thermodynamic equilibrium which alter in particular the role of stimulated emission.

The functional dependence of emission and absorption on physical conditions is quite different in the two limiting situations in which $h\nu$ is either much greater than or much less than kT. In both cases, the emissivity depends only on the population of the upper level, whereas the absorption depends on the population difference between the two levels. For $h\nu/kT$ large, the population of the upper level is small, and most interstellar atoms are in the ground level. The absorption from the ground level is determined by the total number of atoms in the proper stage of ionization and is otherwise independent of temperature, whereas the emission depends sensitively on the temperature and on the detailed mechanisms for populating and depopulating the upper level. For $h\nu/kT$ small, the populations of the upper and lower levels tend to be comparable, and they will be proportional to the statistical weights if the b_k factors are near unity. The emission is proportional to the number of atoms in the proper stage of ionization and is otherwise independent of temperature, and now it is the net absorption that depends sensitively on the detailed mechanisms for populating and depopulating the upper level.

Next we use equations (3-10), (3-13), (3-14), (3-23), (3-24) and also (2-27) to obtain the actual ratio of j_ν to κ_ν. If we make the plausible assumption that j_ν and κ_ν have the same frequency distribution function, $\phi(\Delta\nu)$, we obtain

$$\frac{j_\nu}{\kappa_\nu} = \frac{2h\nu^3}{c^2}\left[\frac{b_j}{b_k}e^{h\nu/kT} - 1\right]^{-1}. \tag{3-31}$$

To express j_ν/κ_ν when $h\nu/kT$ is small and the Rayleigh-Jeans law, equation (3-7), is applicable, we first rewrite equation (3-31) in a form more similar to equation (3-23) and then expand the exponents, obtaining

$$\frac{j_\nu}{\kappa_\nu} = \frac{2h\nu^3 b_k}{c^2 b_j} e^{-h\nu/kT}\left[1 - \frac{b_k}{b_j} e^{-h\nu/kT}\right]^{-1} = \frac{2\nu^2 kT}{c^2} \frac{b_k}{b_j \chi}, \tag{3-32}$$

where χ is defined in equation (3-30) and we have replaced $(1 - h\nu/kT)$ by 1, an approximation similar to that in equation (3-7). From equation (3-3) we see that for large $\tau_{\nu r}$, equations (3-31) and (3-32) give I_ν directly provided that j_ν/κ_ν is constant throughout the emitting-absorbing region.

3.3 EMISSION LINES

The theory in the preceding sections is applied here to the interpretation of those interstellar emission lines for which the emissivity is essentially that in thermodynamic equilibrium or can be computed with a simple correction applied to the ETE value [S2.4]. For several types of emission lines the detailed mechanism of excitation is not so conveniently separated from the discussion of the observations; such lines are treated in Sections 4.2 and 4.3 below. Here we discuss optical recombination lines first; these are theoretically the simplest, since self-absorption can generally be ignored. Next, certain of the data obtained from the 21-cm H emission line are discussed, while finally, maser emission lines are considered and some results derived which are entirely independent of specific excitation mechanisms.

a. Optical Recombination Lines

Faint emission in the Balmer lines is observed from extended regions around O stars [8] and from the galactic disk generally [9] and may be attributed to electron-proton recombination. The emissivity integrated over the line may be obtained from equation (3-10), with n_k expressed in terms of $n_e n_p$ by equation (3-11); for emission by atomic H, r is zero, $f_{r+1} = 1$, and $n(X^{(r+1)}) = n_p$. We consider the radiation emitted in transitions from the level of total quantum number m to the corresponding level n, replacing the subscript k with m. In this way we obtain

$$4\pi \int j_\nu \, d\nu = h\nu \alpha_{mn} n_e n_p, \tag{3-33}$$

where

$$\alpha_{mn} = \frac{b_m g_m A_{mn} e^{-E_1/m^2kT}}{f_e}$$
$$= 4.14 \times 10^{-16} \frac{b_m m^2 A_{mn} e^{158{,}000°/m^2T}}{T^{3/2}} \ \mathrm{cm}^3\,\mathrm{s}^{-1}; \tag{3-34}$$

the electron partition function f_e has been evaluated from equation (2-31). Here E_1/m^2 is the energy of level m relative to zero potential energy at infinity. The "production coefficient" α_{mn} (sometimes called an "effective recombination coefficient") is defined so that $\alpha_{mn} n_e n_p$ equals the total number of photons emitted per second per cubic centimeter in transitions from m to n [10]. Since the correction factor b_k, which allows for departures from ETE [S2.4], depends on the orbital quantum number l, a mean value of $b_k g_k A_{kj}$, averaged over all l for each m, must be used in equation (3-34).

To obtain the surface brightness, or specific intensity I_ν, equation (3-1) must be integrated along the line of sight. The absorption corresponding to the inverse of emission is entirely negligible, a result which follows from the fact that the observed I_ν for these interstellar optical lines is always many orders of magnitude less than $B_\nu(T) b_m/b_n$ [see equations (3-3) and (3-31)]. The absorption produced by grains can be taken into account separately, and the observations corrected to give I_ν in the absence of dust absorption. This corrected I_ν is then equal to the integral of $j_\nu\, ds$ along the line of sight. It is customary to express this integral in terms of the emission measure E_m, defined by

$$E_m = \int_0^L n_e^2\, ds, \tag{3-35}$$

integrated along the line of sight, here assumed to be of length L. In this equation, ds is normally measured in parsecs, with n_e in cm^{-3}. If T and n_p/n_e are taken to be constant along the line of sight, equation (3-33) now yields

$$\int I_\nu\, d\nu = h\nu\alpha_{mn} \frac{n_p}{n_e} \times 2.46 \times 10^{17} E_m. \tag{3-36}$$

Values of α_{mn} computed for recombining atoms are given in Table 4.5 [S4.2a].

These results have been used to obtain E_m from observed intensities of Hα and Hβ. The values so obtained range from 10^7 pc cm^{-6} for a line of sight through the center of the Orion nebula down to values of about 5 pc cm^{-6} (computed with $T = 8000°$K) for the widespread emission extending up to galactic latitudes of about 30° [9]. For visible H II regions around O stars, E_m is typically between 10^3 and 10^4 pc cm^{-6} [8].

The rms electron density can be determined directly from E_m if the extension, L, of the emitting region along the line of sight can be estimated. Thus in the Orion nebula the rms n_e is about 10^3 cm^{-3}, whereas a more typical value for visible H II regions is between 10 and 100 cm^{-3} [8]. For the diffuse galactic Hα emission in the galactic plane, E_m is typically between 5 and 15 pc cm^{-6}; the extension L along the line of sight is about 1 kpc if the extinction at Hα produced by dust is taken to be 1 mag per kpc. The resultant $\langle n_e^2 \rangle$ is between 0.005 and 0.015 cm^{-6}. As shown in the analysis of ionization equilibrium within H II regions [S5.1b], the total number of Balmer photons emitted per second equals the corresponding rate of emission of ultraviolet photons (at $\lambda < 912$ Å) from the stars involved; the observed diffuse Hα emission is roughly consistent with what would be expected from B stars and nuclei of planetary nebulae [11]. The value of E_m for a line of sight extending through the Galaxy, perpendicular to the plane, should equal the mean E_m observed at about 30°, or about 5 pc cm^{-6}; however, this value is uncertain observationally. In Section 3.5b these various results are compared with those obtained from thermal emission and absorption of radio waves.

b. Hydrogen 21-cm Emission Line

The 21-cm emission line of atomic H [12] is produced by radiative transitions between the two hyperfine levels of the ground ($n = 1$) electronic level. In the upper level, the electron and proton spins are parallel and $g_k = 3$, whereas in the lower level, they are antiparallel and $g_j = 1$. The frequency of the emitted radiation is 1420.406 MHz, corresponding to a wavelength of 21.11 cm; the spontaneous transition probability A_{kj} is 2.869×10^{-15} s^{-1}, giving an upward oscillator strength of 5.75×10^{-12}.

The total emissivity in the line is given by equation (3-10). For n_{H} greater than 1 cm^{-3}, the ratio of the b_k factors for these two hyperfine levels is nearly unity [S4.1b]; that is, the ratio of populations is given by the Boltzmann relation, equation (2-26), at the kinetic temperature, and LTE may be assumed. Since the energy difference between the two levels, equal to about 5.9×10^{-6} eV, is much smaller than kT, the levels are populated very nearly according to their statistical weights; as a result, three-quarters of the H atoms are in the upper state, and the integral of $j_\nu \, d\nu$ over the line is independent of T. Since the self-absorption may be important, the value of I_ν must be found from equation (3-8), where we may ignore T_{b0}. The optical thickness $\tau_{\nu r}$ is obtained from equation (3-16), with s substituted from equation (3-29); since $b_k/b_j=1$, $\chi=1$ in this equation. The line profile is entirely determined by the atomic velocity, and we may therefore express $\phi_a(\Delta\nu)$ in terms of the velocity distribution function $P(w)$ [S3.2a]. If we combine equations (3-16), (3-20), (3-25) and (3-29), we obtain

$$\tau_{\nu r}=5.49\times 10^{-14}\frac{N(\mathrm{H\ I})P(w)}{T}, \qquad (3\text{-}37)$$

where $N(\mathrm{H\ I})$ denotes the number of neutral H atoms per cm^2 in the line of sight equal to four times the column density in the lower level. As in equation (3-18), we obtain $N(\mathrm{H\ I})$ by integrating $\tau_{\nu r}T$ in equation (3-37) over all w. If we evaluate T from equation (3-8), with $T_{b0}=0$, and with T_b taken as a function of the velocity w (in cm s^{-1}), we obtain

$$N(\mathrm{H\ I})=1.823\times 10^{13}\int T_b(w)\left[\frac{\tau_{\nu r}}{1-e^{-\tau_{\nu r}}}\right]dw. \qquad (3\text{-}38)$$

For small $\tau_{\nu r}$, the factor in brackets equals unity and $N(\mathrm{H\ I})$ is proportional to the integral of the brightness temperature. Since equation (3-8) applies for a homogeneous medium, equation (3-38) must be modified if T varies with position along the line of sight.

Extensive measures of T_b have been made as a function of w, b, and l. Values of $N(\mathrm{H\ I})$ may be determined from these data if saturation is neglected and $\tau_{\nu r}\ll 1$ in equation (3-38). To determine the H-atom particle density, $n(\mathrm{H\ I})$, requires an estimate of the position of the emitting atoms. Since the interstellar gas participates in the galactic rotation, an estimate of the distance r from the Sun is obtained if w is attributed entirely to differential galactic rotation with the circular velocity computed from a dynamical model of the Galaxy. Since the velocity of the gas is known to have random components, and perhaps also systematic variations from circular velocity resulting from spiral density waves [S13.2], this assumption is certainly only approximate, but should give results that are about correct on the average. The mean hydrogen density found in this way [13] averages about 0.7 cm^{-3} in the galactic plane at values of R, the distance from the galactic center, between 7 and 11 kpc; at smaller or greater R, $n(\mathrm{H\ I})$ decreases. Within spiral arms in the solar neighborhood the mean $n(\mathrm{H\ I})$ is between 1 and 2 cm^{-3}. A galactic thickness of some 250 pc has been determined [14] from the distribution of T_b with galactic latitude, b, at a value of w corresponding to the radial velocity, v_r of a spiral arm; in determining the distance of a spiral arm from the Sun, using v_r and a dynamical model of the galaxy, we take 10 kpc as the distance of the Sun from the galactic center for consistency with most analyses [15].

The more detailed properties of the emitting H gas have been studied by means of high-resolution systematic observations of small regions of the sky [16] and by statistical analyses of available data in all directions [17]. A study of this first type, extending over some 160 square degrees at $b \approx 15°$, $l \approx 120°$, shows a rather smooth background, with n(H I) about equal to 0.2 cm^{-3}. In addition, two emission peaks are present over much of the region, with velocities differing by about 10 km s^{-1}, with a combined additional column density about one-third that of the background, and with many irregularities. The smallest of these, called "cloudlets," have values of N(H I) equal to some 2×10^{19} cm^{-2}, with radii and densities of about 3 pc (comparable with the observing beam width) and 2 cm^{-3}, respectively. Finally, some dozen larger clouds were observed, with N(H I) about 2×10^{20} cm^{-2}, and with radii and densities of some 15 pc and 4 cm^{-3}, respectively. Since all distances were uncertain, the radii and densities of cloudlets and clouds could well be in error by a factor 2.

Since the properties of the interstellar gas, like those of the Milky Way in general, probably vary appreciably from one region to another, the statistical analysis of a wider set of data provides an important addition to the detailed study of small regions. While the data can be fitted with clouds all of one type, assumed to be randomly distributed in the galactic disk, the agreement is appreciably better if two emitting components are assumed: an intercloud background of uniform density, with a temperature sufficiently high so that absorption at 21 cm can be ignored, and separate clouds assumed to be at a temperature of 60°K. The observed 21-cm emission is consistent [17] with a model in which the densities of both components in the galactic plane are constant within 15 kpc of the center and zero outside. For the intercloud medium, n(H I) in the plane is found to equal 0.17 cm^{-3}, whereas for the clouds, the corresponding particle density, averaged over the volume between the clouds as well as in the clouds, is 0.29 cm^{-3}. The former is presumably more reliable than the latter, since cold, opaque gas within clouds, even if atomic, would have relatively little effect on the observed 21-cm emission. The cloud diameters are found to lie between 1 and 13 pc.

c. Radio Maser Lines

When χ in equation (3-29) is negative, s and κ_ν are negative, and any line radiation will be amplified by the emitting atoms. The principles which determine χ are discussed in Chapter 4. Here we take χ to be a known negative number and consider the formation of the resultant emission line. The intensity I_ν in the line will be assumed small enough so that it does not affect the relative population of the upper and lower levels; that is, the maser is then said to be "unsaturated," in which case the assumption of a constant χ is a realistic one. The results of this relatively simple analysis [18] give general qualitative agreement with the data as well as physical insight into the phenomena involved.

The emitting region will be assumed spherical, with radius a. We compute the intensity emitted along a line of sight which passes at a distance d from the center of the sphere. Spontaneous emission will be ignored, and maser amplification of the incident blackbody radiation will be considered. The total pathlength through the sphere along the line of sight equals $2a(1-d^2/a^2)^{1/2}$, and the product of this with κ_ν obtained from equation (3-14) gives the total optical thickness $\tau_{\nu r}$ through the region, which we write in the form

$$\tau_{\nu r} = \left\{1 - \frac{d^2}{a^2}\right\}^{1/2} e^{-(\lambda\Delta\nu/b)^2}\tau_0, \tag{3-39}$$

where τ_0 is the value of $\tau_{\nu r}$ at the line center for a path through the center of the sphere. Thermal broadening has been assumed for the absorption coefficient profile, with equation (3-20) used for $\phi_a(\Delta\nu)$. For amplification of the universal blackbody radiation, we obtain, with use of equation (3-8),

$$T_b = 2.7° e^{-\tau_{\nu r}}. \tag{3-40}$$

For the negative values of τ_0 considered here, equations (3-39) and (3-40) give a narrowing of the outward beam both in frequency and in angle. We denote by $\Delta\nu_e$ the value of $\Delta\nu$ for $d=0$ at which T_b is reduced by a factor $1/e$ below its central value. Similarly, we let d_e be the value of d for which T_b in the line is reduced by a factor $1/e$ below its peak value. Expanding to first order in $1/\tau_0$, we obtain readily for $-\tau_0 \gg 1$,

$$\frac{\Delta\nu_e}{b/\lambda} = \frac{d_e}{2^{1/2}a} = \frac{1}{(-\tau_0)^{1/2}}. \tag{3-41}$$

For the intense and fluctuating emission lines of OH at about 18 cm and of H_2O at 1.35 cm, generally attributed to masers, the brightness temperature is normally in the range 10^{11} to 10^{15}°K [19]. If the equations above are assumed to be applicable, τ_0 should be in the range from -24 to -34. Thus the emergent beam should be narrowed in frequency to about one-fifth of the intrinsic width of $\tau_{\nu r}$ and in apparent size to about one-fourth of the angular diameter of the emitting region. While b is unknown, the observed $\Delta\nu_e$ for these maser lines often corresponds to a velocity of about 0.2 km s^{-1}, or a kinetic temperature of 100°. A small-scale turbulent velocity of about 1 km s^{-1} or a kinetic temperature of at most 2500°K in the source would be not unreasonable. The observed angular diameters are a small fraction of an arcsecond, with some sources unresolved by long-baseline interferometry (combining signals received simultaneously by radio telescopes thousands of kilometers apart) at diameters of 2×10^{-3} arcsecond [19], corresponding to linear dimensions of about an astronomical unit at a distance of 1 kpc.

From a source of given diameter a narrower emission beam results if partial maser saturation is considered [18]. Saturation appears in the outer layer if the radiation which is being amplified reaches the cloud isotropically; the unsaturated inner core then appears as a hot spot with a resultant decrease in angular diameter. An even narrower emitted beam will be produced if the source is directional, which is probably required by the maser polarization measures. The observed OH lines often show one circularly polarized component, with a resultant overall polarization of nearly 100 percent. This may be explained [20] if B is assumed to exceed about 10^{-3} G, and if gradients of $\mathbf{v}$ and $\mathbf{B}$ through the cloud are considered. With this field strength, the Zeeman components will be separated by more than the observed line width and will begin to be somewhat independent of each other. The strongest maser amplification is then achieved along those lines of sight for which the central line frequency is constant over the greatest distance; that is, for which the sum of the Doppler and Zeeman effects is nearly constant. This cancellation of Doppler and Zeeman effects can occur only for one of the Zeeman components; radiation traveling in the opposite direction will be amplified with the opposite polarization and a Zeeman shift in the opposite direction. On this picture, only radiation within a small cone of solid angles will be amplified; for the same peak T_b, the energy density will be much reduced below its value for the isotropically radiating sphere, and much greater amplification is possible before maser saturation sets in. To account for the observed number of sources, the total number of maser sources within a particular region must be correspondingly increased, and the total energy radiated is unchanged. However, the apparent angular width of each source is substantially decreased by this mechanism.

3.4 ABSORPTION LINES

Most of the observed interstellar absorption lines can be analyzed without much reference to conditions of excitation. As we have already noted, the relative population of the two H hyperfine structure levels in H I clouds is believed to obey the Boltzmann equation at the kinetic temperature of the gas. Most of the optical absorption lines are produced by atoms or molecules in their ground states. The observed information on most of these lines is discussed below. Absorption lines from excited atoms or molecules, whose interpretation depends intimately on excitation conditions, are treated in Chapter 4.

a. Hydrogen 21-cm Line

Absorption lines at 21 cm produced by interstellar H have been measured in the spectra of various extragalactic radio sources [21, 22]. Representative data are shown in Fig. 3.1, where the lower plot indicates T_b for the radio source, while the upper plot gives the average T_b for adjacent directions. To obtain the true brightness temperature of the source, which is directly proportional to $\exp(-\tau_{\nu r})$, the observed T_b in the source direction has been diminished by the brightness temperature of the 21-cm galactic emission, shown in the upper plot. The kinetic temperature T may be found directly from $\tau_{\nu r}$ and the value of T_b in adjacent directions with use of equation (3-8); both these quantities are functions of w. The column density N (H I) can then be obtained from equation (3-38).

This determination of N(H I) and T is based on two uniformity assumptions. First, N(H I) is assumed to be reasonably uniform with pointing direction, so that the column density along the line of sight to the source equals the mean value obtained for adjacent regions. Second, T is assumed constant along the line of sight; in fact, it is the harmonic mean temperature, averaged over all the H I atoms in the line of sight, that properly appears in equation (3-37). If the absorption is produced by a single large, relatively uniform cloud, these uniformity assumptions are likely to be realistic; in particular, the emission line profile observed in different regions of the cloud should match rather closely the absorption line profiles.

The values of T obtained in this way, for absorption features observed [21] with a central τ_0 exceeding 0.2, range between about 20 and 110°K, with a mean value of 71°K. In another investigation [22] the mean T found for clouds with $\tau_0 > 0.2$ was 87°. Evidently a mean T of about 80°K for these diffuse clouds seems relatively well established.

The clouds of lower τ_0 appear to be at a relatively higher temperature, and the "uniform" background seen in 21-cm emission, discussed in Section 3.2, appears to be at a higher temperature still. A systematic increase of T from 50 to 300°K has been observed [23], on the average, as τ_0 decreases from 2 to 0.01, with a lower limit of about 1000°K for the

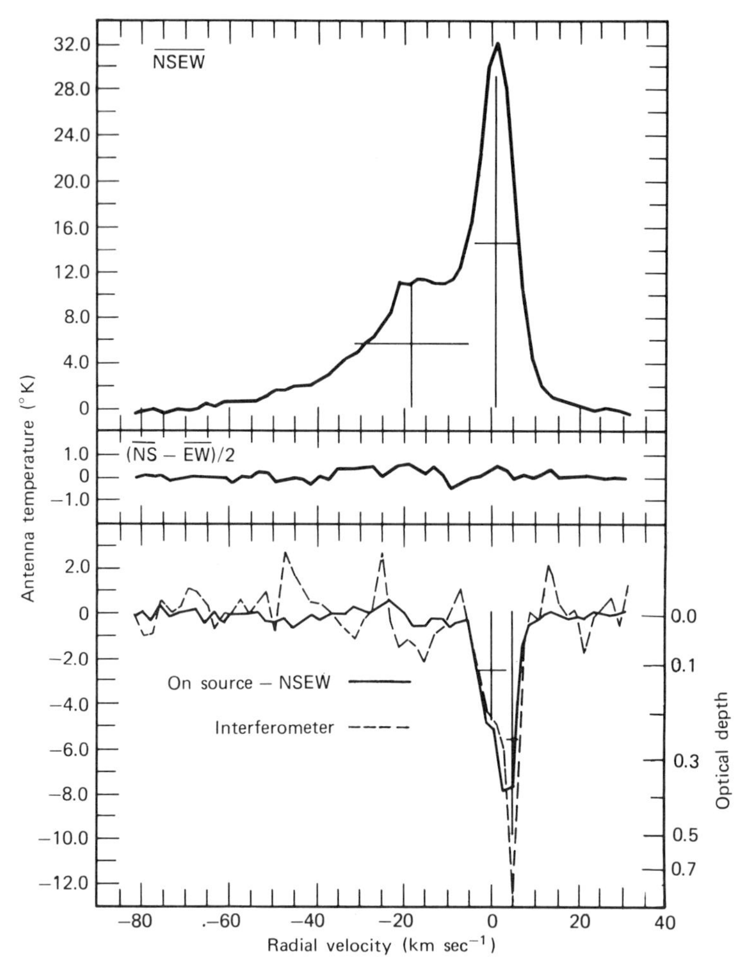

Figure 3.1 Sample 21-cm emission and absorption line profiles [22]. The lower curve gives the difference of measured antenna temperature (equal to 0.8 T_b) between the 21-cm absorption line and the continuous emission in the extragalactic radio source 1610-60, plotted against the radial velocity in the local standard of rest; the dashed curve shows similar data obtained with an interferometer. The right hand vertical scale shows the optical depth computed from the line profiles. The upper plot gives the average antenna temperature of the emission observed one beamwidth (15′) away from the source in the north, south, east, and west directions, while the middle plot gives the difference between the means of the north–south and east–west measures. The line profiles have been fitted by Gaussian components at the indicated velocities; the horizontal lines show for each component the full width at one-half of the maximum τ or antenna temperature.

broad emission which is presumably caused by the uniform background and which shows no associated absorption. Values of T up to 8000°K have been measured [24] in a few weak absorption components seen in the strongest radio sources. These high temperatures are somewhat suspect because of the rapid variation of T_b at each w with direction, which has been observed interferometrically, and which amounts typically to a factor 2 across the 5-arcminute diameter of the Cas A source [25]. A cloud that did not cover the source would produce only weak absorption even though its temperature were low. However, the scale of these fluctuations is apparently greater than 0.3 pc [26], and 21-cm emission and absorption measures with this resolution confirm a large spread of temperatures [27]. Other observations which also suggest a high temperature for the gas outside the observed clouds are summarized together in Section 11.1.

Other physical properties of the relatively opaque clouds can also be determined from the 21-cm data. Again we consider only those absorption components for which τ_0, the optical depth at the line center, exceeds 0.2. The mean value of the column density N(H I) observed for single clouds toward external galaxies is about 3×10^{20} cm^{-2} [22], roughly the same as for the conspicuous emitting clouds discussed in the previous section. If $\langle N(\text{H I})\rangle$, the mean N(H I) for all galactic sources (mostly with $b<2°$), is divided by the mean pathlength $\langle L\rangle$ of 2.6 kpc to these sources, a value of 1.2×10^{21} cm^{-2} kpc^{-1} is found for $T=80°$ [28]; the corresponding n(H I) is 0.4 cm^{-3}, comparable with the value of about 0.3 cm^{-3} for cold clouds obtained from the emission data [S3.3b]. If N(H I) per cloud is again taken to be 3×10^{20} cm^{-2}, then k, the number of clouds per kpc, equals 4.

The internal dispersion of radial velocities within clouds showing appreciable 21-cm absorption (mostly with $\tau_0>0.2$) averages about 1.8 km s^{-1} [22], while the external dispersion of velocities among such clouds is 6.4 km s^{-1} [29]. In contrast, the emission components which show no measurable absorption are substantially wider, with a minimum width corresponding to an internal radial velocity dispersion of about 8 km s^{-1} [30].

The 21-cm absorption components in strong radio sources have been used for measurements of the Zeeman effect and thus of the magnetic field within clouds. For a H atom in the ground state, the splitting of the hyperfine structure corresponds to a normal Zeeman triplet, and $\Delta\nu_B$, the frequency difference between the two components of opposite circular polarization, becomes

$$\Delta\nu_B=\frac{eB_{\parallel}}{2\pi m_e c}=2.80\times10^6 B_{\parallel}\ \text{Hz}, \tag{3-42}$$

where $B_{\parallel}$ denotes the component of **B**, in Gauss, along the line of sight.

This shift is much less than the line width of about 2×10^4 Hz, but it produces slight circular polarization on each side of the absorption component. The measures give mostly upper limits on $B_{\parallel}$, but in a few absorption components, fields in the range from 2 to 70 μG have been measured [31]. Estimates of n(H I) in these clouds, based on the measured angular diameters and on N(H I), give values ranging from about 10 cm^{-3} for the low-B clouds up to nearly 10^3 cm^{-3} for the clouds with the highest field. If $B_{\parallel}$ is assumed to vary as $n(\text{H I})^p$, a value of p about equal to two-thirds would be consistent with the data [31] for n_{H} greater than roughly 5 cm^{-3}, with a much smaller slope for low n_{H}.

b. Wide H and H_2 Optical Lines

When the column density of absorbing atoms is sufficiently great, the width of an absorption line will substantially exceed the value corresponding to the spread of particle velocities in the line of sight. In this case, $\phi_a(\Delta\nu)$ outside the line core is determined by the intrinsic width of the absorption line, and for absorptions from the ground state j (or an excited metastable state) up to level k, we have

$$\phi_a(\Delta\nu) = \phi(\Delta\nu) = \frac{\delta_k/\pi}{\delta_k^2 + (\Delta\nu)^2}, \tag{3-43}$$

where δ_k is given by

$$\delta_k = \frac{1}{4\pi} \sum_m A_{km}, \tag{3-44}$$

summed over transitions to all lower states. The quantity $4\pi\delta_k$ is sometimes called a "radiation damping" constant. The regions of the line profile for which equation (3-43) is applicable are known as "radiation damping" wings of the line. Within the line core, equation (3-43) is inapplicable, but if damping wings are present, I_ν will in any case be essentially zero in this central region. The equation for I_ν in the damping wings now becomes, with use of equations (3-3), (3-16), (3-23), (3-25), and (3-43),

$$\frac{I_\nu}{I_\nu(0)} = \exp\left\{ -\frac{e^2\lambda^4 f_{jk} N_j \delta_k}{m_e c^3 (\Delta\lambda)^2} \right\}, \tag{3-45}$$

where $\Delta\lambda$ is the wavelength distance from the line center. For these interstellar optical lines, stimulated emission is negligible, and s equals s_u in equation (3-23).

Equation (3-45) has been used to determine the column densities of H I and H_2 from the profiles of $L\alpha$ at 1215.7 Å, and of the strongest H_2 lines in the Lyman bands shortward of 1110 Å; overlapping of H_2 line wings must be considered. The $L\alpha$ data [32] give mean particle densities along the line of sight ranging from less than 10^{-2} cm^{-3} up to about 2 cm^{-3}. For 10 unreddened stars, with $E_{B-V} \leqslant 0.01$, the mean n(H I) is about 0.08 cm^{-3}, with some values up to 0.3 cm^{-3}. This same mean is about an upper limit for 11 stars within 100 pc [33]. The correlation between N(H I) found from these $L\alpha$ profiles and E_{B-V} measured photoelectrically is shown in Fig. 1.1, and is discussed in Section 7.2a, where an overall mean $n_{\rm H}$ within 1000 pc from the Sun is found, exceeding by more than a factor 10 the value within 100 pc.

Column densities for H_2 found from corresponding profile measures of strong H_2 lines show [34] that the fraction of hydrogen which is molecular along these lines of sight is between 10 and 70 percent, with a tendency toward higher values in the more reddened clouds. In each of several Lyman bands, the lines from the two lowest rotational levels, $J=0$ and 1 (parahydrogen and orthohydrogen, respectively [S4.3b]) are the ones which show damping wings, since the column densities $N(0)$ and $N(1)$ in these two levels are much the greatest. When these H_2 lines are strong, the ratio of populations in these two levels is believed to follow the Boltzmann formula, equation (2-26) [S5.3b], and the ratio $N(1)/N(0)$ can be used to determine the kinetic temperature. For 13 stars with these strong H_2 lines, the mean temperature obtained is given by

$$T = 81 \pm 13\,^{\circ}\mathrm{K}, \tag{3-46}$$

where the dispersion of individual values is indicated. This result appears to be in satisfactory agreement with the mean temperature of about 80°K found from the comparison of absorption and emission in the 21-cm line [S3.4a]. The mean value of N(H I) for the lines of sight averaged in equation (3-46) is about 5×10^{20} cm^{-2}, not much greater than the mean value of 3×10^{20} cm^{-2} for single absorbing clouds determined from the 21-cm data [S3.4a].

c. Narrow Optical Lines

When the number of absorbing particles is insufficient to produce damping wings, interstellar absorption lines are generally relatively narrow, with widths corresponding to velocity dispersions of only a few km s^{-1}, similar to the 21-cm absorption components. This narrowness is helpful in radial velocity measurements and can also be of decisive importance observationally in distinguishing interstellar lines from the rotationally broadened stellar absorption features. However, the narrow width complicates intensity studies, since the spectroscopic resolution is usually insufficient to measure the line profile. Before describing any of the observational results, we give first the techniques that have been developed for analyzing line intensities in this situation.

When the true line profile cannot be fully resolved it is customary to measure the "equivalent width," given in wavelength units, by

$$W_\lambda = \int \left[1 - \frac{I_\nu}{I_\nu(0)} \right] d\lambda = \frac{\lambda^2}{c} \int \left[1 - e^{-\tau_\nu} \right] d\nu, \tag{3-47}$$

where equation (3-3) has again been used and the integral extends over the line with λ representing the wavelength at the line center. The quantity $I_\nu(0)$, the value which I_ν would have if the line were absent, is the so-called "continuum intensity." It is readily seen that for an isolated line W_λ, which measures the fraction of energy removed from the spectrum by the line, remains unchanged if the apparent I_ν, measured with low resolution, is used instead of the true I_ν.

The relationship between W_λ and $N_j f_{jk}$, the effective number of absorbing atoms in the line of sight, is obtained relatively simply when τ_ν is small across the line. The exponent in equation (3-47) may then be expanded, and from equations (3-16), (3-23), and (3-25) we obtain

$$\frac{W_\lambda}{\lambda} = \frac{\pi e^2}{m_e c^2} N_j \lambda f_{jk} = 8.85 \times 10^{-13} N_j \lambda f_{jk}, \tag{3-48}$$

where $N_j \lambda$ has the dimension cm^{-1}; the stimulated emission factor in equation (3-23) has again been set equal to unity. Lines for which equation (3-48) is applicable are said to be on the "linear" part of the curve of growth. Physically, the curve of growth is linear as long as each atom is exposed to essentially the full continuum intensity, $I_\nu(0)$, so that the total rate of energy absorption varies linearly with the number of atoms.

More generally the curve of growth gives W_λ/λ as a function of $N_j \lambda f_{jk}$. This relationship depends very closely on $\phi_a(\Delta\nu)$ and thus on the velocity distribution function. If equation (3-20) is assumed, equation (3-47) may be written in the form

$$\frac{W_\lambda}{\lambda} = \frac{2bF(\tau_0)}{c}, \tag{3-49}$$

where by definition

$$F(\tau_0)=\int_0^{+\infty}\left\{1-\exp\left(-\tau_0 e^{-x^2}\right)\right\}dx, \tag{3-50}$$

and where τ_0, the value of τ_ν at the line center, is given by

$$\tau_0=\frac{N_j s\lambda}{\pi^{1/2}b}=\frac{1.497\times10^{-2}}{b}N_j\lambda f_{jk}. \tag{3-51}$$

The same equations have been used here as in the derivation of equation (3-48). Series expansions [35] recover equation (3-48) for small τ_0, and give for large τ_0 the asymptotic result

$$F(\tau_0)=(\ln\tau_0)^{1/2}. \tag{3-52}$$

When equation (3-52) is applicable, an equivalent width is said to be on the "flat" part of the curve of growth; that is, $\phi_a(\Delta\nu)$ is decreasing so rapidly at the edges of the line that large increases in $N_j f_{jk}$ increase W_λ only very slighnly. Numerical values of $F(\tau_0)$ are given in Table 3.1.

For sufficiently strong lines the damping wings become important. If the resolution is inadequate to measure the profile of these wings, the equivalent width may again be used. When these wings dominate, we obtain by the same methods as before

$$\frac{W_\lambda}{\lambda}=\frac{2}{c}\left(\lambda^2 N_j s\delta_k\right)^{1/2}. \tag{3-53}$$

Lines for which equation (3-53) is applicable are said to be on the "square-root" section of the curve of growth. Values of W_λ/λ have been

Table 3.1. Curve of Growth for a Maxwellian Velocity Distribution

τ_0	0.000	0.10	0.20	0.30	0.40	0.50	0.60	0.80
$F(\tau_0)$	0.000	0.086	0.165	0.240	0.309	0.374	0.435	0.545
τ_0	1.0	1.2	1.4	1.6	2.0	3.0	4.0	6.0
$F(\tau_0)$	0.643	0.728	0.804	0.872	0.986	1.188	1.320	1.483
τ_0	10	20	30	40	60	100	1,000	10,000
$F(\tau_0)$	1.66	1.86	1.97	2.04	2.14	2.26	2.73	3.12

computed [36] for intermediate situations where both radiation damping and Doppler broadening must be considered, giving a "Voigt profile" instead of the simpler Maxwellian.

We turn now to the observations of narrow interstellar absorption lines. The extensive ground-based data on interstellar Na I and Ca II absorption provide information on the velocity distribution of the interstellar gas [35]. Many separate components are seen in these lines, with radial velocities for some Ca II components as high as 100 km s^{-1}. The strongest components, with velocities up to about 10 km s^{-1} in the local standard of rest, generally correlate reasonably well with 21-cm emission components, but for the clouds of higher velocity the correlation is poor, with often no 21-cm emission observable at the velocity of a Ca II cloud. If a correction is made for blending, the number k of components per kpc has been estimated as between 8 and 10 [37], and with more recent higher resolution [38, 39] the number of components has been somewhat increased. Most of these components have presumably much less material in the line of sight than the absorbing H I clouds with $\tau_0 \geqslant 0.2$, for which a value of 4 per kpc was obtained above for k. If the clouds with weak Na I absorption are excluded, the dispersion σ_e of radial velocities between different clouds is about the same 6 km s^{-1} [29] obtained from the 21-cm data.

The effective thickness $2H$ of the gaseous galactic disk has been obtained [40] from the measured W_λ of the K line of Ca II at 3933.7Å. Since this line is relatively unsaturated, W_λ is nearly proportional to N(Ca II). Dividing N(Ca II) for a line of sight through the Galaxy, pole to pole, by the mean n(Ca II) in the galactic plane gives $2H = 240$ pc, in good agreement with the value obtained from 21-cm emission [S3.3b]. Some clouds producing measurable Ca II absorption are at much greater distances from the galactic plane; on the average, a line perpendicular to this plane intersects from 0.25 to 0.5 such clouds beyond 750 pc and with a radial velocity exceeding 24 km s^{-1} [41].

The composition of several H I clouds has been determined [34] from measures of W_λ for many lines, mostly in the ultraviolet. The most detailed results are for the line of sight to ζ Oph ($E_{B-V} = 0.32$). A few of the interstellar lines in this star, including some lines of the dominant singly ionized atoms of Mg, P, Ni, and Fe, are on the linear part of the curve of growth, giving column densities directly from the adopted f_{jk}. For other lines a curve of growth must be used. Since there is no reason to expect that $\phi_a(\Delta\nu)$ is Maxwellian, the theoretical curve of growth discussed above is not necessarily realistic. It is necessary to construct empirical curves of growth, combining in one group those atoms which might be expected to show the same mean velocity distribution function. For example, atoms of the dominant ionization species in H I regions, such as N I, O I, Mg II, and Si II, would be expected to have the same $\phi_a(\Delta\nu)$, provided that their relative abundances were the same in all clouds. However, neutral atoms such as Mg I and Si I, whose relative abundances should be proportional to the density, will have different abundances relative to their corresponding ions in different clouds and consequently a different $\phi_a(\Delta\nu)$ from that of Mg II and Si II. Within each group the values of $\log(W_\lambda/\lambda)$ from a variety of lines absorbed by a particular atom can be plotted against $\log(f\lambda)$ to give a section of the curve of growth. Different sections are then shifted horizontally to agree with corresponding sections found from other atoms in the same group; N_j is then determined from lines on the linear section. Figure 3.2 shows the empirical curves of growth obtained in this way for ζ Oph, indicating the difference between different groups of elements; for the triangles, representing Mg I, Si I, and neutral atoms of similar elements, the curve of growth clearly does not correspond to the theoretical curves for a Maxwellian $\phi_a(\Delta\nu)$, but it may be reproduced by

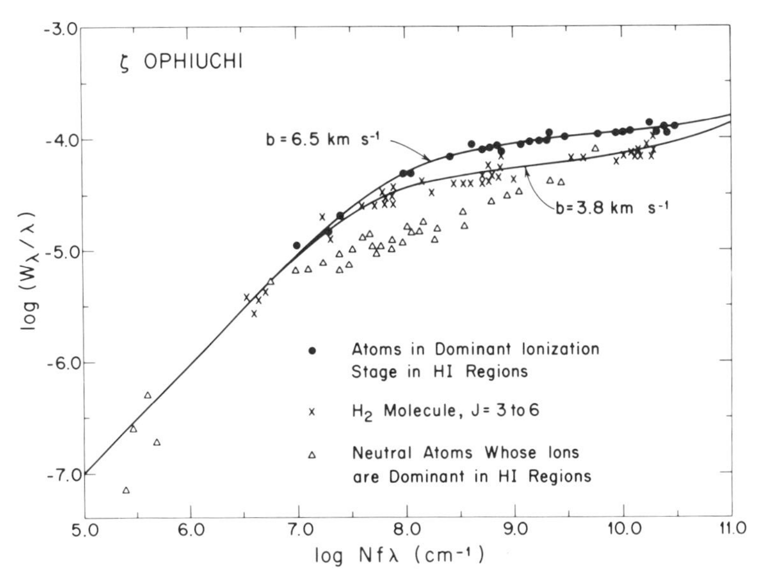

Figure 3.2 Curves of growth for interstellar lines in ζ Oph [34]. The filled circles represent lines produced by Al I, Ar I, Mg II, Si II, S II, and Fe II, representing for each element the dominant stage of ionization in an H I region. The triangles represent C I, Na I, Mg I, S I, K I, and Fe I; these are neutral atoms of elements which are mainly singly ionized in H I regions. The crosses represent H_2 Lyman lines from the levels with rotational quantum number J between 3 and 6. The two solid lines represent theoretical curves for a Maxwellian velocity distribution with the indicated values of b, equal to $2^{1/2}$ times the dispersion of radial velocities.

several clouds with different radial velocities and with widely different column densities.

While some of the observed depletions for ζ Oph in Table 1.1 are uncertain, the general tendency toward selective depletion of some elements seems well established. The atoms missing from the gas constitute about 1.5 percent of the total mass density in H and He. In unreddened stars and in high-velocity clouds there is some indication that the depletion is substantially less. In some six unreddened stars, with $E_{B-V} \leqslant 0.03$, the atoms of N and O were depleted relatively little on the average (at most 0.2 in the logarithm), with the missing atoms of all elements heavier than O amounting to only 0.2 percent of the total mass [42]. Diminished depletions of Mg, Si, and Fe have been observed directly in a few high-velocity clouds [43], with relative abundances about equal to their cosmic values for velocities of about 100 km s^{-1}.

Supporting evidence for a change of relative abundance with cloud velocity is provided by extensive measures of the Na I/Ca II ratio. Average values [44] of N(Na I)/N(Ca II) for components with radial velocities in different groups are given in Table 3.2. The scatter of this ratio for low-velocity clouds is very great, with values up to 90 observed in ζ Oph. On the other hand, this ratio is less than 0.1 for four of the 18 stars with $|v| \geqslant 20$ km s^{-1}. This scatter appears to be correlated with E_{B-V}; in three relatively unreddened stars with $E_{B-V} \leqslant 0.03$, the Na I-to-Ca II ratio in the weak, low-velocity lines seen was found to average less than 0.22 [45], a typical value for the high-velocity clouds in Table 3.2.

Table 3.2. Dependence of Sodium-Calcium Ratio on Cloud Velocity

Velocity range (km s^{-1})	0–9	10–19	20–29	30–39	>40
N(Na I)/N(Ca II)	5.4	1.6	0.32	0.24	0.65
Number of stars	61	16	4	3	11

Attempts have been made to explain these trends with changing ionization equilibria resulting from collisional ionization, with no depletion considered, but the ultraviolet results on the depletion of other elements suggest that in fact Ca is strongly depleted in the more opaque, low-velocity clouds. If photon ionization is assumed, N(Na I)/N(Ca II) should be independent of n_e [S5.2a] and photon flux. Hence the variations shown in Table 3.2 may be attributed to variation of relative depletion with cloud velocity.

Column densities obtained from interstellar lines interpreted by means of curves of growth have been used to discuss rotational excitation of H_2 molecules [S4.3b], the ionization equilibria of H I clouds [S5.2a], and H_2 dissociation equilibria [S5.3a].

The highly ionized O VI atom has absorption lines from the ground state at 1031.9 and 1037.6 Å. Broad interstellar absorption features at these wavelengths, with b in the range from 15 to 50 km s^{-1}, have been observed in most early-type stars [34]. If the line width is attributed to thermal broadening and equations (3-20) and (3-21) are used, values of T between 2×10^5 and 3×10^{6}°K are found for various stars. The mean value of n(O VI), averaged over all lines of sight, is 1.7×10^{-8} cm^{-3}. Only traces of N V absorption have been found for these components; the ratio N(N V)/N(O VI) is less than 2.5×10^{-2} in some cases [46]. These results may be used to set limits on the temperature and density of the collision-ionized gas that presumably produces these lines [S5.2b], and that may occupy much of the volume of the galactic disk [S11.3].

3.5 CONTINUOUS EMISSION AND ABSORPTION BY THERMAL ELECTRONS

At radio wavelengths and, for high temperatures, at optical wavelengths the continuous emission from atoms and molecules is dominated by free–free transitions of electrons in encounters with positive ions; the resulting radiation is known as bremsstrahlung. According to classical theory, in one such encounter an electron emits a single electromagnetic pulse, with no oscillations in E. The Fourier spectrum of such a pulse is independent of frequency for low frequencies. Consequently, the emission coefficient j_ν for this process is nearly independent of frequency except for an exponential cutoff resulting from the Maxwellian distribution of electron velocities. The usual formula for j_ν for free–free transitions, with electrons assumed to interact with ions of charge Z_ie and particle density n_i, is

$$\begin{aligned} j_\nu &= \frac{8}{3}\left(\frac{2\pi}{3}\right)^{1/2} \frac{Z_i^2 e^6}{m_e^{3/2} c^3 (kT)^{1/2}} g_{ff} n_e n_i e^{-h\nu/kT} \\ &= 5.44\times10^{-39} \frac{g_{ff} Z_i^2 n_e n_i}{T^{1/2}} e^{-h\nu/kT} \text{ erg cm}^{-3}\text{ s}^{-1}\text{ sr}^{-1}\text{ Hz}^{-1}. \end{aligned} \tag{3-54}$$

Here g_{ff}, which varies slowly with frequency ν, is the Gaunt factor for free–free transitions. For radio frequencies ν may be assumed to be small compared to kT/h but large compared to the plasma frequency, given in equation (3-59); for these conditions we have [47]

$$g_{ff} = \frac{3^{1/2}}{\pi}\left\{\ln\frac{(2kT)^{3/2}}{\pi e^2 Z \nu m_e^{1/2}} - \frac{5\gamma}{2}\right\}$$
$$= 9.77\left(1 + 0.130\log\frac{T^{3/2}}{Z\nu}\right), \tag{3-55}$$

where γ is Euler's constant, equal to 0.577; in the numerical evaluation, T and ν are in units of degrees Kelvin and Hertz, respectively. For radio wavelengths of practical interest, g_{ff} varies as $T^{0.15}\nu^{-0.1}$ [48].

The rate at which free-bound radiation is emitted, as a result of electron–ion recombination, may be computed from the recombination coefficients [S5.1a]. The corresponding emissivity for such transitions, down to a level of principal quantum number n, is smaller than j_ν above by an approximate factor $-4E_n/kT$, where kT much exceeds the binding energy $-E_n$ of level n. While free-bound transitions contribute importantly to optical spectra from nebulae, they are unimportant for most radio frequencies, since $-E_n$ is less than $h\nu$ by the kinetic energy, $m_e w^2/2$, and $h\nu/kT$ is usually very small.

The total amount of energy radiated in free–free transitions per cm³ per second, which we denote by ε_{ff}, is obtained by multiplying equation (3-54) by $4\pi\, d\nu$ and integrating over all ν. We obtain

$$\varepsilon_{ff} = \left(\frac{2\pi kT}{3m_e}\right)^{1/2}\frac{2^5\pi e^6}{3hm_e c^3} Z_i^2 n_e n_i \langle g_{ff}\rangle$$
$$= 1.426\times 10^{-27} Z_i^2 n_e n_i T^{1/2}\langle g_{ff}\rangle \text{ ergs cm}^{-3}\text{ s}^{-1}. \tag{3-56}$$

The values of the mean Gaunt factor $\langle g_{ff}\rangle$, shown in Table 3.3 [49], are between 1.1 and 1.5 for conditions of interest.

Table 3.3. Mean Gaunt Factor for Free–Free Emission

$\log T\,(^\circ\mathrm{K})/Z_i^2$	4.0	4.25	4.5	4.75	5.0	5.25
$\langle g_{ff}\rangle$	1.26	1.31	1.35	1.39	1.42	1.43
$\log T\,(^\circ\mathrm{K})/Z_i^2$	5.5	5.75	6.0	6.25	6.5	6.75
$\langle g_{ff}\rangle$	1.44	1.43	1.41	1.37	1.32	1.27
$\log T\,(^\circ\mathrm{K})/Z_i^2$	7.0	7.25	7.5	7.75	8.0	8.25
	1.24	1.21	1.19	1.17	1.15	1.14

Corresponding to the emissivity j_ν in free–free transitions, there must be a corresponding absorption coefficient κ_ν according to Kirchhoff's law, equation (3-5). To obtain κ_ν for radio waves, we substitute equations (3-54) and (3-4) into equation (3-5); or κ_ν may be computed directly [50] and multiplied by the stimulated emission factor $h\nu / kT$, giving

$$\begin{aligned}\kappa_\nu &= \frac{4}{3}\left(\frac{2\pi}{3}\right)^{1/2} \frac{Z_i^2 e^6 n_e n_i g_{ff}}{c m_e^{3/2} (kT)^{3/2} \nu^2} \\ &= 0.1731\left\{1 + 0.130 \log \frac{T^{3/2}}{Z\nu}\right\} \frac{Z_i^2 n_e n_i}{T^{3/2} \nu^2} \text{ cm}^{-1},\end{aligned} \tag{3-57}$$

where equation (3-55) for g_{ff} has been used.

a. Free–Free Radio and X-ray Emission

The emissivity for free–free emission depends on n_e^2 in the same way as does the line emission discussed in Section 3.2; hence the observed intensity may be used to determine E_m in equation (3-35) and thus n_e. In the analysis of radio waves from the galactic disk, a primary observational uncertainty is the separation between thermal emission on the one hand and nonthermal synchrotron radiation on the other. The greater steepness of the synchrotron spectrum produced by relativistic electrons, with the resultant j_ν varying about as $\nu^{-0.6}$, is of basic importance in this separation.

The free–free radio emission from individual H II regions can give more accurate values of E_m than do the optical measures of recombination lines, since the former require no correction for extinction by grains. A comparison [8] of the values of n_e obtained from the radio and visual measures shows that the mean ratio of these values, averaged over 19 H II regions, equals 1.4, with an average deviation of 0.5; individual values of n_e range from 5 to 700 cm^{-3}. The agreement between these two sets of values is relatively good, considering the uncertainties.

At low radio frequencies, at wavelengths of about 1 meter, H II regions become opaque according to equation (3-57), and from equation (3-8) we see that the observed T_b should approach the kinetic temperature T. Analysis of the observations is complicated by the relatively poor directivity at these long wavelengths, as well as by the background of galactic synchrotron radiation. The values of T obtained [10] for conspicuous H II regions range from 4000 to 10,000°K, but generally average from 7000 to 8000°K, in good agreement with other determinations [S11.1].

Radio emission is also observed from dense ionized gas in compact H II regions, with E_m in the range from 10^6 to 10^8 pc cm^{-6} [51]. Since the diameters are generally a small fraction of a parsec, the corresponding values of n_e exceed 10^4 cm^{-3}. In accordance with theoretical expectations [S5.1c], the optical thickness of the dust in these regions is high, and with additional obscuration by dust in surrounding H I clouds, these compact H II regions generally cannot be seen in visible light.

The overall distribution of n_e in the galactic plane has been evaluated from the general intensity of the continuous emission at radio frequencies [52, 53] which are high enough so that self-absorption can be ignored. The mean n_e^2 at 10 kpc from the galactic center appears to be about 0.1 cm^{-6}, with local concentrations rising to higher values. In the solar neighborhood the ultraviolet flux from known early-type stars [S5.1b] is only about one-fourth of that needed to keep hydrogen ionized with this value of $\langle n_e^2 \rangle$ [54]; this discrepancy may be attributed in part to heavily obscured stars in compact H II regions and in part to the generally irregular distribution of interstellar matter. The effective thickness $2H$ of the ionized emitting layer appears to be about the same as for neutral H, for which 250 pc was found [S3.3b]. This value presumably refers to the relatively dense H II regions around the bright early-type stars, since these regions contribute predominantly to n_e^2 at low galactic latitude.

Free–free emission is also important in the X-ray region. For H and He, equation (3-54) gives the bulk of the emissivity, but for the other ions, line emission and, to a lesser extent, radiative recombination contribute significantly. Detailed computations [55] of j_ν for a hot gas, in collisional ionization equilibrium, have been used to interpret the X-rays observed from supernova remnants within the spectral range from about 0.2 to 10 keV [56]. Values of the temperature T obtained from the analysis are given in Table 1.2, together with corresponding parameters deduced from the theoretical models of expanding supernova shells [S12.2].

b. Continuous Absorption of Radio Sources

Information on the interstellar n_e^2 is also provided [57] by the absorption of extragalactic nonthermal radiation, whose spectrum at low ν shows a dip at low galactic latitudes, consistent with κ_ν varying as $1/\nu^2$ [58], as indicated in equation (3-57). The pathlength through a disk stratified in layers parallel to the galactic plane varies as H csc b, where b is again the galactic latitude and H is half the effective thickness. Radio absorption measures at $b=60°$ give an emission measure which, for $T=8000°$K, is about 6 cm^{-6} pc for a line of sight extending through the Galaxy, perpendicular to the galactic plane in the solar neighborhood; for $b=5°$, the corresponding E_m is 12 cm^{-6} pc. The value for high b agrees with the corresponding but less certain value of 5 cm^{-6} pc obtained from the diffuse galactic Hα emission

[S3.3a]. Although these two surveys observed different parts of the sky, the agreement is consistent with the assumption that similar emitting regions are involved. Also, since the values of E_m obtained from radio absorption and optical emission vary about as $T^{1.35}$ and $T^{0.85}$, respectively, a temperature not grossly different from the assumed value of 8000°K seems indicated.

Without specifying values for the absorption at radio frequencies per unit distance along the galactic disk, a mean effective thickness $2H$ of this free electron galactic layer cannot be determined directly from these data. If $2H$ is assumed to lie between 250 and 750 pc (see the discussion at the end of next section), an emission measure pole to pole of 5 cm^{-6} pc yields a mean n_e^2 between 0.02 and 0.007 cm^{-6}, about equal to the range of values obtained from Hα emission [S3.3a] and much less than the value 0.1 cm^{-6} obtained from the free–free emission data. This difference is to be expected, in view of the dominant contributions to the observed radio emission made by the dense and conspicuous H II regions, which have been deliberately excluded from the measures of diffuse galactic Hα radiation and which because of their spotty distribution should have less influence on the average absorption of radio waves, especially at high b.

3.6 REFRACTION BY FREE ELECTRONS

When a radio wave passes through a gas containing free electrons, the phase velocity of the wave is altered. Hence variable gradients of the electron density n_e produce irregular refraction and consequent scintillation of pulsar signals. The change of group velocity with frequency produces a dispersion of the radio emission peaks observed from pulsars. In the presence of a magnetic field, double refraction appears, with a resultant rotation of the plane of polarization if the radio source is at least partially plane polarized. These processes [58] are analyzed below; the analysis of the data in terms of physical theory gives a number of important results on the characteristics of the interstellar gas.

a. Dispersion of Pulsar Signals

When electromagnetic waves propagate through a gas of free electrons, the phase velocity V, equal to $\lambda\nu$, is given by [59]

$$m = \frac{c}{V} = \left\{ 1 - \frac{\nu_p^2}{\nu^2} \right\}^{1/2}, \tag{3-58}$$

where m is the index of refraction, and ν_p, the "plasma frequency," is defined by

$$\nu_p = \left\{ \frac{n_e e^2}{\pi m_e} \right\}^{1/2} = 8.97 \times 10^3 n_e^{1/2}\ \text{s}^{-1}. \tag{3-59}$$

The velocity with which a disturbance will propagate is the group velocity V_g, equal in general to

$$\frac{1}{V_g} = \frac{d}{d\nu}\left(\frac{\nu}{V}\right). \tag{3-60}$$

With some algebra, equations (3-58) and (3-60) yield

$$V_g = \frac{c^2}{V} = c\left\{1 - \frac{\nu_p^2}{\nu^2}\right\}^{1/2}. \tag{3-61}$$

If a pulse travels over a distance L, the travel time t will equal approximately

$$t = \frac{L}{c} + \frac{e^2}{2\pi m_e c} \times \frac{D_m}{\nu^2}, \tag{3-62}$$

where the "dispersion measure" D_m is defined by

$$D_m = \int_0^L n_e\, ds. \tag{3-63}$$

In deriving equation (3-62), the square root in equation (3-61) has been expanded, and only the first term in ν_p^2/ν^2 is retained.

The radio signals observed from pulsars show a progressive delay as ν is decreased, and measurement of $dt/d\nu$ gives directly a value for D_m, normally expressed in units cm^{-3} pc. While distances of individual pulsars are known for only a few cases, limits on the distances of about a dozen of these objects with measured D_m have been determined from the presence or absence of 21-cm absorption features produced by known spiral arms; these data indicate [29, 60] that $\langle n_e \rangle$ in the galactic disk, outside conspicuous H II regions, is about 0.03 cm^{-3}.

This result can be reconciled with mean values of n_e^2 of about 0.010 cm^{-6} [S3.5b] if the clumpy distribution of the H II gas is considered. For example, if n_e in the H II regions about B stars is about 0.3 cm^{-3}, and 10 percent of the volume of the galactic disk is occupied by such regions, then $\langle n_e \rangle = 0.03\ \text{cm}^{-3}$, and $\langle n_e^2 \rangle = 0.009\ \text{cm}^{-6}$.

b. Interstellar Scintillation

The refraction of radio waves in accordance with equation (3-58) has been used also to give statistical information on small-scale density inhomogeneities in the ionized gas. We consider the value of θ_s, the rms scattering angle which results when an electromagnetic wave travels a distance z through a region in which the index of refraction m varies spatially with an rms amplitude Δm. We denote by a the mean distance over which m is nearly constant. (More precisely, a may be set equal to the distance at which the autocorrelation coefficient of Δm falls to $1/e$.) Each region of size a will produce a phase change for radiation passing through, with a rms value of about $2\pi a \Delta m/\lambda$. Over a distance z, the number of inhomogeneities will be about z/a. The different phase changes produced by successive inhomogeneities will add quadratically. If we let $\Delta\phi$ denote the resultant rms change of phase along a path of length z,

$$\Delta\phi \approx \left(\frac{z}{a}\right)^{1/2} \frac{2\pi a \Delta m}{\lambda}. \tag{3-64}$$

To compute the rms scattering angle θ_s, we now adopt the "thin phase screen" model [61], in which the scattering is computed as though the phase change $\Delta\phi$ were produced by a thin screen, at a distance $z/2$ from the source. Since the horizontal distance over which $\Delta\phi$ changes is also about equal to a, the inclination of the wave front, after passing through this phase screen, will equal about $\lambda\Delta\phi/2\pi a$. We assume that $\Delta\phi$ is large compared with unity, in which case θ_s much exceeds the diffraction angle for radiation passing through a circular aperture of radius a, and ray optics is valid. Hence θ_s equals the rms inclination of the wave front, and equation (3-64) yields

$$\theta_s = \left(\frac{z}{a}\right)^{1/2} \Delta m = \frac{1}{2\pi} \frac{e^2}{m_e} \left(\frac{z}{a}\right)^{1/2} \frac{\Delta n_e}{\nu^2}, \tag{3-65}$$

where we have made use of equations (3-58), (3-59), and (3-64), again keeping only the first order term in ν_p^2/ν^2.

These results can now be applied to the scintillation of pulsar signals, an observed large change of pulse amplitude over periods of minutes. The physical picture is that many different rays reach the Earth by multiple paths from the diffusing phase screen, and that interference between these rays produces a pattern of marked intensity maxima and minima. Because of the motions of the pulsar and the Earth with respect to this refracting gas, the pattern shifts with time relative to the observer. If the source has a sufficiently small intrinsic diameter, a strongly modulated pattern of interference will exist at any one frequency if two conditions are satisfied. In the first place, z must be sufficiently great so that rays from sections of the phase screen separated by a lateral distance a will actually cross each other; this condition gives

$$\theta_s > \frac{2a}{z}. \tag{3-66}$$

In the second place, the rms difference of phase between the various rays must exceed 1 radian; if ΔL is the dispersion in pathlengths, this condition gives

$$\Delta L = \frac{1}{4} z\theta_s^2 > \frac{\lambda}{2\pi}. \tag{3-67}$$

Finally, if the interference pattern is to be observed, the receiver bandwidth must be sufficiently small so that the phase difference between different rays, about equal to $2\pi\Delta L/\lambda$, changes by at most 1 radian over the range of frequencies measured. If we replace λ by c/ν, we find that this condition gives for $\Delta\nu_s$, the maximum bandwidth for strong scintillation,

$$\Delta\nu_s = \frac{c}{2\pi\Delta L} = \frac{2c}{\pi z\theta_s^2}, \tag{3-68}$$

where equation (3-67), relating ΔL to $z\theta_s^2$, has been used. In deriving equation (3-68) the change of θ_s^2 and thus of L over the bandwidth $\Delta\nu_s$ is ignored, since it is the phase difference between fixed rays, passing through specific regions of the phase screen, that must be constant over the bandwidth.

Equation (3-65) has been compared with the apparent sizes of quasistellar radio sources, determined both from very long baseline interferometry (VLBI) and from measurement of interplanetary scintillation, which disappears for sufficiently extended sources. The measured sizes decrease about as ν^{-2}, as predicted, ranging from about 0.2 arcsecond at 100 MHz to about 0.001 arcsecond at 2000 MHz [62]. If Δn_e is assumed proportional to n_e, equation (3-68) indicates that $\Delta\nu_s$ should vary as ν^4/D_m^2; the observations of $\Delta\nu_s$, the bandwidth at which the scintillation amplitude drops sharply, are consistent with this prediction [63]. For pulsars, comparison of the observed values of θ_s^2 and $\Delta\nu_s$ should give the distance, z, to within the accuracy of the theoretical model used, although the probably clumpy distribution of the refracting gas certainly complicates any detailed analysis of such data. It is not easy to evaluate Δn_e and a separately from the scintillation observations with any precision, but the data are consistent with values of about 3×10^{-5} cm^{-3} and 10^{11} cm for these two quantities. These various values satisfy conditions (3-66) and (3-67). The physical origin of these fluctuations in n_e is not yet known.

The effective thickness of the galactic scattering layer has been determined from a comparison of scintillation observations at high galactic latitudes with corresponding data for pulsars at known distances in the galactic plane; the resultant value for $2H$ is 1100 ± 250 pc [64]. If the relative density fluctuation $\Delta n_e/n_e$ increases with distance from the galactic plane, this result would be consistent with a somewhat smaller thickness for the layer of free electrons, as suggested by the discussion at the end of this section.

c. Faraday Rotation

If a magnetic field is present, equation (3-58) is modified. For radiation propagating parallel to **B**, we have [59]

$$m^2=\frac{c^2}{V^2}=1-\frac{\nu_p^2}{\nu^2}\left\{1\pm\frac{eB}{2\pi m_e c\nu}\right\}^{-1}. \tag{3-69}$$

The minus sign refers to one mode of circular polarization (the one in which the electric vector rotates in the same direction as the electron gyrates around the lines of force), whereas the positive sign refers to the opposite mode. If we make use of equation (3-59), the difference of velocity ΔV between these two modes is

$$\frac{\Delta V}{c}=\frac{n_e e^2}{2\pi^2 m_e}\frac{eB}{m_e c}\frac{1}{\nu^3}. \tag{3-70}$$

For propagation in other directions, equation (3-70) is still valid if we substitute for B its component $B_{\|}$ in the direction of wave propagation, and if $eB/2\pi m_e c$ is very much less than ν.

The influence of the magnetic field on the change of V with ν is very small. However, if the radiation traveling through the medium is plane polarized, the presence of the magnetic field will produce a rotation of the plane of polarization. The plane-polarized wave may be expressed as the sum of two circularly polarized components. In traveling one wavelength these two waves will change their relative phase by an amount $2\pi\Delta V/c$, where ΔV is given in equation (3-70). The plane of vibration (defined by the direction of the electric vector and the direction of propagation) rotates by half this amount, and the total rotation angle, ψ, over the full pathlength L is given by

$$\psi = \frac{\pi}{\lambda} \int_0^L \frac{\Delta V}{c}\, ds = R_m \lambda_m^2, \tag{3-71}$$

where the "rotation measure" R_m is given by

$$R_m = \frac{e^3}{2\pi m_e^2 c^4} \int_0^L n_e B_{\|}\, ds = 8.12 \times 10^5 \int_0^L n_e B_{\|}\, ds \text{ pc m}^{-2}; \tag{3-72}$$

the numerical constant in this equation is obtained with the wavelength λ_m in equation (3-71) expressed in meters, whereas n_e, $B_{\|}$, and ds are in units of cm^{-3}, Gauss, and parsecs, respectively.

Faraday rotation has been measured in the spectra of several hundred extragalactic radio sources [65]. The measured R_m increases with decreasing b, indicating that propagation in a galactic magnetic field must be responsible. However, uncertainties in n_e and at low latitudes the long pathlength through regions of varying B prevent a reliable determination of field strength from these data.

Determination of a mean B within the solar neighborhood is best achieved with pulsar measures both of R_m and of D_m, since according to equations (3-63) and (3-72), $\langle B_{\|} \rangle$ equals R_m / D_m times a known constant. The polarization observed in pulsar signals fluctuates wildly from pulse to pulse, but the mean of 100 pulses seems stable, and it permits a determination of the mean plane of polarization as a function of time during a pulse for each of several wavelengths. The mean values of $B_{\|}$ obtained in this way [66] are shown in Fig. 3.3 for 26 pulsars plotted against galactic longitude l; measures for which b exceeds 35° or the estimated distance exceeds 2500 pc have been omitted, as have six pulsars within the North Polar Spur. The opposite sign of the field for galactic longitudes separated by 180° indicates that $\mathbf{B}$ has a uniform component roughly parallel to the galactic plane. A least-squares solution gives [66] for B_0 and l_0, the amplitude and direction of this field in galactic longitude,

$$B_0 = (2.2 \pm 0.4) \times 10^{-6}\ \text{G}, \qquad l_0 = 94 \pm 11°, \tag{3-73}$$

where the anticipated rms errors are given.

Since the observational errors of $B_{\|}$ are relatively small, the scatter shown in Fig. 3.3 indicates that $\mathbf{B}$ may be expressed as a uniform field $\mathbf{B}_0$ plus an additional irregular component $\mathbf{B}_1$. Analysis of the pulsar data indicates [66] that $\langle B_{1\|}^2 \rangle^{1/2}$ is 1.0×10^{-6} G, corresponding to an rms value

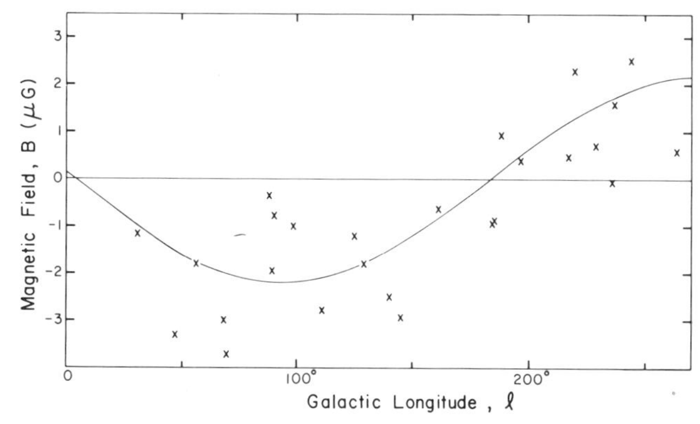

Figure 3.3 Magnetic field found from pulsar observations. For each pulsar the component of **B** parallel to the line of sight is determined from the measures of dispersion and Faraday rotation [66]. Pulsars more distant than 2500 pc or with galactic latitude greater than 35° or in the direction of the North Polar Spur have been omitted. The solid line represents the sinusoidal variation expected for a uniform field with parameters obtained by a least-squares solution and given in equation (3-73).

of 1.7×10^{-6} G for B_1 if this component is statistically isotropic. If the six pulsars in the North Polar Spur (centered at $l\approx24°$) and four more distant than 2000 pc, which were all omitted from the least-squares analysis, are included, the resultant value of B_1 is increased by about 40 percent. Evidently the irregular field is comparable with the uniform field.

The values of R_m for quasars, which apparently do not have as much intrinsic Faraday rotation as other extragalactic radio sources, can be fitted for high latitudes ($|b|>25°$) by the expression [29]

$$R_m=(-18\pm5)|\cot b|\cos(l-l_0). \tag{3-74}$$

The value of l_0 agrees with that in equation (3-73) within the computed error. According to equation (3-72), the constant factor in equation (3-74) is proportional to the mean value of B_0 times the column density of electrons in the z direction (normal to the galactic plane) from z equal to 0 out to ∞. If we take the value of B_0 from equation (3-73), we find that this column density equals 10 cm^{-3}pc. If we multiply this result by two to obtain the column density pole to pole and divide by 0.03 cm^{-3}, the mean interstellar electron density obtained from pulsar dispersion measures [S3.6a], we obtain a value of about 700 pc for $2H$, the effective thickness of the electron layer. This procedure assumes that B is constant with height z above the galactic plane. If B decreases with increasing z at about the same rate as n_e, the value of $2H$ derived refers to the effective thickness of n_e^2 rather than of n_e.

Both this value of 700 pc and the comparable value of 1100 pc obtained for the thickness of the scintillating layer substantially exceed the 250 pc for $2H$ found for the H I clouds [S3.3b] and for the general free–free emission from the Galaxy, resulting presumably from relatively dense H II regions [S3.5a]. The low-density electron–ion gas which is seen at high galactic latitudes evidently extends much farther from the plane than the diffuse clouds, ionized or neutral, which are conspicuous at low values of b.

REFERENCES

1. S. Chandrasekhar, *Radiative Transfer*, Oxford University Press (London), 1950, Chapt. 2.
2. V. V. Sobolev, *A Treatise on Radiative Transfer* (translated by S. Gaposchkin), D. Van Nostrand (Princeton, N.J.), 1963, Chapt. 1.
3. L. C. Green, P. P. Rush, and C. D. Chandler, *Ap. J. Supp.*, **3**, 37, 1957; No. 26.
4. D. H. Menzel, *Ap. J. Supp.*, **18**, 221, 1969; No. 161.
5. D. C. Morton and H. Dinerstein, *Ap. J.*, **204**, 1, 1976.
6. D. C. Morton and W. H. Smith, *Ap. J. Supp.*, **26**, 333, 1973; No. 233.
7. D. R. Johnson, F. J. Lovas, and W. H. Kirchhoff, *Microwave Spectra of Molecules of Astrophysical Interest*, *J. Phys. Chem. Ref. Data*, **1**, 1011, 1972, and subsequent papers in this series.
8. H. M. Johnson, *Stars and Stellar Systems*, Vol. **7**, University of Chicago Press (Chicago), 1968, p. 65.
9. R. J. Reynolds, F. L. Roesler, and F. Scherb, *Ap. J. (Lett.)*, **192**, L53, 1974.
10. D. E. Osterbrock, *Astrophysics of Gaseous Nebulae*, W. H. Freeman (San Francisco), 1974, Chapts. 4 and 5.
11. B. G. Elmergreen, *Ap. J. (Lett.)*, **198**, L31, 1975.
12. F. J. Kerr, *Stars and Stellar Systems*, Vol. **7**, University of Chicago Press (Chicago), 1968, p. 575.
13. F. J. Kerr, *Ann. Rev. Astron. Astroph.*, **7**, 39, 1969.
14. P. D. Jackson and S. A. Kellman, *Ap. J.*, **190**, 53, 1974.
15. J. H. Oort and L. Plaut, *Astron. Astroph.*, **41**, 71, 1975.
16. C. Heiles, *Ap. J. Supp.*, **15**, 97, 1967; No. 136.
17. P. L. Baker and W. B. Burton, *Ap. J.*, **198**, 281, 1975.
18. P. Goldreich and D. A. Keeley, *Ap. J.*, **174**, 517, 1972.
19. M. M. Litvak, *Ann. Rev. Astron. Astroph.*, **12**, 97, 1974.
20. A. H. Cook, *Nature*, **211**, 503, 1966.
21. M. P. Hughes, A. R. Thompson, and R. S. Colvin, *Ap. J. Supp.*, **23**, 323, 1971; No. 200.
22. V. Radhakrishnan, J. D. Murray, P. Lockhart, and R. P. J. Whittle, *Ap. J. Supp.*, **24**, 15, 1972; No. 203.
23. B. Lazareff, *Astron. Astroph.*, **42**, 25, 1975.
24. R. D. Davies and E. R. Cummings, *M. N. Roy. Astr. Soc. London*, **170**, 95, 1975.
25. E. W. Greisen, *Ap. J.*, **184**, 379, 1973.
26. E. W. Greisen, *Ap. J.*, **203**, 371, 1976.
27. J. M. Dickey, E. E. Salpeter, and Y. Terzian, *Ap. J.*, (*Lett.*) **211**, L77, 1977.

28. V. Radhakrishnan and W. M. Goss, *Ap. J. Supp.*, **24**, 161, 1972; No. 203.
29. E. Falgarone and J. Lequeux, *Astron. Astroph.*, **25**, 253, 1973.
30. G. B. Field, *Molecules in the Galactic Environment*, Wiley, (New York), 1973, p. 21.
31. G. L. Verschuur, *Interstellar Gas Dynamics*, IAU Symp. No. 39, edited by H. J. Habing, Reidel Publ., Dordrecht, 1969, p. 150.
32. R. C. Bohlin, *Ap. J.*, **200**, 402, 1975.
33. P. Henry et al., *Ap. J.*, **205**, 426, 1976.
34. L. Spitzer and E. B. Jenkins, *Ann. Rev. Astron. Astroph.*, **13**, 133, 1975.
35. G. Münch, *Stars and Stellar Systems*, Vol. **7**, University of Chicago Press (Chicago), 1968, p. 365.
36. A. Unsöld, *Physik der Sternatmosphären*, 2nd edition, J. Springer, 1955, Fig. 114, p. 292.
37. A. Blaauw, *B.A.N.*, **11**, 405, 1952; No. 433.
38. L. M. Hobbs, *Ap. J.*, **157**, 135, 1969.
39. L. A. Marschall and L. M. Hobbs, *Ap. J.*, **173**, 43, 1972.
40. P. J. Van Rhijn, *Publ. Kapteyn Astr. Lab.* No. 50, Groningen, 1946.
41. G. Münch and H. Zirin, *Ap. J.*, **133**, 11, 1961.
42. D. G. York , in preparation.
43. L. Spitzer, *Comments on Astroph.*, **6**, 177, 1976.
44. R. S. Siluk and J. Silk, *Ap. J.*, **192**, 51, 1974.
45. L. M. Hobbs, *Ap. J. (Lett.)*, **206**, L117, 1976.
46. D. G. York, *Ap. J. (Lett.)*, **193**, L127, 1974.
47. P. A. G. Scheuer, *M.N.R.A.S.*, **120**, 231, 1960.
48. P. G. Mezger and A. P. Henderson, *Ap. J.*, **147**, 471, 1967.
49. W. J. Karzas and R. Latter, *Ap. J. Supp.*, **6**, 167, 1961; No. 55.
50. D. H. Menzel and C. L. Pekeris, *M.N.R.A.S.*, **96**, 77, 1935; eq. (2.24).
51. R. L. Brown and B. Zuckerman, *Ap. J. (Lett.)*, **202**, L125, 1975.
52. G. Westerhout, *B.A.N.*, **14**, 215, 1958; No. 488.
53. R. W. Wilson, *Ap. J.*, **137**, 1038, 1963.
54. D. O. Richstone and K. Davidson, *A. J.*, **77**, 298, 1972.
55. W. H. Tucker and M. Koren, *Ap. J.*, **168**, 283, 1971 and **170**, 621, 1971.
56. P. Gorenstein and W. H. Tucker, *Ann. Rev. Astron. Astroph.*, **14**, 373, 1976.
57. G. R. A. Ellis and P. A. Hamilton, *Ap. J.*, **146**, 78, 1966.
58. V. L. Ginzburg and S. I. Syrovatskii, *Ann. Rev. Astron. Astroph.*, **3**, 297, 1965.
59. L. Spitzer, *Physics of Fully Ionized Gases*, Wiley (New York), 2nd edition, 1962, Chapt. 3.
60. J. Gómez-Gonzáles and M. Guélin, *Astron. Astroph.*, **32**, 441, 1974.
61. P. A. G. Scheuer, *Nature*, **218**, 920, 1968.
62. A. C. S. Readhead and A. Hewish, *Nature Phys. Sci.*, **236**, 440, 1972.
63. K. R. Lang, *Ap. J.*, **164**, 249, 1971.
64. A. C. S. Readhead and P. J. Duffett-Smith, *Astron. Astroph.*, **42**, 151, 1975.
65. F. F. Gardner, D. Morris, and J. B. Whiteoak, *Austral. J. Phys.*, **22**, 813, 1969.
66. R. N. Manchester, *Ap. J.*, **188**, 637, 1974.

4. Excitation

We consider now the processes which determine the population densities of various energy levels under interstellar conditions. Again we let n_j be the number density of particles in a level j, characterized by g_j quantum states all of about the same energy E_j. Under steady-state conditions, n_j is constant with time, and the number of transitions per second per cm^3 into level j must equal the corresponding number out. We denote by $(R_{jk})_Y$ the probability per unit time that a particle in level j undergoes a transition to level k as a result of process Y. Then we have

$$\frac{dn_j}{dt} = -n_j \sum_Y \sum_k (R_{jk})_Y + \sum_Y \sum_k n_k \times (R_{kj})_Y = 0, \tag{4-1}$$

where the sums extend over all physical processes Y and over all other levels k; the convective term $\mathbf{v} \cdot \nabla n_j$ has been ignored and we have momentarily suspended the convention that level j has an energy lower than level k. Equation (4-1) forms the basis for much of the theoretical work on interstellar matter.

In general, if there are M separate levels, the set of equations (4-1) provides $M-1$ linearly independent equations, which determine the $M-1$ ratios of the population densities n_j to each other. If the probability rates $(R_{jk})_Y$ are known, the solution is in principle straightforward and, thanks to modern computers, feasible in practice.

The specific form of equation (4-1) depends, of course, on the types of excitation, deexcitation, ionization, and recombination that are being considered. Successive sections in this chapter discuss the solutions of these equations when the primary sources of excitation are due first to collisions, then to recombination, and then to photons. The equilibrium between recombination and either ionization or dissociation is discussed in Chapter

5. There is some inevitable overlap of subject matter between these various sections. For example, radiative deexcitation must be considered throughout, and ionization equilibrium cannot be altogether ignored in analyses of excitation by recombination.

4.1 EXCITATION BY COLLISIONS

If collisions alone are acting, the value of n_k/n_j will equal n_k^*/n_j^*, since all the probability coefficients will have their values in LTE [S2.4]. For a nontrivial problem we must therefore include radiative transitions as well. We obtain the form of equation (4-1) which includes these two processes.

For the rate of excitation or deexcitation by collisions with electrons, for example, we define γ_{jk} as $(R_{jk})_{\text{coll}}/n_e$. The quantity $(R_{jk})_{\text{phot}}$, the corresponding rate for radiative processes, may be expressed in terms of the three Einstein coefficients introduced in Section 3.2. With these substitutions, the steady-state equation (4-1) now becomes

$$n_j\left\{\sum_k (n_e\gamma_{jk}+B_{jk}U_\nu)+\sum_{k<j} A_{jk}\right\}=\sum_k n_k(n_e\gamma_{kj}+B_{kj}U_\nu)+\sum_{k>j} n_k A_{kj}, \quad (4\text{-}2)$$

where the radiant energy density U_ν per unit frequency interval is related to I_ν by the expression

$$U_\nu=\frac{1}{c}\int I_\nu\,d\omega, \quad (4\text{-}3)$$

the integral extending over all solid angles. For collisions with H atoms, $n(\text{H I})$ replaces n_e in equation (4-2), and the appropriate γ_{jk} for excitation by neutral H must of course be taken.

The three Einstein coefficients are interrelated by equations (3-13). Using the same argument for detailed balancing in thermodynamic equilibrium, with n_k^*/n_j^* given by equation (2-26), we obtain, omitting the subscript r,

$$g_j\gamma_{jk}=g_k\gamma_{kj}e^{-E_{jk}/kT}, \quad (4\text{-}4)$$

where $E_{jk}\equiv E_k-E_j$; with the usual convention that level k lies above level j, E_{jk} is positive.

a. Collisional Rate Coefficients

To evaluate the rate coefficients γ_{jk} and γ_{kj} for inelastic excitation and for superelastic deexcitation we make use of the corresponding cross sections for these processes, in which potential and kinetic energy are interchanged. As in Section 2.1, test particles may be assumed to be moving with a relative velocity u through field particles whose particle density per unit volume is n_f. The probability that a test particle, initially in state j, performs a transition to state k as a result of a collision during the time interval dt is then $n_f u \sigma_{jk}(u)\,dt$. The rate coefficient γ_{jk} is the mean value of this transition probability per unit density of field particles, averaged over a Maxwellian distribution for u [equation (2-17)]. If we replace **du** by $4\pi u^2 du$, we obtain

$$\gamma_{jk} = \langle u\sigma_{jk}(u)\rangle = \frac{4l^3}{\pi^{1/2}}\int_0^\infty u^3\sigma_{jk}(u)e^{-l^2u^2}du. \tag{4-5}$$

Instead of equation (2-18) for l we use the reduced mass for the two interacting particles [equation (2-11)] and write

$$l^2 = \frac{m_r}{2kT} = \frac{m_t m_f}{(m_f + m_t)2kT}. \tag{4-6}$$

With this definition, equation (2-17) gives the distribution of the relative velocity u.

We now use the condition of detailed balancing in thermodynamic equilibrium to derive a relationship between σ_{jk} and σ_{kj}, the cross sections for excitation and deexcitation, respectively. We again consider that the test particles are being excited or deexcited, and denote by n_j^* and n_k^* their particle densities in levels j and k in LTE. If u_j represents the relative velocity before excitation, and u_k represents the corresponding velocity afterward, then in thermodynamic equilibrium the number of upward transitions per cm^3 per second produced by collisions in the relative velocity range between u_j and $u_j + du_j$ must equal the number of deexcitations produced by collisions in the corresponding velocity range between u_k and $u_k + du_k$; hence we have

$$n_j^* n_f f(u_j)u_j\,\mathbf{du}_j\sigma_{jk}(u_j) = n_k^* n_f f(u_k)u_k\,\mathbf{du}_k\sigma_{kj}(u_k). \tag{4-7}$$

If now we substitute from equation (2-26) for the ratio n_j^*/n_k^*, and use equation (2-17) for $f(u)$, we obtain

$$g_j u_j^2\sigma_{jk}(u_j) = g_k u_k^2\sigma_{kj}(u_k), \tag{4-8}$$

where the condition

$$\tfrac{1}{2}m_r u_k^2 = \tfrac{1}{2}m_r u_j^2 - E_{jk} \tag{4-9}$$

has been used, with m_r again the reduced mass appearing in equation (4-6). The quantity E_{jk}, introduced in equation (4-4), is positive. Equation (4-4) may be derived directly from equation (4-8).

Excitation can occur only if the initial kinetic energy exceeds the "threshold energy" E_{jk}; the variation of the excitation cross sections with u near threshold is quite different for collisions between ions and electrons and for collisions in which at least one particle is neutral. In the latter case, $u_k \sigma_{kj}(u_k)$ is constant or decreases slowly as u_k decreases; the excitation cross section $\sigma_{jk}(u_j)$ vanishes at threshold. In the former case of ion excitation by electrons, $\sigma_{kj}(u_k)$ varies as $1/u_k^2$ for small u_k because of the electrostatic attraction, and $\sigma_{jk}(u_j)$, the excitation cross section, is nearly constant at threshold.

We now consider specific values of γ_{jk} for collisions of electrons with ions. For such collisions it is customary [1] to express these cross sections in terms of a "collision strength" $\Omega(j,k)$, defined by

$$\sigma_{jk}(w_j) = \frac{\pi}{g_j}\left(\frac{h}{2\pi m_e w_j}\right)^2 \Omega(j,k). \tag{4-10}$$

Since the ion velocity is generally negligible, the electron velocity w has been substituted for u in this expression. Equation (4-8) indicates that $\Omega(j,k)$ for excitation equals $\Omega(k,j)$ for deexcitation. In accordance with the discussion above, $\Omega(j,k)$ tends to be relatively constant above threshold, although resonances are sometimes present. As may be shown directly from equations (4-10) and (4-5), the rate coefficient γ_{kj} for deexcitation may be expressed in terms of $\Omega(j,k)$ by the expression

$$\gamma_{kj} = \frac{h^2 \Omega(j,k)}{g_k (2\pi m_e)^{3/2} (kT)^{1/2}} = \frac{8.63 \times 10^{-6} \Omega(j,k)}{g_k T^{1/2}} \text{cm}^3 \text{ s}^{-1}, \tag{4-11}$$

where $\Omega(j,k)$ has been assumed constant with w_j.

Values of $\Omega(j,k)$ have been computed from quantum mechanics as a function of w_j. Table 4.1 lists for some of the more important transitions the values of $\Omega(j,k)$ averaged over a Maxwellian distribution of w_j at a temperature of 10,000° [1]. At neighboring temperatures, the mean $\bar{\Omega}(j,k)$ usually changes less rapidly than $T^{1/2}$. The ions included in this table are the astrophysically more abundant ones whose excited levels lie mostly within 4 eV of the ground level. The important O III ion is also included.

We digress momentarily to review the standard spectroscopic notation used in Table 4.1. In atoms with several orbital electrons, each energy level is characterized by a particular value of J, the quantum number for total

Table 4.1. Collision Strengths for Excitation by Electrons

Number of p electrons	Ion	Levels Lower	Levels Upper	E_{jk}(eV)	$\Omega(j,k)$	$\Sigma_j A_{kj}(s^{-1})$
1,5	C II	$^2P_{1/2}$	$^2P_{3/2}$	0.0079	1.33	2.4×10^{-6}
	Ne II	$^2P_{3/2}$	$^2P_{1/2}$	0.097	0.37	8.6×10^{-3}
	Si II	$^2P_{1/2}$	$^2P_{3/2}$	0.036	7.7	2.1×10^{-4}
2	N II	$^3P_0$ —	$^3P_1$	0.0061	0.41	2.1×10^{-6}
		$^3P_0$ —	$^3P_2$	0.0163	0.28	7.5×10^{-6}
		$^3P_1$ —	$^3P_2$	0.0102	1.38	7.5×10^{-6}
		^3P —	$^1D_2$	1.90	2.99	4.0×10^{-3}
		^3P —	$^1S_0$	4.05	0.36	1.1
	O III	$^3P_0$ —	$^3P_1$	0.014	0.39	2.6×10^{-5}
		$^3P_0$ —	$^3P_2$	0.038	0.21	9.8×10^{-5}
		$^3P_1$ —	$^3P_2$	0.024	0.95	9.8×10^{-5}
		^3P —	$^1D_2$	2.51	2.50	2.8×10^{-2}
		^3P —	$^1S_0$	5.35	0.30	1.8
3	O II	$^4S_{3/2}$	$^2D_{5/2}$	3.32	0.88	4.2×10^{-5}
		$^4S_{3/2}$	$^2D_{3/2}$	3.32	0.59	1.8×10^{-4}
		$^2D_{3/2}$	$^2D_{5/2}$	0.0025	1.16	4.2×10^{-5}

angular momentum, including the vector sum of the orbital and spin angular momentum of all the electrons. Each J level includes g_J separate states all with the same energy; in general, $g_J = 2J + 1$. Usually several "fine-structure" levels, with different J values and separated slightly in energy, constitute a "spectroscopic term." For the lighter chemical elements listed in Table 4.1, each term is designated with a letter S, P, D, or F, and so on, referring to a value of 0, 1, 2, or 3, and so on, for L, the quantum number for the total orbital angular momentum (resulting from the vector sum of the orbital angular momenta of all the bound electrons). In the usual notation for an energy level within some spectroscopic term, the normal number of separate J levels, called the "multiplicity" of that term, appears as a superscript in front of S, P, D, or F, and so forth. (The multiplicity equals $2S + 1$, where S is the quantum number for the vector sum of the angular momenta of all the electron spins.) An "electron configuration," characterized by the values of n and l, the total and azimuthal quantum numbers of all the electrons, usually yields a number of different spectroscopic terms. All the energy levels in Table 4.1 correspond to the ground electronic configurations of the atoms listed, and the excitation processes correspond primarily to a change in vector addition of angular momenta of the different electrons rather than a change in the spatial wave functions involved. The subscript number following each letter (S, P, or D) in Table 4.1 is the J value for each particular level.

Energy level diagrams for the ground electron configurations of O II and O III are shown in Fig. 4.1; those for O I and N II are similar to that for O III except that for O I the relative positions of the three fine-structure levels of the ground ^3P term are inverted, with the $J=2$ level ($^3P_2$) the ground state. For excitation of the higher levels of N II and O III, the transitions for all three levels of the ground ^3P term are grouped together to give a total $\Omega(j,k)$ in Table 4.1; each of these three fine-structure levels contributes to $\Omega(j,k)$ in proportion to the statistical weight $2J+1$ for that level.

For each upper level, the sum of the spontaneous radiative transition probabilities, A_{kj}, from that level to all lower levels is given in the last column of Table 4.1 [1,2]. Electric dipole transitions are forbidden for all these transitions, as is generally the case among energy levels of the same electron configuration; hence the radiative transition probabilities given in the last column are all very small.

We consider next the values of γ_{jk} for excitation by H atoms and H_2 molecules. Evidently such collisions are of primary interest in H I regions,

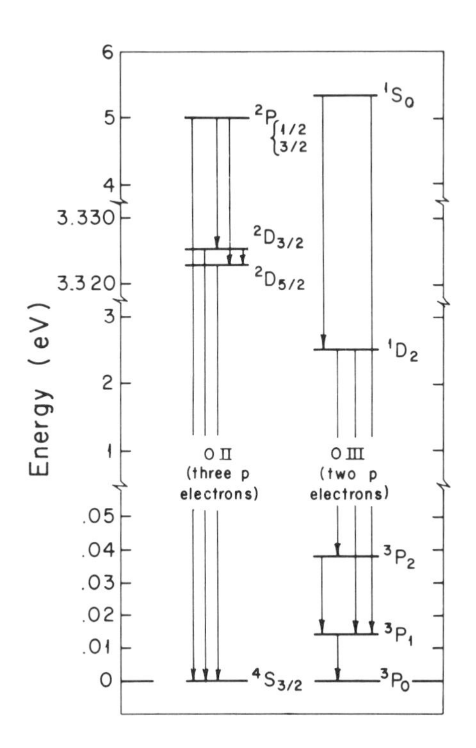

Figure 4.1 Energy level diagram for O II, O III. Each horizontal bar represents an atomic level with an excitation energy shown on the left-hand scale, which changes abruptly to show separation of the fine structure levels in some spectroscopic terms. The forbidden radiative transitions which produce astrophysically important lines are shown by arrows.

where n_e/n_H is relatively small. Hence collisions with the predominant H atoms will be generally much more frequent than those with electrons, and H atoms will usually be responsible for excitation and deexcitation. A variety of such processes are treated here. Some special collisional processes, such as H_2 excitation by proton exchange [S5.3b], are considered subsequently.

Deexcitation rate coefficients for H atoms and H_2 molecules are given in Table 4.2 for the following four types of processes: (a) deexcitation of the hyperfine levels by spin exchange in H–H encounters [3]; (b) deexcitation of the $^2P_{3/2}$ fine-structure level of C II [4]; (c) deexcitation of the $J=2$ rotational level in H_2 [5], with an excitation energy of 0.044 eV; (d) deexcitation of the $J=1$ level in CO [6], with an excitation energy of 4.8×10^{-4} eV. In processes (a) and (b) the electron spins are again changed by the collisions, whereas in process (c) the rotational angular momentum of the molecule is changed. For collisions between neutral particles, the deexcitation cross section tends to be nearly constant at low energies, with γ_{kj} therefore varying about as $T^{1/2}$. For ion-neutral collisions, however, γ_{kj} tends to be constant as with elastic collisions [S2.2]. For processes (c) and (d), the general subscripts k and j are replaced in Table 4.2 by specific values of the rotational quantum number J for the upper and lower levels involved.

Table 4.2. Rate Coefficients (cm^3s^{-1}) for Deexcitation by H and H_2

Temperature T (°K)	10°	30°	100°	300°	1000°
$10^{11}\times\gamma_{kj}$ for H–H	0.23	3.0	9.5	16	25
$10^{10}\times\gamma_{kj}$ for H–C^+	6.9	7.4	8.0	8.4	9.7
$10^{12}\times\gamma_{20}$ for H–H_2	0.96	1.37	3.0	9.1	42
$10^{11}\times\gamma_{10}$ for H_2–CO	1.8	3.2	3.7		

b. Theory for Systems with Two or Three Levels

Most analyses of atomic or molecular populations in statistical equilibrium consider a substantial number of different levels. For each transition the radiative and collisional probabilities must all be determined, and then the simultaneous set of equations (4-2) solved on a computer. Here the methods are illustrated and some of the essential physical principles are emphasized by relatively simple calculations for idealized atoms or molecules with two or three energy levels.

For the two-level system we denote the lower and upper levels by 1 and 2, respectively. Equation (4-2) yields one equation for n_2/n_1, and if equations (2-27), (3-13), and (4-4) are used, we obtain

$$\frac{b_2}{b_1}=\frac{n_e\gamma_{21}/A_{21}+e^{h\nu/kT}c^3U_\nu/(8\pi h\nu^3)}{1+n_e\gamma_{21}/A_{21}+c^3U_\nu/(8\pi h\nu^3)}. \tag{4-12}$$

This result includes specifically collisions with electrons only; a sum over other collisional deexcitation probabilities should be included when appropriate. If we assume that the radiation field is blackbody radiation at the temperature T_R, diluted by a factor W, then use of equations (4-3) and (3-4) gives

$$\frac{c^3U_\nu}{8\pi h\nu^3}=\frac{W}{\exp(h\nu/kT_R)-1}. \tag{4-13}$$

Equation (4-13) is applicable particularly when U_ν results from the cosmic blackbody radiation, for which $W=1$ and $T_R=2.7°$K [S1.1].

For optical transitions under interstellar conditions, U_ν corresponds roughly to B_ν at a radiation temperature of some 10^4°K, with a dilution factor W of order 10^{-14}. Under these conditions the U_ν terms are negligible in equation (4-12), which takes the more transparent form

$$\frac{b_2}{b_1}=\frac{1}{1+A_{21}/n_e\gamma_{21}}. \tag{4-14}$$

When n_e is so large that $n_e\gamma_{21}$ much exceeds A_{21}, collisional deexcitation dominates over radiative deexcitation; under this condition, the ratio of the population of the two levels is essentially determined by collisions, and approaches its value in LTE [S2.4], with $b_2/b_1=1$. When $n_e\gamma_{21}$ is much less than A_{21}, collisional deexcitation can be ignored, and n_2/n_1 now represents a balance between collisional excitation and radiative deexcitation, with every collisional excitation producing one photon radiated; b_2/b_1 is then small and proportional to n_e. For permitted optical transitions A_{21} is of order $10^8\,s^{-1}$, and $A_{21}/n_e\gamma_{21}$ usually exceeds $10^{14}/n_e$, enormously greater than unity. Evidently the upper levels of such lines will be almost completely depopulated under interstellar conditions.

Equation (4-12) may be applied to the excitation of the hyperfine levels in the H atom by cosmic blackbody radiation and by collisions with other such atoms; n_H replaces n_e, of course. As noted earlier [S3.3b], $A_{21}=2.87\times 10^{-15}\,s^{-1}$ for this transition. If we expand all the exponential terms in equation (4-13), we find that b_2/b_1 is nearly unity, while χ, the absorption correction factor in equation (3-30), becomes

$$\chi=\frac{1+(A_{21}/n_H\gamma_{21})(kT/h\nu)}{1+(A_{21}/n_H\gamma_{21})(kT_R/h\nu)}\approx 1+\frac{1}{23n_H}; \tag{4-15}$$

terms of order $h\nu/kT$, comparable to those involved in the expansion of the exponentials, have been ignored compared to unity. A value of 80°K has been used for T in this numerical result, with a value of $8.0 \times 10^{-11}\,\mathrm{cm^3\,s^{-1}}$ found from Table 4.2 for γ_{21} at this temperature. Evidently the absorption is slightly increased above its value in LTE; the effect is small, but according to equation (3-32) it could reduce the saturation T_b somewhat below T for pathlengths through neutral H at low density. If a source of ionizing radiation permeates or is adjacent to an H I region, the resultant $L\alpha$ radiation can exert through photon pumping a dominant influence on the relative population of the hyperfine levels [S4.3a].

The three-level system which we consider next is simplified by the neglect of direct transitions between the two upper levels, designated as 2 and 3. Also we neglect all induced radiative transitions. As a result, the ratios b_2/b_1 and b_3/b_1 are each given separately by equation (4-14). If we neglect H atoms and consider electron collisions only, the ratio of emission intensities, I_{31}/I_{21}, in these two lines becomes

$$\frac{I_{31}}{I_{21}} = \frac{n_3 A_{31} h\nu_{31}}{n_2 A_{21} h\nu_{21}} = \frac{g_3 A_{31}\nu_{31}}{g_2 A_{21}\nu_{21}} \left[\frac{1 + A_{21}/n_e\gamma_{21}}{1 + A_{31}/n_e\gamma_{31}} \right] e^{-E_{23}/kT}. \tag{4-16}$$

If n_e is large, the quantity in brackets in equation (4-16) is unity, and the levels are populated as in thermodynamic equilibrium; as one would expect in this case, the ratio of intensities for these two lines equals n_3^*/n_2^* multiplied by $A_{31}\nu_{31}/A_{21}\nu_{21}$. If n_e is small, the values of A_{kj} cancel out in equation (4-16), and with the use of equation (4-4) we see that in this case the ratio of photon emission rates in the two lines equals γ_{13}/γ_{12}, the ratio of excitation probabilities; in this case every excitation produces one emitted photon.

c. Optical Emission Lines Observed from Heavy Atoms

The preceding analysis may be used to interpret the emission lines observed in the visible from H II regions, as well as from planetary nebulae [1]. The chief emission features observed are those from O II and O III, whose energy level diagrams are shown in Fig. 4.1, and from N II, whose energy level diagram is very similar to that of the isoelectronic ion O III. Lines from S II and Ne III (whose energy level diagrams are similar to those of O II and O III, respectively) are also observed from the denser H II regions, such as the Orion nebula. The theory of relative intensities can be used to determine both n_e and T, whereas the absolute intensities, compared with those of Hα or Hβ, give abundances relative to H.

Electron densities are determined primarily [1] from the ratio of the two O II emission lines at 3728.9 and 3726.2 Å, which are produced by the radiative transitions from the $^2D_{3/2}$ and $^2D_{5/2}$ levels down to the ground $^4S_{3/2}$ level (see Fig. 4.1); here these levels are numbered 3, 2, and 1 in order of decreasing energy. Levels 2 and 3 are so close together in energy that $\nu_{31}/\nu_{21} \approx 1$, and E_{23}/kT is nearly zero. According to equation (4-16) (with $E_{23} \ll kT$ and $\nu_{31} = \nu_{21}$) and the values in Table 4.1, we find that I_{31}/I_{21} equals 2/3 (the ratio of g_3 to g_2) at low n_e (below about 10^2 cm^{-3}), increasing to 2.9 at high n_e (more than about 10^4 cm^{-3}). A more accurate evaluation of I_{31}/I_{21} [1], taking into account collisional transitions between the two upper levels, agrees with equation (4-16) for n_e very small or very large, but the shape of the transition region is somewhat altered.

From the observed values of I_{31}/I_{21} and the computed dependence of this ratio on n_e, values of n_e in the Orion nebula (NGC 1976) have been found [7] to decrease from about 10^4 cm^{-3} in the center to about 10^2 cm^{-3} in the outer parts. In other H II regions typical densities of about 10^2 cm^{-3} are found. Similar results have been obtained for the corresponding S II doublet at 6716.4 and 6730.8 Å.

If the electron densities obtained in this way for the Orion nebula are used to compute the emission measure E_m [S3.3a] to be expected for this object, on the assumption of a smooth spherically symmetrical distribution, the computed E_m is 30 times the value observed from short-wavelength radio data [7]. This discrepancy is attributed to clumpiness in the H II gas; that is, the electron density determined from line ratios refers to dense clumps of gas, with much lower density in between. We adopt a simple theoretical model in which all the gas is concentrated in uniform clumps occupying a fraction F of the volume of the nebula. If n_{ec} is the electron density in each clump, with negligible electron density in between, we obtain

$$\langle n_e^2 \rangle = F n_{ec}^2 = \frac{\langle n_e \rangle^2}{F}. \tag{4-17}$$

Thus the rms n_e is about 5.5 times the mean n_e in Orion. Similar results are found for the H II regions NGC 6523 (M8) and NGC 7000 by comparing the local n_e found from O II line ratios [1] with the rms n_e found from E_m [8]. Clumpiness is probably a general feature of all H II regions.

Electron temperatures are found from the intensity ratio of the O III lines emitted from the $^1S_0$ and $^1D_2$ levels, with a difference of excitation energy, denoted by E_{23}, equal to 2.84 eV. The corresponding N II lines may also be used. To simplify the situation, considering only three electron levels, we group together the three ^3P fine-structure levels of the ground spectroscopic term as level 1, with the $^1D_2$ and $^1S_0$ levels designated as levels 2 and 3, respectively. The situation is now somewhat more complex than envisaged in equation (4-16) in that transitions between levels 2 and 3 must be considered as well as the 12 and 13 transitions. While a general equation for I_{32}/I_{21} is readily derived, we use here the simple physical result that in the low density limit every upward collisional excitation is followed by a downward radiative transition; we assume that $\gamma_{13}/\gamma_{12} \ll 1$, and thus ignore the population of level 2 produced by radiative transitions from level 3. Then the ratio of photons emitted per second from levels 3 and 2 equals γ_{13}/γ_{12}, exactly as found above from equation (4-16). Photons emitted from level 3 are divided between the 32 and 31 lines in accordance with the ratio of A_{32} to A_{31}. Hence, if we again use equation (4-4), we obtain

$$\frac{I_{32}}{I_{21}} = \frac{g_3 \nu_{32} \gamma_{31}}{g_2 \nu_{21} \gamma_{21}} \left[\frac{A_{32}}{A_{32} + A_{31}} \right] e^{-E_{23}/kT}. \tag{4-18}$$

From detailed values of A_{kj} in O III [1] we find that the factor in brackets equals 0.88, and if we insert other numerical values we obtain

$$\frac{I_{32}}{I_{21}} = 0.120 \exp\left(\frac{-32{,}900^\circ}{T} \right). \tag{4-19}$$

Evidently I_{32}/I_{21} and γ_{13}/γ_{12} also are both small, a condition noted above for the validity of these results.

Observations of O III lines in several H II regions have been compared with this theoretical result [1], allowing for the slight variation of $\Omega(1,2)$ and $\Omega(1,3)$ with T. The I_{21} line is actually the sum of three lines, but only two of these, $\lambda 5007$ and $\lambda 4959$, are strong enough to be measured. Since I_{32}, the intensity of the $\lambda 4363$ line, is relatively weak, the measurements are restricted to the brighter emitting regions. The resultant temperatures for three regions in NGC 1976 and for two other H II regions are all between 8000 and 9000°K. Similar measurements for the N II lines show greater variation (from 7000 to 11,000°K), but about the same average value. These data provide probably the most reliable determination of T in H II regions. The agreement with other determinations [S11.1a] leaves little question but that the temperature of such regions averages about 8000°K.

Finally, the intensities of these various emission lines relative to each other can be analyzed to yield the relative abundances of elements. In the low density limit, where collisional deexcitation is unimportant, equations (3-10) and (4-2) indicate that the total emissivity in each line is proportional to $n_e n_j(X^{(r)})\gamma_{jk}$; hence the intensity integrated over the line varies as $\gamma_{jk} E_m n_j(X^{(r)})/n_e$, where E_m is again the emission measure defined in equation (3-35). When a ratio of two intensities is taken, E_m cancels out, but T must be known accurately to compute γ_{jk} and thus to determine the relative abundances. In addition, the ratio between $n_j(X^{(r)})$, the particle density in level j for stage of ionization r, and the total particle density n for atoms of type X, in all stages of excitation and ionization, must be estimated. Evidently the final results are not very certain, but the abundances of N, O, Ne, and S relative to H in H II regions appear to agree with the cosmic abundances shown in Table 1.1 to within about a factor 2 [1], comparable with the errors in the determination.

d. Molecular Radio Lines

At radio wavelengths between 0.1 and 50 cm many types of interstellar molecules have been observed [9, 10], mostly in emission from relatively dense interstellar clouds. Interpretation of these observations is complicated, since a wide variety of excitation processes must be considered. In particular, radiative excitation and deexcitation can be of dominant importance, and in regions of high line opacity these rates can be evaluated only with a full solution of the coupled equations of transfer (3-1) and of statistical equilibrium (4-1). Here we derive a few simple results, first with $\tau_{\nu r}$, the optical thickness of the emitting region, and $h\nu/kT$ both assumed small, and second with $n_H\gamma_{kj}/A_{kj}$ so large that LTE may be assumed. These two limiting cases have the advantage that the equations of statistical equilibrium can be solved independently of the radiative transfer equations, greatly simplifying the problem. The effects to be expected with more general assumptions are briefly discussed. These results are then used to obtain general values for cloud densities and temperatures. To illustrate how maser lines might be excited, a collisional pumping mechanism proposed for H_2CO is described.

In computing τ_ν for a molecular radio line, the matrix element μ_{jk} [S3.2a] is normally used, related to B_{jk} and A_{kj} by equations (3-13), (3-24) and (3-28). We consider here pure rotational transitions within a diatomic or linear molecule which is in the ground vibrational level and which has no net electronic angular momentum (electron configuration denoted by $^1\Sigma$). In this simple case the square of μ_{jk}, summed over all the degenerate states of the upper level, is given by [11]

$$|\mu_{J,J+1}|^2 = \mu^2 \frac{J+1}{2J+1}, \tag{4-20}$$

where J is the rotational quantum number of the lower level and μ is the permanent dipole moment of the molecule. Values of μ [11, 12], in Debyes (10^{-18} cgs), are given in Table 4.3 for several such molecules in their ground vibrational levels.

We now derive quantitative relations for the brightness temperature T_b (a function of frequency ν) in a radio emission line. First we consider the simple case where $\tau_{\nu r}$ is small and where only two levels need be considered, with equation (4-12) applicable for b_2/b_1. With small $\tau_{\nu r}$, I_ν is very

Table 4.3. Permanent Dipole Moments of Diatomic and Linear Molecules

Molecule	CO	CS	SiS	SiO
μ (Debyes)*	0.112	1.96	1.73	3.10

Molecule	N_2O	HCN	OCS	HC_3N
μ (Debyes)*	0.166	2.98	0.71	3.6

*In $v=0$ vibrational level.

nearly equal to the cosmic radiation intensity $B_\nu(T_R)$, and we obtain a first approximation for b_2/b_1 with the assumption $I_\nu = B_\nu(T_R)$. To simplify the equations we take $h\nu/kT$ small also, computing j_ν/κ_ν from equations (3-32) and (3-30), and using the Rayleigh-Jeans law, equation (3-7), to replace $B_\nu(T_R)$ and I_ν by T_R and T_b, respectively. In obtaining T_b from equation (3-3) we must retain the $I_\nu(0)\exp(-\tau_{\nu r})$ term, leading to a corresponding $T_R \exp(-\tau_{\nu r})$ term. Expanding all exponentials we finally obtain to first order in $\tau_{\nu r}$

$$T_b - T_R = \tau_{\nu r} \frac{T - T_R}{1 + (A_{21}/n_{\rm H}\gamma_{21})(kT/h\nu)}. \tag{4-21}$$

The column density N_1, for molecules in the lower level, may be found from equation (3-18), substituting for $\tau_{\nu r}$ from equation (4-21) and obtaining s from equation (3-29) with χ again found from equation (3-30). If we express s_u in terms of $|\mu_{12}|^2$, from equation (3-28) we have

$$\int (T_b - T_R)\,d\nu = \left[\frac{1 - T_R/T}{1 + (A_{21}/n_{\rm H}\gamma_{21})(kT_R/h\nu)} \right] \frac{8\pi^3\nu^2|\mu_{12}|^2}{3kc} N_1. \tag{4-22}$$

As a result of the factor χ in $\tau_{\nu r}$, the correction factor in brackets in equation (4-22) differs slightly from that in equation (4-21), with kT_R replacing kT. Apart from this factor, which corrects for physical effects ignored in Chapter 3, equation (4-22) is equivalent to equation (3-38), derived for 21-cm radiation, except that in the earlier equation the total number of H atoms is computed rather than the column density in level 1 only, and $\tau_{\nu r}$ is assumed finite. If the correction factor in equation (4-22) is set equal to unity, a minimum value of N_1 is obtained.

The physical significance of the bracketed correction factor in equation (4-22) is easily understood. If $T = T_R$ the situation is indistinguishable from complete thermodynamic equilibrium. In this case $T_b = T_R$ at all frequencies and no line can appear; the $1 - T_R/T$ factor allows for this effect. In the denominator the product of the two factors each in parentheses is the ratio of radiative deexcitation to collisional deexcitation; the factor $kT_R/h\nu$ is the ratio of stimulated-to-spontaneous emission in thermodynamic equilibrium at the radiation temperature T_R. To the extent that radiative transitions dominate over collisional ones, the excitation of the levels approaches that in thermodynamic equilibrium at the radiation temperature T_R, rather than at the gas kinetic temperature. It is sometimes helpful in discussing this topic to introduce the excitation temperature T_{ex}, defined by the relation

$$\frac{n_2}{n_1} = \frac{g_2}{g_1} \exp\left(- \frac{h\nu}{kT_{ex}} \right). \tag{4-23}$$

From equation (4-12) it follows directly that if all exponents are small,

$$\frac{1}{T_{ex}} = \left[\frac{n_{\mathrm{H}}\gamma_{21}}{T} + A_{21} \frac{kT_R}{h\nu} \frac{1}{T_R} \right] \times \left[n_{\mathrm{H}}\gamma_{21} + A_{21} \frac{kT_R}{h\nu} \right]^{-1}. \tag{4-24}$$

Evidently T_{ex} is a harmonic mean of T and T_R, with each temperature weighted by the corresponding transition probability for collisional and radiative deexcitation, respectively. The correction factor in brackets in equation (4-22) is simply $1 - T_R/T_{ex}$. As T_{ex} approaches T_R, the excitation approaches that in thermodynamic equilibrium at the temperature characterizing the incident radiation, and the line disappears.

These results are clearly very idealized, and much more detailed models are required for comparison with observations. In particular, both $h\nu/kT_R$ and $\tau_{\nu r}$ must be assumed comparable with unity. Equation (3-9) shows that $h\nu/kT_R = 2.1$ for the ^{12}CO $J=0$ to $J=1$ line at 2.60 mm, with $T_R = 2.7$°K. The value of $\tau_{\nu r}$ at the center of the CO and other lines much exceeds unity in the larger, denser clouds.

The analysis for general values of $h\nu/kT_R$ and $\tau_{\nu r}$ can be carried through if we impose the restriction that collisional excitation and deexcitation dominate over the corresponding radiative processes; that is, that $n_{\mathrm{H}}\gamma_{21}/A_{21}$ is large compared either to $kT/h\nu$ or unity. In this "collisionally dominated" case, T_{ex} equals the kinetic temperature T, the conditions for LTE are fulfilled, and the computation of I_ν is straightforward. Since Kirchhoff's law is valid in LTE, equation (3-6) now applies for arbitrary $\tau_{\nu r}$, provided that T is constant in the cloud. If we replace $I_\nu(0)$ by $B_\nu(T_R)$, where T_R again equals 2.7°K, we obtain

$$I_\nu - B_\nu(T_R) = \{ B_\nu(T) - B_\nu(T_R) \} (1 - e^{-\tau_{\nu r}}). \tag{4-25}$$

If T is known and I_ν is measured, $\tau_{\nu r}$ may be found from equation (4-25). If $h\nu/kT_R \ll 1$, equation (3-7) is applicable, and we have the physically plausible result that with increasing $\tau_{\nu r}$, T_b within the line increases from T_R to T.

To determine I_ν when A_{21} exceeds $n_H\gamma_{21}$, the equation of radiation transfer must be solved with n_2/n_1 determined at each point by the joint action of radiative and collisional transitions. The photons now tend to be trapped within the cloud, increasing the radiant intensity I_ν significantly above $B_\nu(T_R)$. The determination of I_ν now involves complications which have not been solved in any general way. In one idealized case, where a linear gradient of velocity exists within the cloud, solutions have been obtained for this radiative-trapping case [13, 14]. The results show that the intensity emitted in the line can approach the Planck value at the kinetic temperature provided that $\tau_{\nu r}$ exceeds $A_{21}/n_H\gamma_{21}$. One must also consider the excitation of higher rotational levels, which by downward cascading significantly increases T_b for the CO $J=0$ to $J=1$ line above the values computed for a two-level molecule.

These various results above have been applied to molecular emission lines observed from dark visible clouds [15, 10], from prominent infrared sources [10], including compact H II sources, and from the galactic center [10]. In interpreting the observations an important datum is the value of n_H at which the collisional and spontaneous radiative transition rates are equal; that is, at which $n_H\gamma_{21}$ equals A_{21}; we denote this quantity by n_{Hc}. For a line with an optical depth about equal to unity, we have seen above that $T_b - T_R$ decreases when n_H/n_{Hc} falls below unity. We disregard here the reduction factor for stimulated emission, which usually does not alter the order of magnitude of $T_b - T_R$ for the CO lines. However, for a line of large optical thickness $\tau_{\nu r}$, n_H/n_{Hc} can be as small as $1/\tau_{\nu r}$ without much decrease in $T_b - T_R$, which will not be far below $T - T_R$. Detailed computations for CO and other strong molecular emission lines indicate that these have measurable strengths only if n_H/n_{Hc} in the source region is not less than about 10^{-2}.

Values of n_{Hc} for a few of the better observed molecules, together with computed values of A_{21} [10] for the indicated transitions, are given in Table 4.4. The rate coefficient γ_{21} for collisions of H_2 and CO molecules equals $3.2 \times 10^{-11}\,\mathrm{cm^3\,s^{-1}}$ at 30°K, according to Table 4.2; since $n_H = 2n(H_2)$ if all hydrogen is molecular, $\gamma_{21}/2$ has been used in computing n_{Hc}. Since γ_{21} for molecules other than CO may differ substantially from the values assumed and since $h\nu/kT$, the stimulated emission correction factor, is appreciably less than unity for the lines of longer wavelength, the values of n_{Hc} are very approximate for molecules other than CO. However, these values are consistent with the view that in visible dark clouds, where

Table 4.4. Hydrogen Densities giving Equal Radiation and Collisional Transition Rates

Molecule	CO	NH_3	CS	HCN
J_j, J_k	0, 1	1, 1	0, 1	0, 1
λ(cm)	0.260	1.265	0.612	0.338
$A_{21}(s^{-1})$	6×10^{-8}	1.7×10^{-7}	1.8×10^{-6}	2.5×10^{-5}
$n_{Hc}(cm^{-3})$	4×10^{3}	1.1×10^{4}	1.1×10^{5}	1.6×10^{6}

mainly CO (and also some OH) emission is observed, n_H is at least about $10^2\ cm^{-3}$, whereas in the smaller and more opaque dense clouds surrounding H II regions and infrared sources in general, where all the lines listed in Table 4.4 are often seen, the minimum n_H is higher by some two orders of magnitude. Within 3 pc from the galactic center a wide variety of molecular lines are emitted, some with A_{21} as large as 10^{-2}, suggesting that n_H in this region exceeds $10^7\ cm^{-3}$.

These ranges for n_H are consistent with some other evidence. For example, the densities estimated in this way for dark clouds are consistent [15] with those found from the total extinction through these regions, together with the average dust-to-gas ratio (by mass) of about 10^{-2}, typical of less dense regions [S7.3a, c]. If the values of n_H found from these rather imprecise arguments are accepted, the corresponding masses of the observed molecular clouds lie within the range from $10^2\ M_\odot$ for small, very dense configurations up to values as great as $10^5\ M_\odot$ for the extended clouds seen primarily in CO. The large "giant molecular clouds," in which most of the CO mass appears, are much more abundant within 4 to 8 kpc of the galactic center than they are in the solar neighborhood; observations of H II regions, supernova remnants, γ radiation, and presumably interstellar matter generally show this same galactic distribution, although n(H I), in contrast, is nearly constant at distances from the galactic center between 4 and 13 kpc [16]. The value of the effective thickness, $2H$, of the galactic CO layer is about 120 pc [17], roughly half that observed for atomic hydrogen.

Values of the gas kinetic temperature T can be found from the values of T_b for the strongest molecular lines. Particularly for the ^{12}CO lines, the optical depth is believed to be large, and T_b should approach closely some average T for the emitting regions. In dark visible clouds, the mean T_b averages about 10°K, with individual values ranging from 6 to 18°K [18]. Other molecular measures on dark clouds, including emission by NH_3 and absorption by OH, are consistent with temperatures in this range [10]. For CO emission from the vicinity of H II regions [19], T_b generally averages about 30°K, with a value of 70°K observed for the Orion A source, close to the Orion nebula.

While precise determination of molecular column densities from the observed emission lines is complicated by problems of excitation and radiation transfer, approximate results have been obtained by the use of relatively weak lines and the assumption of LTE. Thus measures of I_ν in the ^{13}CO $J=0$ to 1 line at 2.72 mm have been used [20] to determine $N(^{13}\mathrm{CO})$, with T obtained from the strong ^{12}CO line. In LTE, at a known T, $\tau_{\nu r}$ can be determined from equation (4-25) and N_j from equation (3-18), if we substitute for s from equation (3-23) with $b_k/b_j=1$. Since $\tau_{\nu r}$ computed in this way is about unity, this procedure determines N_j with reasonable accuracy, subject to an error of perhaps a factor 2 introduced by the assumption of LTE. Multiplication of N_j by $f_{\mathrm{CO}}/g_j e^{-E_j/kT}$, where f_{CO} is the CO partition function given in equation (2-29), then yields $N(^{13}\mathrm{CO})$ in each of some 50 dark clouds. Determination of N_{H} from the measured visual extinctions of these clouds gives abundances relative to H [S7.3c].

Evidently the cloud properties deduced as yet from molecular line data are all rather approximate. In principle, much more definite results could be obtained by analysis of several different emission lines from each molecule, fitting the observations with rather detailed physical models.

We turn now to a brief discussion of more complex excitation mechanisms. When transitions among many energy levels are considered, in conditions far from thermodynamic equilibrium, the ratio of populations between two adjacent levels can be entirely different from those in LTE, permitting either maser amplification or enhanced antimaser absorption, depending on whether χ in equation (3-30) is negative or positive. One of the simple situations considered is that for H_2CO (formaldehyde), some of whose lines are widely seen in absorption against the blackbody radiation, indicating that the upper states are more depopulated compared with the lower ones than in LTE at a temperature of 2.7°K; i.e., $T_{ex} < T_R$.

A possible mechanism [21] for explaining these observations involves collisional excitation of orthoformaldehyde (H_2CO), of which the four lowest levels are shown in Fig. 4.2. Dipole radiation can occur only in transitions between states of opposite parity, a designation which depends on whether the total wave function changes sign on inversion through the origin. For "even" states, the wave function is unchanged when the radius vector r_i to each electron and ion is replaced by $-r_i$, whereas for "odd" states, this operation changes the sign of the wave function. Atomic states resulting from the same electron configuration [S4.1a] all have the same parity. In the ortho-H_2CO molecule, levels 1 and 4 are odd, whereas 2 and 3 are even. The transition 12 produces the absorption line seen at 4.83

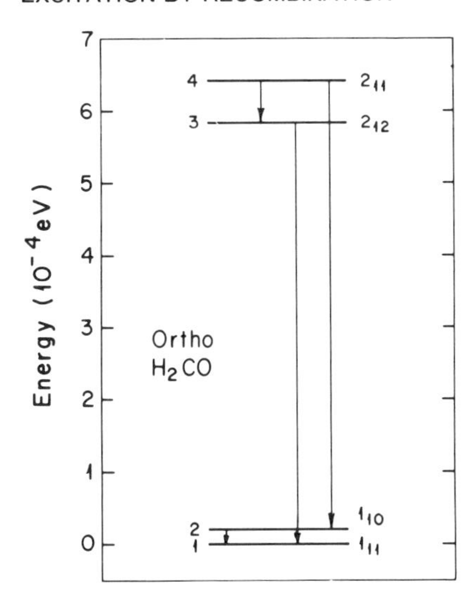

Figure 4.2 Energy level diagram for H_2CO. Energy levels are shown for ortho-formaldehyde, with permitted radiative transitions shown by arrows. Collisional transitions are possible between any two levels.

GHz (6.2 cm); transition 34 produces a line at 14.5 GHz (2.07 cm), also usually seen in absorption. The 13 and 24 transitions produce emission lines at about 2 mm. If γ_{13} and γ_{23}, the collisional excitation rate coefficients for collisions with H_2 molecules, are both larger than γ_{14} and γ_{24}, level 3 will be populated more frequently than level 4. If A_{31} much exceeds the collisional deexcitation rate, an excess population will result in level 1 as compared with level 2, which is populated by radiative transitions downward from level 4. Quantum-mechanical calculations of these coefficients show [22] that if the two next higher levels (3_{11} and 3_{12}) are also taken into account, this pumping mechanism can in fact explain the observed antimaser effects seen in the longer wavelength H_2CO lines. The particle density of H_2 required is somewhat greater than $10^2\,cm^{-3}$.

4.2 EXCITATION BY RECOMBINATION

When an electron recombines with an atom, or two atoms combine to form a molecule, the system is generally in an excited state. The excited system then decays in subsequent downward radiative transitions, cascading in due course down toward the ground state, usually passing through various other excited levels. Collisionally induced transitions may also be of some importance during this sequence of events. At some stage in this process, perhaps after a long wait in the ground state, the atoms will be ionized again, or the molecule dissociated and the process repeated. Within a volume element of the gas a steady population of excited levels will result. The relative population densities may be determined from equation (4-2), provided that additional terms are included to allow for the rate of ionization (or dissociation) from each level and the corresponding rate of recombination.

Computations of the population densities to be expected in atomic H have been carried out in considerable detail [1]; some of the main assumptions and results are summarized here. The lower and higher quantum levels (n less than or greater than about 40) are discussed separately below, partly because the relevant processes are different in these two cases and partly because the results are applied differently, with the former regions relevant to optical recombination lines and the latter relevant to radio lines.

a. Lower Quantum Levels

For bound hydrogenic states with total quantum number n less than about 40 (more specifically, less than 45 to 30 for n_e between 1 and $100\,\mathrm{cm}^{-3}$ [1]), it is necessary to consider the distribution of atoms in levels of different orbital angular momentum; that is, to compute $n_{n,l}$ for each n and l. Collisions with protons are the most effective for producing transitions between states of the same n but different l, but for these low values of the total quantum number and for densities of interest are generally less rapid than downward radiative transitions. Electron collisions are of primary importance in producing collisional changes of the total quantum number.

The relative populations in equilibrium are appreciably influenced by whether or not the Lyman photons, emitted in transitions to the $n=1$ level, escape from the emitting region [S3.3a]. Instead of solving the equation of transfer to determine $U(\nu)$ in equation (4-2), the computations have considered two limiting cases, with $U(\nu)=0$ in both cases. In Case A, all photons in the Lyman series, emitted when an electron jumps down to the $n=1$ level, are assumed to escape with no reabsorption. In Case B, complete reabsorption of all Lyman photons is assumed with no reabsorption at longer wavelengths. In this case, each higher Lyman photon (emitted from a level with $n \geqslant 3$) will be converted after at most a few absorptions and reemissions into photons of lower frequency and a Lyα photon, which either escapes in the line wings or is absorbed by dust [S5.1c]; hence in equation (4-2) all downward transitions to the $n=1$ level are ignored. Because of conversion of Lyman photons into Balmer ones, the total Hβ emission in Case B is about 1.7 times greater than in Case A, but the relative intensities of the various emission lines in the two cases differ by less than 10 percent.

In Section 3.3, production coefficients α_{mn} for photon emission were defined in equation (3-34). Values of these quantities for the more realistic Case B [1, 23] are given in Table 4.5 for $n_e \approx 1\,\mathrm{cm}^{-3}$. If $T \geqslant 10{,}000°\mathrm{K}$, these production coefficients increase by less than 4 percent as n_e increases to $10^6\,\mathrm{cm}^{-3}$. These emission rates vary about as $T^{-0.8}$. The values of $b_{n,l}$ are between about 0.1 and 1 at 10^4°K for $n=3$ or 4, and increase faster than proportionally with increasing T. At each T, α_{32} is about half the corresponding recombination coefficient $\alpha^{(2)}$ [equation (5-14)]. The values given in Table 4.5 have been used in Section 3.3 in the interpretation of H emission observations. General confirmation of this theory is provided by the excellent agreement [24] between the computed values of α_{n2}/α_{42} and the relative intensities of the Balmer lines in planetary nebulae, where the lines are strong enough to permit accurate measures.

Table 4.5. Photon Production Coefficients for Hα and Hβ

T(°K)	1,250	2,500	5,000	10,000	20,000
$10^{14} \times \alpha_{42}(\mathrm{cm}^3\,\mathrm{s}^{-1})$	14.8	9.1	5.35	3.01	1.61
$10^{14} \times \alpha_{32}(\mathrm{cm}^3\,\mathrm{s}^{-1})$	73	41	22.1	11.7	5.97

b. Higher Quantum Levels

For values of n greater than about 40, the computation of equilibrium populations is simplified, since levels of different orbital quantum number l are populated in accordance with their statistical weight. This results from collisions with protons, which become more rapid for higher n, while the spontaneous downward radiative probabilities diminish markedly. As a result, an atom will change its l value many times before n is changed, a sufficient condition for statistical equilibrium among these levels. Hence all states with the same total quantum number n may be grouped together in the calculations.

Ionization by electron collisions and recombination by three-body encounters now play a dominant role in the calculations. Approximate results [25] for b_n are given in Fig. 4.3 as a function of n for $T=10^4$°K and $n_e = 10^4\,\mathrm{cm}^{-3}$; Case B, described above, has been assumed. Computations [26, 27] with improved values of the various rate coefficients give similar results.

The dashed lines (a) and (b) in the plot for b_n show the solution when collisional ionizations (and recombinations) and collisional transitions between adjacent n levels, respectively, are each added separately to purely radiative processes, shown by the dotted line. With larger n the atomic

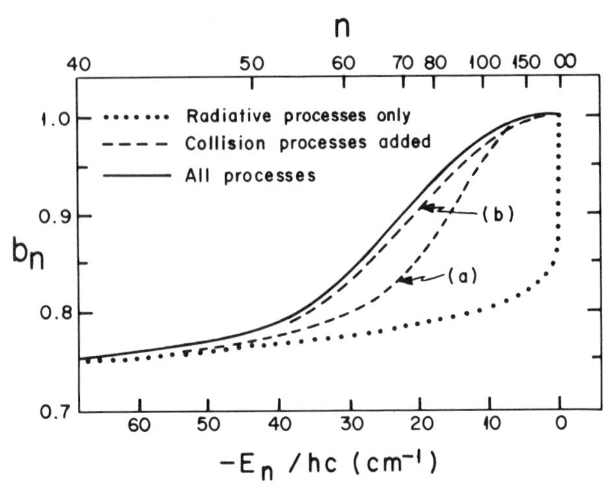

Figure 4.3 Values of b_n for recombining H atoms [25]. The quantity b_n gives the population of the H-atom level with total quantum number equal to n, relative to the population in thermodynamic equilibrium. Different curves show the effects produced by different processes. The dotted line is based on electron-ion radiative recombination followed by radiative downward cascading. In dashed curve (*a*) collisional ionization and three-body recombination are also included, while curve (*b*) includes instead collisional transitions from n to $n \pm 1$. The solid curve includes all these processes.

radius increases, collisions become more frequent, and b_n approaches unity more rapidly than in the radiative case. The solid curve shows b_n when all effects are included.

Such results make possible the determination of χ, the stimulated emission correction factor given in equation (3-30). For these high n levels, where b changes smoothly with n, we may write approximately

$$\chi = 1 - \frac{kT}{h\nu}\frac{d\ln b}{dn}\Delta n, \tag{4-26}$$

where Δn is the decrease in n for the emission line under consideration, and equals 1, 2, 3, and so on for the lines designated as $n\alpha$, $n\beta$, $n\gamma$, and so on. The frequency ν is given by

$$\nu = \nu_1\left[\frac{1}{n^2} - \frac{1}{(n+\Delta n)^2}\right] = 6.58\times 10^9 \frac{\Delta n}{n^3}\left[1 - \frac{1.5\Delta n}{n}\cdots\right]\text{MHz}, \tag{4-27}$$

where ν_1 is the frequency at the Lyman limit. Values [26] of χ are given in Table 4.6 for the $n\alpha$ line (i.e., for the transition from $n+1$ to n) for

Table 4.6. Absorption Correction Factor for $n\alpha$ Recombination Lines

Parameter		Value of $-\chi$ for Indicated Quantum Number n						
n_e (cm^{-3})	T (°K)	40	50	60	70	80	90	100
10	5×10^3	5.3	7.8	10.7	15.3	24	44	77
10	10^4	6.7	10.0	13.4	18.8	29	51	90
10^3	5×10^3	8.0	22.0	42	47	42	35	27
10^3	10^4	9.5	25.4	54	69	66	56	45
n_e	T	120	140	160	180	200	240	280
10	5×10^3	133	142	124	100	79	49	31
10	10^4	180	211	196	164	132	84	55
10^3	5×10^3	16.5	10.2	6.5	4.1	2.7	1.03	0.22
10^3	10^4	28.7	18.3	12.0	8.0	5.5	2.61	1.18

different values of n. Since ν is nearly proportional to Δn, χ will be about constant for lines of the same n but different Δn.

These results may be used to express the line emission and absorption coefficients, j_ν and κ_ν, in terms of their values j_ν^* and κ_ν^* in the ETE system [S2.4]. From equations (2-25), (3-14), and (3-29) we find

$$\frac{\kappa_\nu}{\kappa_\nu^*} = b_n\chi, \tag{4-28}$$

while from equation (3-10)

$$\frac{j_\nu}{j_\nu^*} = b_m. \tag{4-29}$$

In these equations the subscripts n and m denote the total quantum numbers of the lower and upper level, respectively. Since χ is negative over the entire range of values in Table 4.6, maser amplification is generally possible.

c. Radio Recombination Lines

The discussion above will now be applied to radio recombination lines, for which n generally exceeds 40. The emitted intensity is obtained in general by substituting j_ν/κ_ν from equation (3-32) into equation (3-3), the general solution of the equation of radiative transfer. The optical thickness $\tau_{\nu r}$

includes both continuum and line absorption. We evaluate $\tau_{\nu r}$ at the center of the recombination line, denoting by τ_L the contribution of line absorption to this quantity; we omit the subscript r. Similarly, we let τ_C denote the optical thickness of the region for continuous absorption.

The value of τ_L^*, equal to τ_L in the ETE system, is obtained from equation (3-16), with s found from equation (3-29), and with $\chi = 1$. We take equation (3-20) for $\phi_a(\Delta\nu)$, with $w=0$ at the line center. If we use equation (3-11) to relate n_n, the particle density in the lower level, to $n_e n_p$, as in the derivation of equation (3-34), we obtain, setting $g_n = 2n^2$,

$$\tau_L^* = \frac{\pi^{1/2}e^2h}{m_e}\frac{g_n f_{nm}}{kTf_e b}\frac{n_p}{n_e}E_m\,(\mathrm{cm}^{-5})$$
$$= 27.53\frac{n_p}{n_e}\frac{n^2 f_{nm}}{bT^{5/2}}E_m, \tag{4-30}$$

where we have substituted from equations (2-31) and (3-35) for f_e and the emission measure E_m, respectively, with n_e assumed constant and E_m in the second expression in pc cm^{-6}. The factor $\exp(-E_1/n^2kT)$ has been set equal to unity, a valid approximation for $n \geqslant 40$ and $T \gg 100°$.

We now evaluate the oscillator strength f_{nm} and the Doppler parameter b in equation (4-30). For large n the total upward oscillator strength is given by [28]

$$f_{n,n+\Delta n} = nM(\Delta n), \tag{4-31}$$

where $M(\Delta n)$ decreases from 0.191 for $\Delta n = 1$ to 0.0263, 0.0081, and 0.0034 for $\Delta n = 2$, 3, and 4, respectively. The velocity-spread parameter b obtained from equation (3-21), on the assumption that thermal motions dominate, is about 10^6 cm s^{-1} at 8000°K. [The frequency width at half maximum intensity, often used by observers, is related to b by equation (3-22).] From equation (4-30) we see that for this same T, for $\Delta n = 1$ and $n = 200$, $\tau_L^* \approx 10^{-8}E_m$. Inspection of the detailed values of χ in Table 4.6, taking into account that b_n is generally between 0.75 and 1 for these higher quantum levels (see Fig. 4.3), indicates that τ_L may be comparable with unity, giving some maser amplification, if τ_L^* exceeds 10^{-2} [see equation (4-28)]. Thus maser amplification is possible if E_m is as great as 10^6 cm^{-6} pc, a value surpassed in the densest visible H II regions [S3.3a] and in many of the compact H II regions [S3.5a], although at very high n_e, χ approaches unity.

We derive now a simple formula for the strength of a radio recombination line in the limit of small τ_L. In the radio region $h\nu/kT$ is small; the brightness temperature T_b can be substituted in place of I_ν, and equation

(3-7) can be used for $B_\nu(T)$. We define r as the excess relative intensity at the center of the recombination line, giving

$$r=\frac{T_{bL}-T_{bC}}{T_{bC}}, \tag{4-32}$$

where again subscripts L and C refer to the line center and the continuum, respectively. If we consider the case where τ_L is small, the absorption or maser amplification may be ignored. The emitted intensity may then be obtained most simply by integrating j_ν along the line of sight, neglecting the $\kappa_\nu I_\nu$ term in equation (3-1) and also $I_\nu(0)$, the intensity of the radiation incident on the far side of the region. We express the integral of $j_\nu ds$ for the line in terms of τ_L^* by means of equations (4-29), (3-5), and (3-2), with equation (3-7) used to express I_ν in terms of T_b. For the continuum radiation, $\kappa_\nu=\kappa_C=\kappa_C^*$, while at the line center, $\kappa_\nu^*=\kappa_C+\kappa_L^*$. Equation (4-32) now becomes

$$r=\frac{b_m\tau_L^*}{\tau_C}. \tag{4-33}$$

If we substitute τ_L^* from equation (4–30), and compute τ_C from equation (3-57) (with $n_p=n_e$ and $Z_i=1$), we find after some algebra and with the substitution of $\nu_1=2\pi^2 m_e e^4 h^{-3}$ in equation (4-27)

$$\frac{rb}{c}=\frac{3(3\pi)^{1/2}}{4}\,\frac{b_m\Delta n\,M(\Delta n)}{g_{ff}}\left(\frac{h\nu}{kT}\right). \tag{4-34}$$

The quantity rb/c is about equal to W_λ/λ, the dimensionless equivalent width of the emission line referred to the adjacent continuum. Equation (4-34) gives in principle a simple method for determining T, valid for H II regions with E_m not more than about $10^5\,\mathrm{cm}^{-6}$ pc. While the recombination lines which are strong enough to be measured accurately arise in regions of larger E_m, where maser amplification is significant, equation (4-34) usually gives the correct order of magnitude for rb/c.

More detailed computations of T_{bL}, based on solutions of equation (3-3) for negative $\tau_{\nu r}$ and on the detailed values of χ in Table 4.6, have been used to interpret the recombination line measurements. If absorption effects were entirely negligible, the quantity rb would vary as $1/T$, independent of n_e or E_m. Thus different lines would give the same information on T, but E_m or n_e could not be determined. When maser amplification is present, the situation alters. From equations (4-30) and (4-31) we see that τ_L^* varies as n^3; hence maser action tends to be more pronounced for higher n, despite some decrease of $|\chi|$ at larger values of n. From an analysis of several lines in the same source, it is possible to determine E_m, n_e, and T separately. The resultant values of T for seven H II regions [1] range from 7200 to 12,000°, averaging 9100°K, in good agreement with the results obtained in other ways [S11.1a]. The values of n_e range from 5×10^3 up to $7\times10^4\,\text{cm}^{-3}$; the corresponding values of E_m yield pathlengths between 0.008 and 0.1 pc, indicating that relatively compact, dense regions are being observed.

4.3 PHOTON PUMPING

After absorption of a photon an atom or molecule will usually cascade back through a variety of states, reaching levels which could not be populated by direct upward radiative transitions from the ground state. Examples of this process, which are considered below, are the excitation of the hyperfine levels of the H I ground state, excitation of the fine structure levels in O I, N II, and similar atoms, and the excitation of H_2 rotational levels. This indirect excitation of low-lying levels by photon absorption is called "photon pumping."

In this situation all the relevant energy levels may be divided into two groups, upper levels and lower levels. The upper levels are those from which downward radiative transitions to one or more of the lower levels occur so rapidly that the fraction of particles in these levels at any one time is negligibly small. Moreover, the amount of time spent in each upper level is so short that collisionally induced transitions out of each such level can be ignored. Under these conditions, the values of n_j for the lower levels only need be considered directly in equation (4-1). The upper levels can be treated indirectly as channels by which a particle initially in level j of the lower group can be excited to another level k in this same group through the mechanism of photon absorption in level j followed by radiative cascading down to level k. The rate coefficient $(R_{jk})_{\text{ph pump}}$ for this process, which must be included in equation (4-1), we denote by P_{jk}. Since collisional processes are assumed negligible for the upper levels, P_{jk} is independent of density.

We give now the basic equations for P_{jk}. If m represents all the quantum numbers for one of the upper levels, reached by photon absorption, then from the equations in Section 3.2 we have

$$P_{jk} = \sum_m \beta_{jm}\epsilon_{mk}, \tag{4-35}$$

where in general β_{jm}, the probability of an upward-induced transition from level j to level m per second, is related to $U(\nu)$, the energy density of radiation per cm^3 per Hz, by the expression

$$\beta_{jm} = \int s_\nu \frac{cU(\nu)}{h\nu} d\nu, \tag{4-36}$$

integrated over the absorption line. As in equation (3-12), the integral in equation (4-36) is normally entirely determined by the radiation at the line center $\nu = \nu_{jm}$, and if we ignore stimulated emission for these optical transitions, produced by dilute radiation, combining equations (4-36), (3-15), (3-23), and (3-24) gives the usual result that $\beta_{jm} = B_{jm} U(\nu_{jm})$. However, when very strong absorption lines are present, $U(\nu_{jm})$ is so small that the line wings must be considered in evaluating equation (4-36) [S4.3b].

The quantity ϵ_{mk} in equation (4-35) is the fraction of downward transitions from level m that populate level k when the atom or molecule first reaches the group of lower levels. In the simple case where no radiative cascading occurs among intermediate upper levels, lower level k must be reached in a single downward jump or not at all; ϵ_{mk} is then the branching ratio,

$$\epsilon_{mk} = \frac{A_{mk}}{\sum_j A_{mj}}. \tag{4-37}$$

From equations (4-37) and (3-13) we obtain

$$g_j \nu_{jm}{}^3 B_{jm} \epsilon_{mk} = g_k \nu_{km}{}^3 B_{km} \epsilon_{mj}, \tag{4-38}$$

giving detailed balancing in thermodynamic equlibrium at very high temperatures. For transitions within a multiplet, values of ϵ_{mj} are readily found from tables of relative line strengths [29].

a. Atomic Levels

Transitions between the two hyperfine structure states of the H I $n=1$ level (which we denote again by subscripts 1 and 2) can be excited by Lα radiation, which produces transitions up to the four hyperfine states of the $n=2$ level. Because of the very large optical depth in Lα, the energy density of this radiation is relatively high in regions where at least partial ionization of H is maintained by some process. Hence the effective photon pumping rate P_{12} for transitions between the two ground hyperfine states will usually be much larger than the collisional rate $n_H\gamma_{12}$ [30]. In such a case, n_2/n_1 will be dominated by the ratio of energy densities producing the transitions $1m$ and $2m$; that is, by $d\log U(\nu)/d\nu$ in the center of Lα. However, in a homogeneous medium the chief process that can modify this spectral slope significantly is the frequent scattering of Lα photons. Because of photon energy losses resulting from H-atom recoil in each collision, this spectral slope approaches an equilibrium value which gives $\ln(g_1n_2/g_2n_1)$ equal to $-E_{21}/kT$ after all [31], exactly as in LTE; again T is the kinetic temperature. In a medium with velocity gradients this result is less clear, and the possibility of maser amplification of the 21-cm line in some situations has not been entirely excluded.

Similarly, the N II fine-structure levels may be excited by absorption of photons at 1084 Å, producing transitions to the $^3D_1$ level with an excitation energy of 11.43 eV; we designate this level by the subscript 3. If subscripts 0, 1, and 2 denote the ^3P levels with the corresponding values of J, then we find that $\epsilon_{31}=15/36$, while $\epsilon_{32}=1/36$. Excitation of the $J=2$ or 3 levels in the ^3D term can be ignored, since most interstellar N^+ atoms will be in the $^3P_0$ level, from which transitions are permitted only to $J=1$ levels. Calculations show [32] that close to the exciting star in an H II region this process alters somewhat the values of n_1/n_0 and n_2/n_0 from those obtained with equation (4-14). Throughout most of the H II region this effect is relatively small and both n_1 and n_2 vary mainly as n_0n_e. Since the ratio of n_0 to n_p is about equal to the relative abundance of N with respect to H, the column density of excited N II ions, determined from ultraviolet absorption line measures, is proportional to $E_m \times N_N/N_H$. Separate estimates of E_m give nitrogen abundances about equal to the cosmic abundances in Table 1.1.

b. H_2 Rotational Levels

Another example of photon excitation occurs in the H_2 molecule [33]. In the H_2 energy level diagram, shown in Fig. 4.4, the solid curves show the potential energy $V(r)$ governing the motion of the two protons. The quantity $V(r)$ is the sum of the Coulomb energy of the two protons plus the quantized binding energy of the two electrons, depending on their quantum state; the curves are drawn for the ground state $^1\Sigma_g$, denoted by X, and the first two excited singlet states, $^1\Sigma_u$ and $^1\Pi_u$, denoted by B and C, respectively. Vibrational energy levels are shown by thin horizontal lines. The absorption and emission features produced by the various transitions between X and either B or C are called "Lyman or Werner bands," respectively. The symbols $R(J)$ or $P(J)$ denote lines produced in such a band by transitions in which the rotational quantum number equals J in the lower level and $J+1$ or $J-1$, respectively, in the upper level.

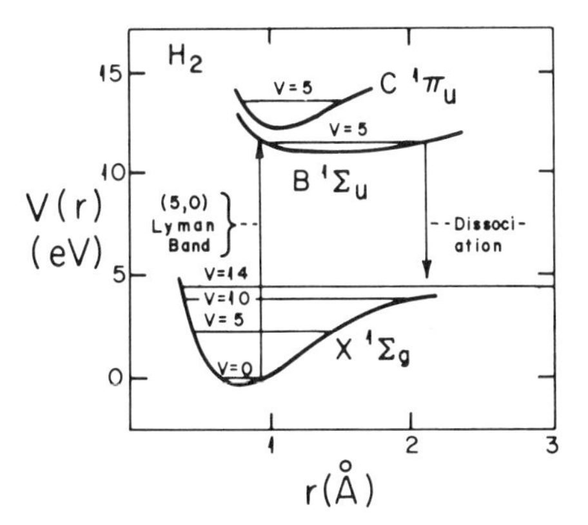

Figure 4.4 Energy level diagram for H_2 molecule [33]. The heavy curves represent the potential energy, $V(r)$ of the system as a function of the proton separation r in three electronic states, including the ground state X. The light horizontal lines represent energy levels for different values of v, the vibrational quantum number.

Under interstellar conditions H_2 molecules are almost entirely concentrated in the lowest $(v=0)$ vibrational level of electron state X. We denote by $n(J)$ the particle density of molecules in this ground vibrational level with a rotational quantum number equal to J; $N(J)$ denotes the corresponding column density per cm^2. Molecules in such a J level can absorb an ultraviolet photon and jump to any vibrational level of state B, with $\Delta J = \pm 1$. They then jump back down to any vibrational level v of the ground electronic state X. Transitions to levels with $v > 14$ (or with $J > 5$ for $v = 14$) lead to dissociation of the molecule [33, 34].

Once the molecule has reached a bound level for electron state X, it will continue to cascade down in a series of rotational-vibrational transitions. The spontaneous transition probabilities for such transitions in H_2 are all very small, since quadrupole radiation is required. In this homonuclear molecule, the center of mass of the two nuclei always coincides with the center of charge, and no electric dipole moment can arise. Thus for any excited vibrational level the sum of the A_{kj} values to all lower levels is between about $0.7 \times 10^{-6} s^{-1}$ and $5.2 \times 10^{-6} s^{-1}$ [35]. Once a molecule has reached the $v=0$ level, it will cascade down through the rotational levels; the corresponding values of A_{kj} for these quadrupole radiation transitions, shown in Table 4.7 [36], are considerably smaller, and the mean radiative

Table 4.7. Spontaneous Radiation Probabilities $A_{J,J-2}$ for H_2 in Ground Vibrational Level

J of Upper Level	2	3	4	5	6
$A_{J,J-2}(s^{-1})$	2.95×10^{-11}	4.77×10^{-10}	2.76×10^{-9}	9.85×10^{-9}	2.65×10^{-8}

life of the levels with $J=2$ or 3 exceeds the mean time to collisional deexcitation, if n_H exceeds about 10 or 300 cm^{-3}, respectively.

An important characteristic of the H_2 molecule which must be considered in such analyses results from the quantum mechanical requirement that the wave function be antisymmetrical in the positions of the two protons. As a result, when the proton spins are parallel, the electron wave function must be invariant except for a change of sign on rotation of π radians about the axis of symmetry, since this rotation interchanges the two protons. When the proton spins are antiparallel, on the other hand, the electron wave function must be completely unchanged on rotation through an angle π, since reversing the spin direction for each of the two protons changes the sign of their contribution to the wave function. As a result, H_2 molecules (like H_2CO molecules [S4.1d]) are divided into two types of states, orthohydrogen with proton spins parallel and parahydrogen with proton spins antiparallel. For electron state X all states of odd J are orthohydrogen and those of even J are parahydrogen. Radiative transitions between these two types of states occur at a negligible rate [33], and as a result, P_{JK}, the rate coefficient for optical pumping between levels of rotational quantum numbers J and K (both with $v=0$), will be effectively nonzero only if J and K are either both even or odd. Under interstellar conditions encounters with protons, which can change the nuclear spins by an exchange process, provide the chief collisional process for converting ortho- to parahydrogen [S5.3b].

Equation (4-35) may now be applied to this cascade process. The lower group of levels, among which collisional as well as radiative transitions must be considered, is limited to the different rotational states of the $v=0$ level, with the electrons in state X. A large variety of upper levels in electron states B and C can be reached by absorption. We replace j and k, the general quantum numbers of specific lower levels, by J and K, the initial and final rotational quantum numbers. The cascade probabilities ϵ_{mK} needed in equation (4-35) for P_{JK} may be computed from the transition probabilities down to various v'',J'' levels of state X [37], and the subsequent cascade efficiency factors $a(v''J'';K)$, defined as the fraction of molecules initially in level v'',J'', of state X, which will be in a level with rotational quantum number K when the vibrational quantum number reaches zero [35].

In the final equation for P_{JK}, we must take into account that a certain fraction k_m of transitions to an upper level will produce cascades to the $v'' \geqslant 14$ level of state X and thus produce dissociation rather than cascade down to the $v''=0$ level. Hence the sum of ϵ_{mK} over all K will equal $1-k_m$ rather than unity. The value of k_m [34] increases sharply with v', and is nearly zero for the Werner lines longward of 912 Å, including only through the (5, 0) band. In equilibrium the number of new molecules formed per cm^3 per second must just equal the number destroyed [S5.3a], which equals a mean value of k_m, denoted simply by $\langle k \rangle$, times the sum of all the upward transition probabilities times the number of H_2 molecules per cm^3 [see equation (5-42)]. If we let $G(K)$, the "formation distribution function," be the fraction of new molecules that are in rotational level K when they first reach the ground vibrational level, we find that equation (4-35) becomes

$$P_{JK} = \beta(J)\left[\epsilon_{JK} + \langle k \rangle G(K)\right], \tag{4-39}$$

where $\beta(J)$, the total probability per second of an upward radiative transition from level J is given by

$$\beta(J) = \sum_m \beta_{Jm} \tag{4-40}$$

and

$$\epsilon_{JK} = \sum_m \frac{\beta_{Jm}\epsilon_{mK}}{\beta(J)}. \tag{4-41}$$

Evidently all the quantities in equation (4-39) except $G(K)$ depend on the spectrum of the exciting radiation. In clouds with high H_2 column densities, self-absorption may sharply reduce $U(\nu)$ in the central region of the strong $R(0)$, $R(1)$, and $P(1)$ lines, reducing $\beta(0)$ and $\beta(1)$ markedly.

The values of P_{JK} for optical pumping obtained from equation (4-39) may be inserted into equation (4-1) and the relative populations for the "lower levels" (the rotational levels of the ground vibrational level) determined. The terms for collisional and spontaneous radiative transitions must be introduced as in equation (4-2), with the appropriate numerical values for γ_{JK} [5] and A_{KJ} (see Table 4.7). For the sum of P_{JK} over all K, which must be included on the left-hand side of equation (4-2), we find

$$\sum_K P_{JK} = \beta(J). \tag{4-42}$$

Outside the H_2 absorbing clouds, $\beta(J)$ may be set equal to β_0, with a dependence on J so slight that it may be ignored.

The quantities P_{JK} and $\langle k \rangle$ have been computed and the resultant equations of equilibrium have been solved [38] for $n(J)/n(0)$ in relatively transparent clouds, in which self-absorption of the H_2 lines is negligibly small. The results have been fitted to the column densities $N(J)$ obtained from ultraviolet H_2 absorption measures to determine β_0 and n_H, the overall hydrogen density [assumed about equal to n(H I), the density of neutral hydrogen] which multiplies γ_{KJ}, the collisional rate coefficient in $(R_{KJ})_{coll}$. The computed value found for $\langle k \rangle$, averaged over the Werner and Lyman bands with the interstellar radiation field in Table 5.5, is 0.11. While $G(K)$ is essentially unknown, comparison of predicted and observed $N(J)$ values suggest that $G(K)$ is peaked at values of K well above 0 or 1, as would be expected if an appreciable fraction of the energy released in H_2 molecule formation [S6.2b] goes into vibrational and rotational excitation. Theoretical calculations have been made also for the relatively thick clouds [39] in which the strong $R(0)$, $R(1)$, and $P(1)$ lines are on the square-root section of the curve of growth [S3.4c], with each β_{Jm} reduced by the appropriate K_{Jm} factor [S5.3a] [see equations (5-43) and (5-44)]. Again, values of n_H and β_0 were determined by fitting the observed column densities.

The distribution of β_0 values found in these studies is shown in Fig. 4.5. For comparison, the anticipated value of β_0 computed for mean conditions

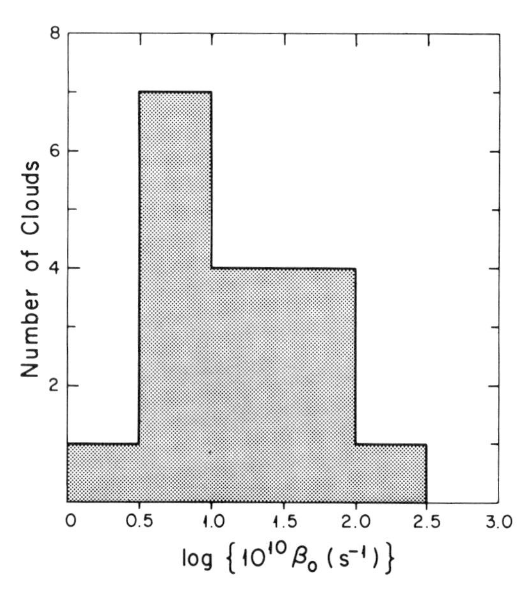

Figure 4.5 Distribution of β_0 values. Plotted in the histogram are the number of β_0 values within each interval of 0.5 in $\log \beta_0$. Individual values were determined [38, 39] by applying optical pumping theory [35] to the observed H_2 column densities in excited rotational levels. The expected mean value of β_0 for interstellar starlight is about $5 \times 10^{-10} s^{-1}$, in agreement with the observed peak at $\log(10^{10}\beta_0) = 0.7$.

in interstellar space, using the radiation field near the Sun given in Table 5.5 [S5.2a], is $5 \times 10^{-10} s^{-1}$ [40]. Evidently the observed peak in the β_0 distribution agrees well with this theoretical computation. However, about half the observed clouds show much higher values of β_0 and are apparently rather close to some early-type star; a distance of some 20 to 30 pc from the O9 stars that are typical of those being observed would be consistent with these high values of β_0.

These clouds of relatively high β_0 tend to show higher n_H than do the others, with values mostly between 100 and $1000 cm^{-3}$, as compared to a range of n_H between 10 and $100 cm^{-3}$ for clouds of average β_0. A few of these high β_0 clouds, studied in more detail [41], show velocities of approach exceeding $10 km\ s^{-1}$ with respect to the clouds of greatest H I column density N(H I), and have relatively low N(H I) combined with high n_H. These absorbing regions must be thin sheets of material, less than about 0.1 pc thick, such as would be produced by the outward moving shocks discussed in Chapter 12.

Excitation of the $J=1$ rotational level of H_2 by proton collision with a molecule in the $J=0$ level is considered subsequently [S5.3b]. For conversion of para- into orthohydrogen, this process competes with the formation of a new molecule following photodissociation.

REFERENCES

1. D. E. Osterbrock, *Astrophysics of Gaseous Nebulae,* W. H. Freeman (San Francisco), 1974, Chapts. 3 and 4.
2. W. L. Wiese, M. W. Smith, and B. M. Glennon, *Atomic Transition Probabilities I,* National Standard Reference Data Series, National Bureau of Standards 4, U. S. Gov't. Printing Office, 1966.
3. A. C. Allison and A. Dalgarno. *Ap. J.,* **158**, 423, 1969.
4. J-M. Launay and E. Roueff, *J. Phys. B* (*Atom. Mol. Phys.*), **10**, 879, 1977.
5. S. Nishimura, *Ann. Tokyo Astr. Obs.,* Ser. 2, **11**, 33, 1968.
6. S. Green and P. Thaddeus, *Ap. J.,* **205**, 766, 1976.
7. D. Osterbrock and E. Flather, *Ap. J.,* **129**, 26, 1959.
8. H. M. Johnson, *Stars and Stellar Systems,* Vol. 7, University of Chicago Press (Chicago), 1968, Table 7.
9. D. M. Rank, C. H. Townes, and W. J. Welch, *Science,* **174**, 1083, 1971.
10. B. Zuckerman and P. Palmer, *Ann. Rev. Astron. Astroph.,* **12**, 279, 1974.
11. C. H. Townes and A. L. Schawlow, *Microwave Spectroscopy,* McGraw Hill (New York) 1955, Chapter 1 and Appendix VI.
12. F. J. Lovas and E. Tiemann, *J. Phys. Chem. Ref. Data,* **3**, 609, 1974.
13. P. Goldreich and J. Kwan, *Ap. J.,* **189**, 441, 1974.
14. N. S. Scoville and P. M. Solomon, *Ap. J.* (*Lett.*), **187**, L67, 1974.
15. C. Heiles, *Ann. Rev. Astron. Astroph.,* **9**, 293, 1971.
16. W. B. Burton, *Ann. Rev. Astron. Astroph.,* **14**, 275, 1976.
17. W. B. Burton and M. A. Gordon, *Ap. J.* (*Lett.*), **207**, L189, 1976.

18. A. A. Penzias, P. M. Solomon, K. B. Jefferts, and R. W. Wilson, *Ap. J.* (*Lett.*), **174**, L43, 1972.
19. A. A. Penzias, Lectures at Les Houches Summer School, 1975.
20. R. L. Dickman, *Ap. J.,* **202**, 50, 1975.
21. C. H. Townes and A. C. Cheung, *Ap. J.* (*Lett.*), **157**, L103, 1969.
22. B. J. Garrison, W. A. Lester, W. H. Miller, and S. Green, *Ap. J.* (*Lett.*), **200**, L175, 1975.
23. R. M. Pengelly, *M.N.R.A.S.,* **127**, 145, 1964.
24. J. S. Miller, *Ann. Rev. Astron. Astroph.,* **12**, 331, 1974.
25. M. J. Seaton, *M.N.R.A.S.,* **127**, 177, 1964.
26. M. Brocklehurst, *M.N.R.A.S.,* **148**, 417, 1970.
27. A. Burgess and H. P. Summers, *M.N.R.A.S.,* **174**, 345, 1976.
28. D. H. Menzel, *Ap. J. Supp.,* 18, **221**, 1969; No. 161.
29. C. W. Allen, *Astrophysical Quantities,* Athlone Press (London), 1963, p. 56.
30. G. B. Field, *Proc. Inst. of Radio Engineers,* **46**, 240, 1958.
31. G. B. Field, *Ap. J.,* **129**, 551, 1959.
32. L. Spitzer and E. B. Jenkins, *Ann. Rev. Astron. Astroph.,* **13**, 133, 1975.
33. G. B. Field, W. B. Somerville, and K. Dressler, *Ann. Rev. Astron. Astroph.,* **4**, 207, 1966.
34. A. Dalgarno and T. L. Stephens, *Ap. J.* (*Lett.*), **160**, L107, 1970.
35. J. H. Black and A. Dalgarno, *Ap. J.,* **203**, 132, 1976.
36. A. Dalgarno and E. L. Wright, *Ap. J. (Lett.),* **174**, L49, 1972.
37. A. C. Allison and A. Dalgarno, *Atomic Data,* **1**, 289, 1970.
38. M. Jura, *Ap. J.,* **197**, 575, 1975.
39. M. Jura, *Ap. J.,* **197**, 581, 1975.
40. M. Jura, *Ap. J.,* **191**, 375, 1974.
41. L. Spitzer and W. A. Morton, *Ap. J.,* **204**, 731, 1976.

5. Ionization and Dissociation

The ionization and dissociation equilibrium of interstellar atoms and molecules is again governed by the steady-state equation (4-1). Thus to compute the ratio of ionized to neutral atoms or of molecules to separate atoms is, in principle, a straightforward problem of enumerating all the relevant physical processes and determining the probability rate $(R_{jk})_Y$ for each of these. In general we shall use a subscript f to denote a free state, either for an electron or an atom.

We give first the equations for ionization equilibrium when ionization by photon absorption is balanced by radiative recombination; this important situation has received the most theoretical study. The probability rate $(R_{jf})_{\text{phot}}$ for radiative transitions from level j to the free state we denote by β_{jf}. The corresponding radiative recombination probability $(R_{fj})_{\text{phot}}$, resulting from collisions of electrons with ions, we set equal to $\alpha_j n_e$, where α_j is the "recombination coefficient" for this process, and n_e is the electron density. In terms of these quantities equation (4-1) gives the simple equation of ionization equilibrium:

$$\sum_j n_j(\mathrm{X}^{(r)})\,\beta_{jf} = \sum_j n_e n(\mathrm{X}^{(r+1)})\alpha_j. \qquad (5\text{-}1)$$

This equation may also be derived by summing equation (4-1) over all bound levels j, since the transitions between such bound levels cancel out in the sum. The equation for dissociative equilibrium of molecules is similar to equation (5-1) if radiative attachment and radiative dissociation are the two primary processes involved.

Under interstellar conditions, only the ground level is usually populated, although the excited fine-structure levels of the ground term must sometimes be taken into account. Here we shall ignore on the left-hand side of equation (5-1) all levels other than the ground level ($j=1$), although recombinations in various excited levels must be considered on the right-hand side. If we denote by x the fraction of atoms of element X that are ionized $r+1$ times, and assume that all other atoms of this element are ionized r times, equation (5-1) then assumes the simple form

$$(1-x)\,\beta_{1f}=xn_e\sum_j \alpha_j = xn_e\alpha, \tag{5-2}$$

where α denotes the total recombination coefficient.

The ionization probability β_{jf} is related by equation (4-36) to the absorption coefficient $s_{f\nu}$ for a transition leaving the electron in a free state; the integral in equation (4-36) now extends over the entire range of frequencies in the continuous absorption spectrum (from threshold frequency to infinity). The recombination coefficient α_j in equation (5-2) is a collisional rate coefficient and with the use of an equation similar to (4-5) may be expressed in terms of σ_{cj}, the cross section for capture of an electron (or of an atom in case of molecule formation) in state j. Both β_{jf} and α_j depend on the type of atom or molecule under consideration and also on the stage of ionization. For atoms in diffuse interstellar clouds, $s_{f\nu}$ is usually important only for transitions from the ground ($j=1$) level.

The cross sections $s_{f\nu}$ and σ_{cj} may be related to each other through the usual condition of detailed balance in thermodynamic equilibrium. As with the derivation of equation (4-8) we obtain the expression (sometimes referred to as the "Milne relation")

$$\sigma_{cj}=\frac{g_{r,j}}{g_{r+1,1}}\,\frac{h^2\nu^2}{m_e^2c^2w^2}\,s_{f\nu}, \tag{5-3}$$

where equations (2-17), (2-28), (2-32), (3-4), and (4-9) have been used. As in Section 2.4, $g_{r,j}$ is the statistical weight of the level which is doing the absorbing and into which the electron recombination is considered, whereas $g_{r+1,1}$ is the statistical weight of the ion in its ground level.

The continuous absorption coefficient $s_{f\nu}$ may be expressed in terms of the oscillator strength for the ionizing transition, with equations (3-14), (3-23), and (3-25). The stimulated emission correction in equation (3-23) is generally negligible for interstellar ionizing radiation. Since the variation of $s_{f\nu}$ with frequency is determined by atomic structure, we introduce this variation into the definition of oscillator strength, defining f_ν as the cumulative oscillator strength for all frequencies less than ν. Then we obtain

$$s_{f\nu}=\frac{\pi e^2}{m_e c}\frac{df_\nu}{d\nu}=\frac{8.07\times10^{-18}}{Z^2}\,\frac{df_\nu}{d(\nu/\nu_1)}\,\text{cm}^2, \tag{5-4}$$

where ν_1 is a reference frequency, which we take to be Z^2 times the frequency at the H I Lyman limit, corresponding to one Rydberg. With this definition equation (5-4) and subsequent equations (S5.1a) are directly applicable to all one-electron atoms of nuclear charge Z.

5.1 IONIZATION OF HYDROGEN

Because of the overwhelming abundance of H, the ionization of this element is of great importance in determining the physical state of the interstellar gas, as discussed in Chapter 1. In addition, the principal ionization and recombination probabilities for H may be computed. In this section we discuss the ionization equilibrium of H, with some brief consideration of He also, since this latter atom may contribute appreciably to the electron density near the hotter stars.

a. Absorption and Recombination Coefficients

For atomic hydrogen and other one-electron atoms we replace the subscript j by n, the total quantum number. The value of $df_\nu/d\nu$ for such atoms may be expressed as

$$\frac{df_\nu}{d(\nu/\nu_1)}=\frac{16}{3^{3/2}\pi}\left(\frac{\nu_1}{\nu}\right)^3\frac{g_{nf}}{n^5}, \tag{5-5}$$

where g_{nf} is the "Gaunt factor" for the bound-free transition, which is generally about unity near threshold, and which for absorption from $l=0$ levels decreases as $(\nu_1/\nu)^{1/2}$ for large ν/ν_1 [1]. Equations (5-4) and (5-5) yield for the absorption coefficient $s_{f\nu}$ for a hydrogenic atom in the $n=1$ level,

$$s_{f\nu}=\frac{7.91\times10^{-18}}{Z^2}\left(\frac{\nu_1}{\nu}\right)^3 g_{1f}\ \text{cm}^2. \tag{5-6}$$

The factor g_{1f} is given by

$$g_{1f}=8\pi 3^{1/2}\frac{\nu_1}{\nu}\frac{e^{-4z\cot^{-1}z}}{1-e^{-2\pi z}}, \tag{5-7}$$

where

$$z^2=\frac{\nu_1}{\nu-\nu_1}. \tag{5-8}$$

Values of g_{1f} are given in Table 5.1.

Table 5.1. Gaunt Factor for Photoionization from Ground Level

$Z^2\lambda(\text{Å})$	912	760	651	570	507	456
ν/ν_1	1.0	1.2	1.4	1.6	1.8	2.0
g_{1f}	0.797	0.844	0.878	0.905	0.926	0.942
$Z^2\lambda(\text{Å})$	304	182	91.2	45.6	22.8	9.12
ν/ν_1	3	5	10	20	40	100
g_{1f}	0.985	0.994	0.939	0.830	0.694	0.515

For recombination, σ_{cn} is given by equations (5-3), (5-4), and (5-5) and becomes

$$\sigma_{cn} = A_r \frac{\nu_1}{\nu} \frac{h\nu_1}{\frac{1}{2}m_e w^2} \frac{g_{nf}}{n^3}, \tag{5-9}$$

where the "recapture constant" A_r is given by

$$A_r = \frac{2^4}{3^{3/2}} \frac{he^2}{m_e^2 c^3} = 2.105 \times 10^{-22}\ \text{cm}^2. \tag{5-10}$$

The kinetic energy of the electron before capture is $\frac{1}{2}m_e w^2$, and $h\nu$ is the energy of the emitted photon, equal to $\frac{1}{2}m_e w^2 + h\nu_1/n^2$. The recombination coefficient α_n to level n equals $\langle w\sigma_{cn}\rangle$ [see equation (4-5)]. Here we shall be interested in the coefficients $\alpha^{(1)}$ and $\alpha^{(2)}$, where

$$\alpha^{(n)} = \sum_{n}^{\infty} \alpha_m. \tag{5-11}$$

Thus $\alpha^{(1)}$ is the total recombination coefficient, summed over all levels, denoted in equation (5-2) simply by α, whereas $\alpha^{(2)}$ is the recombination coefficient excluding captures to the $n=1$ level. To express $\alpha^{(1)}$ and $\alpha^{(2)}$ in a convenient form we introduce the function $\phi_n(\beta)$ defined by the relationship

$$\alpha^{(n)} = 2A_r \left(\frac{2kT}{\pi m_e}\right)^{1/2} \beta\phi_n(\beta), \tag{5-12}$$

where

$$\beta = \frac{h\nu_1}{kT} = \frac{158{,}000 Z^2}{T}. \tag{5-13}$$

Table 5.2. Recombination Coefficient Functions ϕ_1 and ϕ_2

$T\,(°\mathrm{K})/Z^2$	31.3	62.5	125	250	500	1,000
ϕ_1	4.68	4.36	4.04	3.71	3.38	3.05
ϕ_2	3.89	3.57	3.25	2.92	2.60	2.27
$T\,(°\mathrm{K})/Z^2$	2,000	4,000	8,000	16,000	32,000	64,000
ϕ_1	2.73	2.40	2.09	1.79	1.50	1.22
ϕ_2	1.96	1.64	1.34	1.06	0.80	0.59
$T\,(10^6\,°\mathrm{K})/Z^2$	0.128	0.256	0.512	1.0	10	100
ϕ_1	0.96	0.73	0.52	0.36	0.074	0.011

Values of $\phi_1(\beta)$ and $\phi_2(\beta)$, computed with a series expansion for g_{nf} [2], are given in Table 5.2. For temperatures above $10^5\,°\mathrm{K}$, values of ϕ_1 only are listed, since ϕ_2 is generally not relevant for such hot regions; for T equal to 10^7 and $10^8\,°\mathrm{K}$, g_{nf} was set equal to unity, about the effective value computed at $10^6\,°\mathrm{K}$. The errors in this table resulting from the approximate evaluation of g_{nf} are believed to be at most 2 percent for $T/Z^2 \leqslant 16{,}000\,°\mathrm{K}$, with a possible error as great as 10 percent for $T/Z^2 = 10^6\,°\mathrm{K}$. Inserting numerical values from equations (5-10) and (5-13) into equation (5-12), we obtain

$$\alpha^{(n)} = \frac{2.06 \times 10^{-11} Z^2}{T^{1/2}} \phi_n(\beta)\ \mathrm{cm}^3\ \mathrm{s}^{-1}. \tag{5-14}$$

b. H II Regions without Dust

The equations above make it possible to compute the extent of the photon-ionized gas around an early-type star. One fundamental characteristic of the problem is the very large value of τ_ν for ionizing radiation if the hydrogen is neutral. Equation (5-6) together with Table 5.1 gives a value of 6.3×10^{-18} cm^2 for $s_{f\nu}$ at threshold ($\nu = \nu_1$); we ignore absorption from excited levels, and we drop the subscript f from $s_{f\nu}$ in this section. If $n(\mathrm{H\ I})$ is 1 per cm^3, it follows that the mean free path for an ionizing photon is only 0.05 pc at threshold, increasing to 5 pc only for λ as short as about 180 Å, where presumably little energy is radiated except by the very hottest stars. As a consequence, if much of the hydrogen is neutral outside the regions ionized by a hot star, the opacity of the interstellar gas will be so great that none of the hydrogen will be ionized by photons. Unless the hydrogen density is very low indeed, the transition zones between H I and

H II regions must be very sharp, less than 1 parsec in thickness except for the very hottest stars.

We compute now from equation (5-2) the "Strömgren radius" r_S of the ionized H II region [3] surrounding a star with a high ultraviolet luminosity. It should be noted that n_e may slightly exceed n_p, the proton density, because of He ionization.

The energy density U_ν that must be used in equation (4-36) for β_{1f} includes two contributions, the stellar radiation field, of density $U_{s\nu}$, and photons emitted by electron captures in the $n=1$ level; these latter photons produce diffuse radiation in the Lyman continuum, whose energy density we denote by $U_{D\nu}$. For the stellar radiation at a distance r from the central star we have

$$U_{s\nu} = \frac{L_\nu e^{-\tau_\nu}}{4\pi r^2 c}, \tag{5-15}$$

where $L_\nu d\nu$ is the stellar luminosity in the interval $d\nu$, whereas τ_ν is the optical thickness from the star to the distance r, assumed much greater than the stellar radius. From equations (3-2) and (3-14) we have

$$d\tau_\nu = (1-x) n_{\mathrm{H}} s_\nu \, dr, \tag{5-16}$$

where x again denotes the fractional ionization. Absorption by dust particles and He atoms are considered subsequently. If we multiply equation (5-15) by $4\pi r^2 c$ and differentiate, we obtain

$$\frac{1}{r^2}\frac{d}{dr}(r^2 c U_{s\nu}) = -(1-x) n_{\mathrm{H}} s_\nu c U_{s\nu}. \tag{5-17}$$

Next we must consider the diffuse radiation field, which contributes to the rate of ionization at any point. If $\mathfrak{F}_{D\nu}$ is the net flux of this radiation, per cm^2 per second per frequency interval, then equation (3-1) of radiative transfer, expressed in spherical coordinates and integrated over all solid angles, gives

$$\frac{1}{r^2}\frac{d}{dr}(r^2 \mathfrak{F}_{D\nu}) = -(1-x) n_{\mathrm{H}} s_\nu c U_{D\nu} + 4\pi j_{D\nu}, \tag{5-18}$$

where $j_{D\nu}$ is the emissivity for this diffuse radiation in the Lyman continuum. This relation simply equates the divergence of the diffuse flux to the source term minus the sink. If we now add equations (5-17) and (5-18), $U_{s\nu}$ and $U_{D\nu}$ on the right-hand side combine to give the total energy density U_ν at frequency ν. If we substitute from equation (5-2) for $1-x$, and multiply the resultant equation by $d\nu/h\nu$, integrating over all frequency, the integral of $cs_\nu U_\nu d\nu/h\nu$ in the numerator cancels β_{1f} in the denominator [see equation (4-36)], and we obtain on multiplying through by $4\pi r^2$

$$\frac{d}{dr}\left\{4\pi r^2\int_{\nu_1}^{\infty}(cU_{s\nu}+\mathcal{F}_{D\nu})\frac{d\nu}{h\nu}\right\}=-4\pi r^2\left\{xn_e n_{\rm H}\alpha-4\pi\int_{\nu_1}^{\infty}\frac{j_{D\nu}\,d\nu}{h\nu}\right\}. \tag{5-19}$$

The second of the two terms in the brackets on the right-hand side is simply the total number of electron captures in the ground state per cm^3 per second, which equals $xn_e n_{\rm H}\alpha_1$, and may be combined with the first term. If we define $S_u(r)$ as the total number of photons flowing through a shell of radius r per second with a frequency greater than ν_1 (the frequency at the Lyman limit), the quantity in brackets on the left-hand side equals $S_u(r)$, and we obtain

$$\frac{dS_u(r)}{dr}=-4\pi r^2 xn_e n_{\rm H}\alpha^{(2)}, \tag{5-20}$$

where $\alpha^{(2)}$ has been defined in equation (5-11).

Equation (5-20) gives the physically reasonable result that the net number of ultraviolet photons streaming outward from the star is decreased by the number of recombinations to excited levels taking place within the region. Such captures in the second quantum level or higher produce radiation longward of the Lyman limit, which yields no further ionization, and promptly escapes from the H II region. Electrons captured in the ground level, on the other hand, emit an ultraviolet photon, which will promptly be absorbed again before it has gone very far. Not until the electron is captured in a high quantum level will at least some of the resultant photons escape. Thus for computing the reduction of $S_u(r)$, captures in the ground level may be ignored.

Equation (5-20) may now be integrated over dr out to the radius r_S, defined as the value of r at which $S_u(r)$ has decreased to zero. If $n_{\rm H}$ and n_e are assumed constant and x is set equal to unity throughout the region, one finds

$$\frac{4\pi}{3}r_S^3 n_e n_{\rm H}\alpha^{(2)}=S_u(0)\equiv N_u. \tag{5-21}$$

At small r the diffuse flux $\mathcal{F}_{D\nu}$ is negligible, and $S_u(r)$ equals the number of ultraviolet photons radiated by the central star. The assumptions that x is nearly unity and that s_ν is nearly constant with ν are both realistic outside the transition layer, but the assumption of uniform n_H neglects the inhomogeneities that are undoubtedly present.

Values of r_S determined [4] from theoretical models of main-sequence early-type stars are given in Table 5.3, together with the assumed radii in solar units, effective temperatures in degrees Kelvin, and the total rate of photon emission shortward of the Lyman limit. The values of $r_S(n_e n_H)^{1/3}$ have been computed for a kinetic temperature of 8000°K, for which $\alpha^{(2)}$ in equation (5-14) equals 3.09×10^{-13} cm^3 s^{-1}. The increased radii for stars of luminosity class I yield values of r_S greater than in Table 5.3 by factors of 1.5 to 2.0 for spectral types from O7 to B0, respectively [4]. The values of the dust optical thickness τ_{Sd} in the last column are discussed below [S5.1c]. Regardless of the detailed variation of s_ν with ν, or of the resultant structure of the transition layer, equation (5-21) and Table 5.3 give, with a relative accuracy of about $(n_H s_\nu r_s)^{-1}$, the radius of the shell at which x falls to about 0.5. These idealized spherical H II regions are sometimes referred to as "Strömgren spheres."

Table 5.3. Properties of H II Regions

Spectral Type	T_{eff} (°K)	$R/R_\odot$	$N_u \times 10^{-48}$ (s^{-1})	$r_S(n_e n_H)^{1/3}$ (pc cm^{-2})	$\tau_{Sd} n_H^{-1/3}$ (cm)
O5	47,000	13.8	51	110	0.69
O6	42,000	11.5	17.4	77	0.48
O7	38,500	9.6	7.2	57	0.36
O8	36,500	8.5	3.9	47	0.29
O9	34,500	7.9	2.1	38	0.24
B0	30,900	7.6	0.43	22	0.14
B1	22,600	6.2	0.0033	4.4	0.028

We now show that the Balmer radiation from H atoms within an H II region is directly related to the stellar ionizing flux. Virtually every recombining electron will reach the ground state before it absorbs another ionizing photon. The only way such an electron can avoid passing through the $n=2$ level, and emitting one Balmer photon, is to jump from a level of higher n directly to the $n=1$ level. However, the Lyman photon emitted in such a jump will be strongly absorbed, giving an electron in some other H atom the opportunity to cascade down, with this process continuing until the electron reaches the $n=2$ level, emitting a Balmer photon en route. About two-thirds of the time [S9.1b; Table 9.1] an electron which reaches the $n=2$ level will find itself in a state with $l=1$ and will promptly jump down to the $n=1$ level, emitting a Lyman α (Lα) photon; the remaining one-third of the recombinations will lead to an electron in the $n=2$, $l=0$

level, in which case two photons will be emitted simultaneously within less than 1 second; the transition from the $2s$ down to the $1s$ level is forbidden for all one-photon processes. Except in compact H II regions, where extinction by dust is appreciable [S5.1c], the Balmer photons will all escape, and the number of such photons leaving the H II region per second, including the Balmer continuum as well as the emission lines, must equal N_u. The Lα photons will usually be absorbed [S5.1c] by dust grains within the H II region, even for relatively low values of the dust optical thickness.

The effect of helium atoms and ions has been ignored in the above discussion of hydrogen ionization. Just as hydrogen degrades an ultraviolet photon into a photon of Lyman α and other photons, so neutral helium converts most photons of ultraviolet radiation into photons whose wavelength is either 584 or 626 Å, corresponding to the transitions from the 2^1P and 2^3S levels, respectively, down to the ground 1^1S level, in addition to producing one or more photons of greater wavelength [5]. Similarly He II produces a photon at 304 Å plus less energetic photons. If $n_e < 10^3$ cm^{-3}, so that collisional transitions can be ignored, the flux of photons capable of ionizing hydrogen is essentially unaltered by the presence of helium [5], and the radius r_S of the ionized hydrogen region is still given by equation (5-21), with $S_u(0)$ unchanged but with n_e slightly increased. Since any radiation which can ionize helium can also ionize hydrogen, the radius of the region in which helium atoms will be ionized cannot much exceed r_S computed for hydrogen alone, even for the hotter stars. For stars of spectral type later than about O8, the relative number of stellar photons capable of ionizing helium will be much less than the corresponding number for hydrogen, and the helium will be ionized only in a small region close to the central star, with a radius much less than r_S for hydrogen [6].

c. Effect of Dust on H II Regions

Since dust grains are known to be present in H II regions, from the infrared radiation emitted, dust absorption of ionizing radiation must be considered. We define τ_{Sd} as the optical thickness of the dust just shortward of the H ionization limit and over a pathlength equal to r_S found from equation (5-21). Since r_i, the actual radius of the ionized region, will be reduced below r_S because of dust absorption, the actual optical thickness τ_{id} of the dust within the ionized zone will be less than τ_{Sd}. We also define τ_{SH} as the corresponding optical thickness of atomic H at the same wavelength if all the hydrogen were neutral. Evidently

$$\tau_{SH} = n_H s_\nu r_S, \tag{5-22}$$

where s_ν is evaluated from equation (5-6) and Table 5.1 for $\nu = \nu_1$.

We now evaluate the ratio of τ_{Sd} to τ_{SH} together with $\tau_{Sd}/n_H^{1/3}$. In this analysis we take the ratio of dust to gas to be constant; the data suggest that this is a plausible assumption in most H I regions [S7.2a] and perhaps in many H II regions also [S7.4b, 7.5]. The extinction at 912 Å produced by grains is taken to be $13E_{B-V}$ [S7.3b]. We then obtain from equations (5-6), (5-22), and (7-18)

$$\frac{\tau_{Sd}}{\tau_{SH}} = \frac{1}{3100}. \tag{5-23}$$

Values of $\tau_{Sd}/n_H^{1/3}$ computed from equations (5-22) and (5-23) are given in Table 5.3; since equation (5-23) is quite approximate, n_e/n_H has been set equal to unity. Evidently if $n_H = 1$ cm^{-3}, τ_{Sd} is less than unity for all spectral types, whereas if $n_H = 10^3$ cm^{-3}, τ_{Sd} is between 2.4 and 6.9 for main sequence O stars.

We consider now the radius r_i of the ionized gas when τ_{Sd} is appreciable. To solve this problem analytically [7] we make two additional approximations. First we neglect the variation of τ_ν with ν, evaluating the absorption coefficients at $\nu = \nu_1$. Second we assume that any Lyman continuum photons emitted as a result of electron captures in the $n = 1$ level are absorbed locally; this "on-the-spot" approximation, which involves replacing $\alpha^{(1)}$ by $\alpha^{(2)}$ throughout, ignores the transport of energy by the diffuse radiation field, and somewhat underestimates the absorption produced by grains. With these assumptions the ionization probability β_{1f} per H atom in the ground state per second may be written [equations (4-36) and (5-15)]

$$\beta_{1f} = \frac{s_\nu N_u e^{-\tau}}{4\pi r^2}. \tag{5-24}$$

Equation (5-24) may be substituted into equation (5-2) to give the steady-state equation for ionization equilibrium in this situation. To simplify this resultant equation, we define a dimensionless distance y by the relation

$$y = \frac{r}{r_S}, \tag{5-25}$$

where r_S is determined from equation (5-21), and n_e is set equal to n_p. After some algebra we obtain, with use of equations (5-21) and (5-22),

$$\frac{1-x}{x^2} = \frac{3y^2 e^\tau}{\tau_{SH}}, \tag{5-26}$$

where the optical depth τ from the star out to the distance r is the sum of the corresponding optical depths τ_H and τ_d due to neutral H atoms and dust. Since the distribution of dust is assumed uniform, τ_d is given by

$$\tau_d = y\tau_{Sd}, \tag{5-27}$$

whereas for $d\tau_H/dy$, equal to $(1-x)\tau_{SH}$, we obtain, substituting for $1-x$ from equation (5-26),

$$\frac{d\tau_H}{dy} = 3x^2y^2e^{\tau_H+\tau_d}. \tag{5-28}$$

Equation (5-28) can be put in integrable form if we multiply both sides by $e^{-\tau_H}$ and set $x=1$ as before, a very good approximation for r appreciably less than r_i. If we extend the integral out to $y=y_i$, corresponding to the radius r_i of the ionized zone, the left-hand side will be very nearly unity, since τ_H out to the ionization boundary $r=r_i$ is generally very large. Hence if we use equation (5-27), the right-hand side gives

$$3\int_0^{y_i} y^2e^{y\tau_{Sd}}\,dy = 1. \tag{5-29}$$

The integral can be evaluated directly; the values of $y_i = r_i/r_S$ obtained from this equation are shown in Table 5.4. The optical thickness of the dust out to the radius r_i, which we denote as τ_{id}, equals $y_i\tau_{Sd}$. The fraction f of stellar photons shortward of the Lyman limit which are absorbed by H atoms rather than by dust in the entire ionized region equals y_i^3, decreasing from 0.53 to 0.176 as τ_{Sd} increases from 1 to 4.

Table 5.4. Reduction Factor r_i/r_S for Radius of Ionized Hydrogen

τ_{Sd}	0.10	0.20	0.4	0.6	0.8	1.0	1.5	2.0
$y_i=r_i/r_S$	0.98	0.96	0.91	0.87	0.84	0.81	0.75	0.70
τ_{Sd}	3.0	4.0	6.0	8.0	10	15	20	40
$y_i=r_i/r_S$	0.62	0.56	0.47	0.42	0.37	0.30	0.25	0.15

These results are generally consistent with the comparison of infrared and radio emission from H II regions [8] plotted in Fig. 7.3 of Section 7.5. The fact that the observed points tend to lie not far from the dashed curve computed for $f=1$ indicates that f is not much less than unity for many such regions.

Finally it should be noted that in all likelihood the Lα radiation which is produced within H II regions is all absorbed within such regions. In general, about one such photon will be produced for every 1.5 stellar ultraviolet photons absorbed [S9.1b]. For a Lα photon in the center of the line profile [for b corresponding to thermal motions at 8000°—see equation (3-21)], the absorption coefficient equals 10^4 times the continuous absorption coefficient at threshold; the resultant mean free path for such a photon at a distance $r = r_S/2$ from the central star may be shown to be about $10^{-4} r_S$. The escape of these photons from the H II region is made possible by the Doppler effect, which in each scattering shifts the frequency by an amount dependent both on the atomic velocity and on the difference in direction between incident and scattered photons. Photons with strongly shifted frequencies will have a mean free path comparable with r_S and can escape the ionized zone [9]. However, the surrounding H I region, with a much higher density of neutral H atoms, acts as a scattering layer which will reflect the Lα photons back, and they will traverse the H II region again several times until they are absorbed by the grains; the values of $\tau_{id} = y_i \tau_{Sd}$ obtained from Tables 5.3 and 5.4 indicate that only a few crossings of an H II region will be required to absorb a Lα photon. Some absorption of Lα radiation by dust will also occur in the H I region, but this tends to be less important because of the much higher ratio of neutral H to dust. We assume in subsequent chapters that every photon of Lα produced is absorbed by grains within the H II region.

d. Ionization by Energetic Particles

Well outside H II regions the ultraviolet stellar photons that can ionize H are entirely absent. However, H atoms can be ionized by energetic particles of various types, including electrons, protons, and photons. At temperatures of about 10,000°K ionization by energetic thermal electrons is important. Cosmic-ray protons, including those at energies down to a few million eV, if present, can penetrate diffuse clouds and ionize the H atoms there, as can also any X-rays with energies of 200 eV ($\lambda \approx 60$ Å) or more. These effects are discussed briefly.

Ionization by thermal electrons is evidently important in the coronal gas [S5.2b]. In most interstellar clouds three-body recombination is unimportant in overall ionization equilibrium, which at high temperatures is determined by the balance between collisional ionization and radiative recombination (including dielectronic recombination, discussed below). If we denote by $\gamma_{jf}(X^{(r)})$ the rate coefficient for ionization by electron collision of an atom of type X which is in the stage of ionization r and in level j, equation (4-1) now gives, in place of equation (5-1),

$$n_e \sum_j n_j(X^{(r)}) \gamma_{jf} = n_e n(X^{(r+1)}) \sum_j \alpha_j, \qquad (5\text{-}30)$$

where the sum extends over all the excited states of the relevant atom. Evidently n_e cancels out from equation (5-30), and the ionization level depends primarily on T rather than on n_e, although the relative populations of the low-lying levels of the atom $X^{(r)}$ may depend on n_e.

For hydrogen, $n_j(X^{(r)}) = n_n$, the particle density of H atoms with total quantum number n, while $n(X^{(r+1)}) = n_p$, the proton density. The sum on the right-hand side of equation (5-30) must include the ground $n=1$ level, since the Lyman continuum radiation will presumably escape at the low values of n(H I) relevant in collisional ionization generally and in the coronal gas particularly. If we again let x be the fraction of H nuclei that are ionized, and neglect ionization from excited levels, we obtain

$$1 - x = \frac{1}{1 + \gamma_{1f}/\alpha^{(1)}}. \tag{5-31}$$

Values of γ_{1f} for various atoms have been determined both from observation and theory [10]. At a temperature of 10^6 degrees, γ_{1f} exceeds $\alpha^{(1)}$ by more than 10^6, giving an extremely small value of the neutral fraction, $1 - x$, for H in so hot a gas.

Ionization by cosmic-ray particles will produce some ionization of H atoms even in cool H I clouds. We let ζ_H be the probability per second that an H atom will be ionized by such energetic particles, including ionization produced by the secondary electrons. The ionization equilibrium is determined by an equation similar to equation (5-2), but with ζ_H replacing β_{1f}. If we neglect helium ionization and set n_e equal to $n_i + n_p$, where n_i is the particle density of positive ions heavier than helium, we find

$$n_p = \frac{1}{2} n_i \left\{ \left(1 + \frac{4\zeta_H n(\mathrm{H\ I})}{\alpha^{(2)} n_i^2} \right)^{1/2} - 1 \right\}, \tag{5-32}$$

where we have used $\alpha^{(2)}$ on the assumption that all Lyman photons emitted are promptly reabsorbed.

Equation (5-32) is not valid if much of the hydrogen is molecular, since in this case the electron–ion recombination proceeds by a series of exchange reactions, leading finally to dissociative recombination of a molecular ion [S5.3b]. In the lower density clouds, with $n(\mathrm{H\ I}) \leqslant 10\ \mathrm{cm}^{-3}$, the fraction of H_2 is generally small, and we may replace n(H I) by n_H. We shall evaluate equation (5-32) for the warm H I clouds, for which T may be set equal to 6000°K [S11.1a], giving $\alpha^{(2)} = 4.0 \times 10^{-13}\ \mathrm{cm}^3\ \mathrm{s}^{-1}$. The value of ζ_H is somewhat uncertain, since the solar wind shields the solar system from the lower energy particles (less than about 10^9 eV per nucleon) with the greatest cross sections for H ionization. From the observed flux of the more energetic particles reaching the Earth, with a conservative extrapolation to lower energies, a value of 7×10^{-18} s^{-1} is obtained for ζ_H [11]. For sufficiently small n_H, less than about 100 cm^{-3} if n_i/n_H equals 5×10^{-4}, we obtain

$$n_p = 4.2\times10^{-3} n_H^{1/2}. \tag{5-33}$$

Thus if n_H is between 10^{-1} and 10^{-2} cm^{-3}, the fractional ionization of H will be between about 1 and 4 percent, provided that T has the high value assumed.

A value of about 10^{-15} s^{-1} has been proposed for ζ_H [12], corresponding to large fluxes of cosmic-ray protons and other positive ions at energies between 10 and 100 MeV per nucleon. However, there is little direct astronomical evidence for so high a value of ζ_H [13], and analysis of molecule formation in H I regions suggests that in a significant fraction of clouds, ζ_H is nearer to the lower value of about 10^{-17} s^{-1} given above [S5.3b].

Appreciable ionization of H by soft X-rays, with energies between 100 and 250 eV, has also been suggested, with values of ζ_H again about 10^{-15} s^{-1} [12]. The observed fluxes of this radiation yield values of ζ_H less by several orders of magnitude [13], but ionization of interstellar gas by transient bursts of soft X-rays cannot be excluded observationally. While such photons should produce relatively large numbers of highly ionized heavy atoms, charge exchange with H atoms would reduce the numbers of such ions in H I regions below the limit of observability [S5.2b].

5.2 IONIZATION OF HEAVY ATOMS

The same physical processes which affect the ionization of atomic hydrogen also influence the ionization of heavier atoms. In addition, a number of other processes become important. Radiative recombination can be augmented by dielectronic recombination [14], a process in which an electron is captured in some level j, while the energy which it loses goes into exciting an inner-bound electron to some highly excited state k. Some fraction of the time this doubly excited atom will lose its energy radiatively rather than by ejecting the captured electron and returning to its initial state. Because of the large excitation energy required for the inner electron [14], dielectronic recombination is rarely important at $T < 10{,}000°$K.

Other processes are associated with collisions of heavy ions with atoms and molecules of various types. Ion-atom encounters can lead to an exchange of one or more electrons, a process known as "charge exchange"; processes in which no radiation is emitted or absorbed can have reaction cross sections equal to or even greater than the geometrical values, and can therefore dominate over the much lower radiative capture cross sections given in equation (5-9). Collisions with molecules can lead to a complex array of successive chemical reactions, which determine the relative abundances of different molecular species [S5.3c], both neutral and ionized.

The following discussion gives examples of how these processes are used in analyzing the ionization equilibrium of various interstellar atoms other than hydrogen. Equilibrium between photon ionization and radiative recombination is treated first, with collisional ionization at high temperatures following. Finally, some examples of charge exchange are considered together with charge neutralization through reactions with molecules.

a. Photon Ionization

In H II regions and for some conditions in H I regions also, the ionization of atoms heavier than H is again determined by equation (5-2), representing the balance between photon absorption and radiative recombination.  Values of the recombination coefficient α have been calculated [15] for various atoms in different stages of ionization. For direct electron capture in excited levels, σ_{cj} is essentially hydrogenic, and the results in the previous section are applicable. Rate coefficients for dielectronic recombination must be computed theoretically. To compute α_1 for direct radiative capture in the ground level, the corresponding value of $s_{f\nu}$ in equation (5-3) is obtained either from quantum mechanical calculations or, in the case of several neutral atoms, directly from laboratory measurements.

Values of the ionization probability β_{1f} in the average H I region, exposed to the mean interstellar radiation field, have been computed [16] from these same values of $s_{f\nu}$, using equation (4-36), integrated over all frequencies above the ionization threshold. If we set $U(\nu)d\nu/d\lambda$ equal to U_λ, the energy density of interstellar radiation per wavelength interval, the values of U_λ in the computation are about those in Table 5.5, obtained by summing the direct and scattered light from different stars in the solar neighborhood [17, 18]. In the wavelength ranges 2000 to 3000 Å [17] and 1350 to 1480 Å [19] these values of U_λ are in reasonable agreement with direct measurements, but could be in error by as much as a factor 2 [20].

Table 5.5. Energy Density of Ultraviolet Radiation

λ (1000 Å)	2–3	1.7	1.5	1–1.4	0.98	0.93
U_λ (10^{-17} erg cm^{-3} Å^{-1})	3.0	7.1	11.	8.5	7.5	2.9

These results have been used to determine from equation (5-2) the electron density n_e in a variety of H I clouds. Since the recombination coefficient α depends on T, the variation of T along the line of sight must be considered. The most detailed analysis has been for the line of sight to ζ Oph [21]; a two-component model was assumed, with $n_H = 500$ cm^{-3}, $T = 110$°K, and 62 percent of the observed N_H in an outer region and with the remaining hydrogen gas in a dense inner region characterized by five times the density and one-fifth the temperature. With depletions of heavy elements about equal to those in Table 1.1, this model predicts column densities for most atoms and molecules [S5.3b, c] in agreement with the observations. The values of n_e obtained are about 0.06 and 0.25 cm^{-3} in the outer and inner regions, respectively, corresponding to electrons primarily from C.

For other stars the analyses have been based on single-component models, with clouds taken to be at a uniform temperature. In 10 stars showing detectable Ca I absorption, with a mean E_{B-V} of 0.23 mag, and presumably with n_H and n_e above the average for diffuse clouds generally, the mean n_e found for $T = 70$°K is about 0.1 cm^{-3} [22], corresponding to $n_H = 10^3$ cm^{-3} on the same assumptions as above. As in ζ Oph, n_e and the resultant n_H would presumably be decreased somewhat if a lower temperature in the inner region of each cloud had been considered.

An additional application of equation (5-2) and Table 5.5 is to the abundance of Na and Ca in H I regions. In most of the interstellar gas these atoms are predominantly ionized to Na II and Ca III. Thus both N(Na I)/N(Na II) and N(Ca II)/N(Ca III) vary linearly with n_e. The ratio N(Na I)/N(Ca II) is independent both of n_e and of the overall radiation density; this ratio equals 0.013 for T between 50 and 100°K, for the cosmic abundances in Table 1.1, and for the shape of the radiation spectrum in Table 5.5 [23]. Since the observed value for this ratio is between 1 and 10 for low-velocity diffuse clouds [S3.4c], Ca is depleted in these clouds by a factor of 100 to 1000 relative to Na, consistent with the results for ζ Oph shown in Table 1.1. As noted earlier, the observed mean ratios in Table 3.2 indicate a variation of Ca depletion with cloud velocity.

b. Collisional Ionization

At temperatures significantly above 10,000°, collisional ionization becomes important for heavy atoms just as for H and He [S5.1d]. In collisions at such high random velocities, molecules will dissociate, eliminating one source of complication in the ionization equilibrium. If charge exchange between ions is ignored, equation (5-30) is satisfied, and the fraction of atoms in different charge states becomes independent of n_e. Values of this fraction, $n(X^{(r)})/n_X$, for N and O in ionization equilibrium at temperatures between 10^5 and 10^6°K are listed in Table 5.6 [24]. These values are computed with dielectronic recombination taken into account, since this effect can be of substantial importance for some atoms at these high temperatures.

Such values have been used for interpreting the column densities of O VI ions determined from the equivalent widths of the two interstellar absorption features [S3.4c]. For example, the upper limit of 2.5×10^{-2} observed for N(N V)/N(O VI) together with the cosmic abundances for N and O and the ionization fractions in Table 5.6 require that T exceed about 4×10^5°K. Comparable lower limits on T are obtained from similar upper limits on N(S IV) and N(Si IV). According to the values of n(O VI)/n_O in Table 5.6, the mean H density n_H in the collision-ionized gas has a minimum value at $T = 3 \times 10^5$°K; if n_O/n_H is taken from Table 1.1, and n(O VI) is set equal to its observed mean value of 1.7×10^{-8} cm^{-3} [S3.4c], the minimum value of n_H is about 1×10^{-4} cm^{-3}. Since T likely differs from the value for minimum n_H, and since this hot "coronal" gas cannot fill all of interstellar space, $\langle n_H \rangle$ must somewhat exceed this minimum [S11.3].

Table 5.6. Ionization Fractions in N and O in Collision Equilibrium

log T	$-\log[n(X^{(r)})/n_X]$ for Ion:					
	N III	N IV	N V	O IV	O V	O VI
5.1	0.39	0.26	1.65	0.35	1.84	5.06
5.2	0.77	0.19	0.84	0.20	1.02	3.21
5.3	1.49	0.53	0.57	0.24	0.51	1.84
5.4	2.65	1.36	0.89	0.50	0.30	0.93
5.5	3.82	2.24	1.36	1.11	0.52	0.59
5.6	4.88	3.03	1.79	2.16	1.23	0.83
5.7	5.82	3.71	2.16	3.25	2.02	1.24
5.8	6.63	4.29	2.46	4.24	2.75	1.63
5.9	7.32	4.76	2.67	5.12	3.37	1.96

c. Charge Exchange and Reactions with Molecules

Among the various charge-exchange processes [25] that may occur in H I regions, one of the most important and best established is the reaction [26]

$$O^+(^4S_{3/2}) + H \rightarrow O(^3P_2) + H^+ + 0.020 \text{ eV}, \qquad (5\text{-}34)$$

which has a high measured rate coefficient of 0.4×10^{-9} cm^3 s^{-1} [27] for

$T = 300°$K. The symbol k will be used to denote the rate coefficient $\langle \sigma u \rangle$ for these chemical processes. For reactions of ions with neutrals, k for an exothermic reaction is often nearly independent of T over a wide range of temperature [Tables 2.1 and 4.2]; at interstellar cloud temperatures a k value of 0.5×10^{-9} cm^3 s^{-1} for reaction (5-34) is indicated by the molecular line data [21].

A charge-exchange reaction, like many atomic and molecular processes, can occur through several energy levels; these different reaction routes are sometimes referred to as "channels." For example, the final state in which the neutral O atom is in its ground level (and the H atom ionized), which we denote by subscript 1, is the "exit channel" for reaction (5-34), whereas the initial state for this reaction, with the O^+ ion in the $^4S_{3/2}$ level (and the H atom in its ground neutral state), is the corresponding "entrance channel," here denoted by the subscript 3. The rate coefficient for reaction (5-34), from channel 3 to channel 1, we denote by k_{31}. In addition, one must consider the reaction leading to exit channel 2, in which the charge transfer leaves the neutral O atom in its excited $^3P_1$ level, with a negligible energy loss of about 7×10^{-5} eV; if n(H I) is less than about 10^5 cm^{-3}, reactions to this channel are normally followed by spontaneous radiative transitions down to the $^3P_2$ level constituting channel 1. The channel in which the charge transfer leaves the O atoms in the $^3P_0$ level is endothermic by 0.009 eV, and may be ignored at temperatures less than 100°K.

If no reactions other than charge transfer and spontaneous radiation occur, equation (4-1) for statistical equilibrium may readily be solved, using equation (4-4) to relate k_{13} and k_{31}. Transitions from channel 2 back to 3 may also be ignored, since $A_{21} \gg k_{23} n(\text{H II})$, and we find

$$\frac{n(\text{O II})}{n(\text{O I})} = \frac{g_3}{g_1 + g_2} e^{-E_{13}/kT} \frac{n(\text{H II})}{n(\text{H I})} = 10^{-100/T} \frac{n(\text{H II})}{n(\text{H I})}, \tag{5-35}$$

where g_1, g_2, and g_3, the total weights for the three levels described above, are equal to 5, 3, and 8, respectively. Since A_{21} equals 9.0×10^{-5} s^{-1}, and $k_{31} + k_{32}$ is about 0.6×10^{-9} cm^3 s^{-1}, equation (5-35) is valid in detail only if n(H II) is substantially less than 10^5 cm^{-3}.

As we have seen [S5.1d] a small residual ionization of H is produced by cosmic rays in H I regions. If the density of H_2 molecules is relatively low so that reactions such as (5-37) below can be ignored, then equation (5-35) indicates that the relative ionization of O will be maintained at about 5 percent of the level for H in a typical diffuse cloud at $T \approx 80°$K.

Other charge-exchange reactions have been proposed [25, 28] as possibly important in H I regions, but the reaction rates for some of these are uncertain. Detailed calculations for the reaction

$$C^{(3)} + H \rightarrow C^{(2)} + H^+ + 34.3 \text{ eV} \tag{5-36}$$

indicate a large rate coefficient, about 10^{-9} cm^3 s^{-1}, at T between 10^3 and 2×10^4°K [29]. From a semiclassical viewpoint, an electron can jump from one atom (or molecule) to another if no change in energy is involved; the kinetic energy of the atomic nuclei must be the same, of course, immediately before and after the abrupt jump. Thus one would expect charge transfer to be possible at a particular internuclear distance if the change in electron binding energy is equal and opposite to the change in nuclear potential energy at that distance. Between a neutral and an ionized atom the potential energy is relatively small; hence charge exchange in this case has an appreciable cross section only if the binding energy of the electron being exchanged is about the same in the two neutral atoms. For reaction (5-36), however, the transition much increases the positive potential energy of the two colliding atoms, offsetting the equally large increase in the negative energy of the bound electron. After the electron jump, the nuclear potential energy is transformed into kinetic, as the proton is accelerated away from the $C^{(2)}$ ion. As a result of reaction (5-36), any C IV population produced by absorption of X-rays, even at a hypothetical level corresponding to $\zeta_H = 10^{-15}$ s^{-1} [S5.1d], would be reduced [29] below the present upper limits on N(C IV) set by ultraviolet absorption measures.

Atoms may also be either ionized or neutralized by collisions with molecules. A full discussion of such reactions must form part of molecular equilibrium analysis [S5.3b, c]. Here we indicate two of the chief types of reactions which are believed to be important, exchange reactions and direct radiative attachment, followed in either case by dissociative recombination of a more complex molecule.

An important exchange reaction if much H_2 is present is

$$O^+ + H_2 \rightarrow OH^+ + H \tag{5-37}$$

for which k is about 10^{-9}cm^{3}s^{-1} [30]. Reaction (5-37) may be followed by dissociative recombination

$$OH^+ + e \rightarrow O + H. \tag{5-38}$$

Alternatively, further exchange reactons with H_2 can lead to more complex ions, such as H_2O^+ or H_3O^+. These ions in turn will be subject to dissociative recombination, for which the reaction cross section may somewhat exceed the geometrical value, giving a k value perhaps as great as 10^{-7}cm^{3}s^{-1} at 30 to 100°K [30]. Through the charge-exchange reaction

(5-34), the interaction of O^+ with H_2 can affect the proton density as well as the OH abundance [S5.3b].

The atomic ion N^+ interacts with H_2 in much the same manner as does O^+. However, the corresponding reaction with C^+ is endothermic, and instead the following process occurs:

$$C^+ + H_2 \rightarrow CH_2^+ + h\nu. \tag{5-39}$$

The rate coefficient for this radiative attachment reaction at 80°K may be as great as $10^{-14} cm^3 s^{-1}$ [31], although the detailed modeling of molecular column densities in the line of sight to ζ Oph [21] indicates the much lower value of $5 \times 10^{-16} cm^3 s^{-1}$. Again, charge neutralization may occur by dissociative recombination either of the initial molecular ion, CH_2^+, or of some subsequent more complex ion such as CH_3^+ [reaction (5-50)]. Similar reactions occur with Si^+. In the centers of dense clouds, where much of the hydrogen must be molecular, $n(H_2)$ is likely to exceed $10^3 n_e$, and reactions such as (5-39) may outweigh electron-ion radiative recombination and have an important influence on the population densities of some atomic ions.

In addition, production of atomic ions may result from a variety of processes, including charge-exchange reactions between neutral atoms and complex molecular ions [30]. Thus ions of Mg, Ca, Na, and Fe can likely arise from collisions with O_2^+, HCO^+, H_3O^+, and CH_3^+ [32]. Much more detailed information on rate coefficients would be required to evaluate all these processes in interstellar clouds, and fuller observational information would be needed to make sure that all the most important processes had been taken into account.

5.3 FORMATION AND DISSOCIATION OF MOLECULES

The principal methods normally considered for the formation of interstellar molecules from individual atoms [25] are direct radiative attachment and catalytic formation on the surfaces of dust grains. In the first process, two atoms, for example, can stick together to form a molecule, with their binding energy going off as a photon. In the second process, successive atoms hit a grain, stick to the surface momentarily, and then combine to form a molecule, giving off their binding energy as heat transmitted to the grain; the molecule then either escapes by evaporation or by some other process, perhaps involving photon excitation of the molecule to a level which is not bound to the grain. Once the molecule is formed, two of the primary mechanisms for dissociation are believed to be photon absorption to a level in which the molecule as a unit becomes unbound, and dissociative recombination [see equation (5-38)]. Both formation and disruption can be modified by exchange interactions between molecules and either atoms or other molecules.

For the H_2 molecule, with its high abundance, these exchange interactions with other molecules provide only a small perturbation on the H_2 abundance, simplifying the analysis. In addition, the ultraviolet absorption measurements have yielded observational data on H_2 not available for other molecules. The discussion in this section is therefore concentrated on the equilibrium of H_2, with only a brief treatment of other molecules.

a. Equilibrium Abundance of H_2

Since vibration–rotation transitions in the H_2 molecule are forbidden for electric-dipole radiation [S4.3b], radiative attachment of two neutral H atoms to form a molecule is also forbidden. Formation of H_2 molecules is believed to occur on the surfaces of dust grains, which provide a suitable catalyst for this reaction. To apply the steady-state equation (4-1) for computing $n(H_2)$, the particle density of H_2 molecules, we now evaluate the rate of this formation process. We let Σ_d denote the projected area of the dust grains per H nucleus in the gas [S7.3b], so that the total projected area per cm^3 equals $\Sigma_d n_H$; we denote by γ the fraction of H atoms striking the grain which come off attached to other H atoms to form H_2 molecules. Theoretical studies [33] indicate that if the gas is too hot, the H atoms will rebound elastically without sticking momentarily, whereas if the grains are too hot, the atoms will evaporate again before they have had time to migrate over the surface and combine with another adsorbed H atom; a value of about 0.3 for γ seems likely for interstellar conditions.

We denote by $(R_{fj})_d$ the probability rate per neutral H atom for the formation of molecules in state j as a result of collisions with dust grains. We have

$$n(\mathrm{H\ I})\sum_j (R_{fj})_d = \tfrac{1}{2}\langle \gamma w_\mathrm{H}\rangle \Sigma_d n_H n(\mathrm{H\ I}) \equiv R n_\mathrm{H} n(\mathrm{H\ I}), \qquad (5\text{-}40)$$

where γw_H is averaged over all types of grains as well as over all H-atom velocities. If we set Σ_d equal to $1.0 \times 10^{-21}\,\mathrm{cm}^2$ from equation (7-23), assume $\gamma = 0.3$, and take $\langle w_\mathrm{H}\rangle$ at 80°K from equation (2-19) (with $A \gg 1$), we obtain for R, the H_2 formation constant in equation (5-40),

$$R \equiv \tfrac{1}{2}\langle \gamma w_\mathrm{H}\rangle \Sigma_d = 2.0 \times 10^{-17}\,\mathrm{cm}^3\,\mathrm{s}^{-1}. \qquad (5\text{-}41)$$

Dissociation of H_2 is produced by photon absorption, followed by a spontaneous transition to an excited vibrational level ($v \geqslant 14$) of the ground electronic state [S4.3b]. We compute now the total probability per second, $(R_{Jf})_{ph}$, that a H_2 molecule in rotational state J of the ground electronic state and ground vibrational level be dissociated by photon absorption. As in Section 4.3b we denote by β_{Jm} the probability per second of a radiative transition upward from level J to an excited level whose rotational, vibrational, and electron quantum numbers are denoted by m; equation (4-36) evaluates β_{Jm} in terms of $U(\nu)$. The quantity k_m denotes the probability that a molecule initially in level m will cascade down to an unbound level and dissociate; actually k_m does not depend significantly on the upper rotational quantum number, but rather on the upper vibrational quantum number v'. We obtain simply

$$(R_{Jf})_{ph} = \sum_m k_m \beta_{Jm} \equiv \langle k \rangle \beta(J), \tag{5-42}$$

where $\beta(J)$, the total upward transition probability from level J, is given in equation (4-40). In this sum, electron state C may be ignored, since as we have seen [S4.3b] $k_m = 0$ for all Werner bands longward of 912 Å; however, Werner bands contribute to $\beta(J)$.

In evaluating this expression we must take the absorption of radiation into account. If we introduce the usual equation (3-3) for exponential absorption (with j_ν set equal to zero), and assume that $U_0(\nu)$, the energy density outside the cloud, is nearly constant over each absorption line, equation (4-36) for β_{Jm} becomes

$$\beta_{Jm} = (\beta_{Jm})_0 K_{Jm} = \frac{\pi e^2}{m_e} \frac{U_0(\nu_{Jm}) f_{Jm}}{h\nu_{Jm}} K_{Jm}, \tag{5-43}$$

where for each line $(\beta_{Jm})_0$ represents the transition probability outside the cloud, and $U(\nu) = U_0(\nu)$ in the absence of shielding; the absorption correction factor K_{Jm} is given by

$$K_{Jm} = \int \phi(\Delta\nu) e^{-\tau_\nu} d\nu, \tag{5-44}$$

integrated over the line. For a strong H_2 line on the square-root section of the curve of growth, $\phi(\Delta\nu)$ is given by equation (3-43); if the column density down to a particular region of the cloud is $N(J)$, K_{Jm} in that region is given by

$$K_{Jm} = \left(\frac{\delta_m}{N(J) s_{Jm}} \right)^{1/2}; \tag{5-45}$$

s_{Jm} is the integrated absorption coefficient per H_2 molecule for the transition J to m [equation (3-15)]. According to equations (3-43) and (5-45), πK_{Jm}^2 is the value that $1/\tau_\nu$ would have at the line center if there were no Doppler broadening. The opacity produced by grains, which has been ignored in equation (5-45), has been included in more detailed numerical computations [34].

The equation of dissociation equilibrium, obtained by substituting equations (5-40) and (5-42) into equation (4-1), and making use of equation (5-43), may now be written

$$\frac{n(\text{H I})}{n(\text{H}_2)} = \frac{\langle \sum_m k_m (\beta_{Jm})_0 K_{Jm} \rangle}{R n_{\text{H}}}, \tag{5-46}$$

where the quantity in brackets is averaged over all J, weighted by $n(J)$. In relatively transparent clouds, in which all H_2 lines are on the linear part of the curve of growth, all K_{Jm} factors equal unity. Following Section 4.3 we denote by β_0 the sum of $(\beta_{Jm})_0$ over all m, neglecting the slight dependence of β_0 on J for the interstellar radiation field. Equation (5-46) then assumes the particularly simple form

$$\frac{n(\text{H I})}{n(\text{H}_2)} = \frac{\langle k \rangle \beta_0}{R n_{\text{H}}}. \tag{5-47}$$

Equation (5-47) is particularly useful for determining the H_2 formation constant R from observations of transparent H_2 clouds. Analysis of the relative intensities of H_2 lines from different rotational levels [S4.3b] gives values of β_0 and of $n(\text{H I})$. For relatively transparent clouds, where $n(\text{H}_2)/n(\text{H I})$ is very small, $n(\text{H I}) \approx n_{\text{H}}$. The measured ratio of column densities $N(\text{H}_2)/N(\text{H I})$, together with the computed value of 0.11 for $\langle k \rangle$ [S4.3b], then gives directly the value of R in equation (5-47). The values obtained [35] show considerable scatter, ranging from about 1 to 4×10^{-17} cm^3 s^{-1}; a mean of about 3×10^{-17} cm^3 s^{-1} is indicated, in general agreement with equation (5-41). If this value of R is assumed and β_0 is again set equal to its mean interstellar value, then for the clouds producing very weak H_2 lines, in four stars within about 100 pc of the Sun, values of n_{H} as low as about 1 cm^{-3} are obtained from equation (5-47) [36], with greater values if much of the observed H is outside the H_2 clouds.

Detailed computations have also been carried out for $n(\text{H}_2)/n(\text{H I})$ for more opaque clouds, taking the K_{Jm} factors into account with various approximations [34, 13]. The results reproduce the chief features of the observations very well, showing $n(\text{H}_2)/n(\text{H I})$ increasing very rapidly with increasing total thickness of the cloud, as measured either by N(H I) or by $E_{B\text{-}V}$. The physical explanation of this behavior is that when the column density of H I through a cloud is so great that stellar radiation is not able to dissociate all the H_2 molecules formed, then an outer shielding layer develops, in which the Lyman lines are absorbed, and inside which the very low dissociation rate permits much of the hydrogen to accumulate in molecular form. Analyses of opaque clouds, based on the value of R found above, indicate that a spread of n_H values from about 10 to $10^3\,\text{cm}^{-3}$ is required to fit the data to the theory [13]. This range of densities is in general agreement with that obtained from ratios of intensities of different rotational H_2 lines [S4.3b].

b. Equilibrium of HD

The Lyman lines of the HD molecule are observed in most of the stars showing strong H_2 lines. The equilibrium particle density, n(HD), is determined by rather different considerations than that for H_2. In the first place, HD is photodissociated much more rapidly than H_2, since most of the absorption lines produced by these two molecules are quite separate, those of HD are weak, and most of the K_{Jm} factors equal unity; for the H_2 molecules, most of the molecules are in the $J=0$ or 1 levels, the Lyman lines from these levels are strong, and K_{0m} and K_{1m} found from equation (5-45) are very low, typically about 3×10^{-4} for moderately reddened stars ($0.1<E_{B\text{-}V}<0.3$). In the second place, HD is formed primarily by the rapid exchange reaction [37, 38]

$$D^+ + H_2 \rightarrow H^+ + HD + 0.039\ \text{eV}. \tag{5-48}$$

This energy excess results from the difference of vibrational zero-point energy, and is therefore nearly independent of J. The rate coefficient k_{HD} for reaction (5-48) is about $1.0\times10^{-9}\,\text{cm}^3\,\text{s}^{-1}$ [39], a typical value for reactions between neutral molecules and positive ions [30].

We compute the proton density $n(H^+)=n_p$ required to explain the observed column densities of HD. The steady-state equation is obtained if n(HD) times the right-hand side of equation (5-42), with $\beta(J)=\beta_0$, is equated to the formation rate $k_{HD}n(D^+)n(H_2)$. Since charge transfer tends to keep the fractional ionization of H and D equal at $T\geqslant 80°$K, we obtain

$$n_p = \frac{n_H}{n_D}\frac{N(\text{HD})}{N(H_2)}\frac{\beta_0\langle k\rangle}{k_{HD}}, \tag{5-49}$$

provided that $n(H_2)\ll n(\text{H I})$ so that reaction (5-48) does not affect n(D I). The deuterium-to-hydrogen abundance ratio n_D/n_H, found from the ratio of the interstellar ultraviolet atomic absorption lines, is 1.8×10^{-5} [40]. From the measured column densities and the values of β_0 obtained from relative Lyman line strengths (see Section 4.3), values between 10^{-1} and $10^{-3}\,cm^{-3}$ are found for n_p in six stars [41]. The more detailed analysis [21] of the two-component cloud model for ζ Oph [S5.2a] gives $n_p\approx3\times10^{-3}\,cm^{-3}$ in the outer zone, where $n_H=500\,cm^{-3}$.

The proton exchange reaction in which H^+ interacts with H_2 can lead to conversion of orthohydrogen to parahydrogen and vice versa [42]. It is this process, whose rate coefficient k_{pH_2} is about the same as for reaction (5-48), which is responsible for maintaining $n(J=1)/n(J=0)$ at its Boltzmann value at the kinetic temperature [S3.4b]. The condition for this process to be dominant in opaque clouds is that $k_{pH_2}n_p$ much exceed $\Sigma_m k_m \beta_{Jm}$ for $J=0$ or 1, since most of the molecules will be in these two states and photodissociation followed by formation of new molecules is the only other process that can convert ortho- into parahydrogen at an appreciable rate. These two rates are comparable for β_{Jm} equal to 0.01 times its mean interstellar value for each m and J if n_p about equals $10^{-3}\,cm^{-3}$, the lowest value found above. For $J=0$ and 1, the values of the absorption correction factor K_{Jm} computed from equation (5-45) are between 10^{-3} and 10^{-4} for stars with saturated H_2 absorption lines. Hence β_{Jm} for these J values will be much less than 0.01 times the mean interstellar value, and proton collisions will safely dominate over photodissociation in determining the relative population of the $J=0$ and 1 levels. Evidently the marked reduction of the photodissociation rate by self-shielding in these relatively opaque clouds justifies the use of the Boltzmann equation in obtaining equation (3-46).

The values of n_p found from the HD abundances in six stars, noted above, are consistent with values of ζ_H [S5.1d] ranging between 10^{-17} and $10^{-15}\,s^{-1}$ for the clouds in the lines of sight to these stars [41]. The more precise results for ζ Oph give $\zeta_H=1.6\times10^{-17}\,s^{-1}$, a value consistent also with the analysis of OH and CO data for this same line of sight [21]. The steady-state equations used in these analyses take into account electron-ion recombination through a complex series of processes, starting with the inverse of reaction (5-34), followed by reaction (5-37). This low value of ζ_H in the ζ Oph cloud is about the ionization rate produced by cosmic rays reaching the Earth [11], and it provides a strong argument against a significant flux of cosmic rays at energies below 100 MeV [S6.2b], at least in this conspicuous cloud.

c. Other Molecules

While direct radiative attachment of atoms is a quite possible formation process for some molecules (for example, CH^+) formation on grains, which is dominant for H_2 [S5.3a], may be important also for other molecules, either directly [43] or indirectly [31, 44]. On the direct formation theory saturated hydride molecules, such as CH_4, H_2O, SiH_4, and so forth, are formed on grain surfaces. In the gas, these molecules are subject to photodissociation, which produces diatomic hydride molecules. Exchange reactions with C^+ ions produce such diatomic molecules as CO, CN, and CS, which then form more complex molecules on collisions with grains; ions are important because they produce more rapid exchange processes than do neutral atoms or molecules. On this theory, ultraviolet light is required both to liberate the molecules from the grains, since thermal evaporation is insufficient for molecules much heavier than H_2, and to ionize the C atoms.

On the indirect theories, the grains are needed to produce H_2 molecules, which in a series of successive chemical reactions serve as a source for other molecules. We have seen [S5.3b] that HD is formed in this way. In addition, the CH_2^+ molecule ion produced in reaction (5-39) can serve as a source both of CH, produced by dissociative recombination of CH_2^+ [45] as in reaction (5-38), and of CH^+, produced by photodissociation of CH_2^+ [31]. If much of the hydrogen is molecular, the most likely reaction for CH_2^+ is

$$CH_2^+ + H_2 \rightarrow CH_3^+ + H. \tag{5-50}$$

The CH_3^+ ion is then a principal source of CH and CH^+ through dissociative recombination and photodissociation [45, 31], respectively.

Another molecular species formed from processes involving other molecules is the tightly bound and abundant CO molecule, which can be produced in several ways [44]. Formation from CH^+ occurs by the exchange reaction

$$CH^+ + O \rightarrow CO + H^+, \tag{5-51}$$

with a similar reaction involving CH. Alternatively, the OH radical formed following reaction (5-37) can lead to CO by the process

$$OH + C^+ \rightarrow CO + H^+. \tag{5-52}$$

In diffuse clouds where most of the hydrogen is atomic, analysis of CO formation by reaction (5-51), taking into account the optically determined ratio of 10^{-7} for $n(CH^+)/n_H$, predicts that $n(CO)/n_H$ is also about 10^{-7}, in rough agreement with the values determined from ultraviolet absorption lines. If a cosmic C/H ratio of 3×10^{-4} is assumed [Table 1.1], about one C atom in every 3000 within such clouds is in a CO molecule.

In the more opaque clouds a much greater fraction of carbon is apparently in CO in the gas [S7.3c]. It is not clear whether some additional process must be invoked in such clouds to explain why all the molecules have not condensed on the grains; the theoretical condensation time at H densities greater than 10^4 cm^{-3} may be less than about 10^6 years [S9.4a].

REFERENCES

1. W. J. Karzas and R. Latter, *Ap. J. Suppl.,* **6**, 167, 1961.
2. M. J. Seaton, *M.N.R.A.S.,* **119**, 81, 1959.
3. B. Strömgren, *Ap. J.,* **89**, 526, 1939.
4. N. Panagia, *A. J.,* **78**, 929, 1973.
5. D. E. Osterbrock, *Astrophysics of Gaseous Nebulae,* W. H. Freeman (San Francisco), 1974, Section 2.4.
6. R. H. Rubin, *A. J.,* **74**, 994, 1969.
7. V. Petrosian, J. Silk, and G. B. Field, *Ap. J. (Lett.),* **177**, L69, 1972.
8. C. G. Wynn-Williams and E. E. Becklin, *P.A.S.P.,* **86**, 5, 1974.
9. D. E. Osterbrock, *J. Quant. Spectrosc. Radiat. Transfer,* **11**, 623, 1971.
10. W. Lotz, *Ap. J. Suppl.,* **14**, 207, 1967; No. 128.
11. L. Spitzer and M. G. Tomasko, *Ap. J.,* **152**, 971, 1968.
12. A. Dalgarno and R. A. McCray, *Ann. Rev. Astron. Astroph.,* **10**, 375, 1972.
13. L. Spitzer and E. B. Jenkins, *Ann. Rev. Astron. Astroph.,* **13**, 133, 1975.
14. A. Burgess, *Ap. J.,* **139**, 776, 1964 and **141**, 1588, 1965.
15. S.M.V. Aldrovandi and D. Pequignot, *Astron. Astroph.,* **25**, 137, 1973 and **47**, 321, 1976.
16. K. S. deBoer, K. Koppenaal, and S. R. Pottasch, *Astron. Astroph.,* **28**, 145 and **29**, 453, 1973.
17. A. N. Witt and M. W. Johnson, *Ap. J.,* **181**, 363, 1973.
18. M. Jura, *Ap. J.,* **191**, 375, 1974.
19. S. Hayakawa, K. Yamashita, and S. Yoshioka, *Astroph. Space Sci.,* **5**, 493, 1969.
20. P. M. Gondhalekar and R. Wilson, *Astron. Astroph.,* **38**, 329, 1975.
21. J. H. Black and A. Dalgarno, *Ap. J. Supp.,* **34**, 405, 1977.
22. R. E. White, *Ap. J.,* **183**, 81, 1973.
23. M. Jura, *Ap. J.,* **206**, 691, 1976.
24. P. H. Shapiro and R. T. Moore, *Ap. J.,* **207**, 460, 1976.
25. W. D. Watson, *Rev. Mod. Phys.,* **48**, 513, 1976.
26. G. B. Field and G. Steigman, *Ap. J.,* **166**, 59, 1971.
27. F. C. Fehsenfeld and E. E. Ferguson, *J. Chem. Phys.,* **56**, 3066, 1972.
28. G. Steigman, *Ap. J.*, **199**, 642, 1975.
29. R. J. Blint, W. D. Watson, and R. B. Christensen, *Ap. J.*, **205**, 634, 1976.
30. E. Herbst and W. Klemperer, *Ap. J.*, **185**, 505, 1973.
31. J. H. Black, A. Dalgarno, and M. Oppenheimer, *Ap. J.*, **199**, 633, 1975.
32. M. Oppenheimer and A. Dalgarno, *Ap. J.*, **192**, 29, 1974.
33. D. J. Hollenbach and E. E. Salpeter, *Ap. J.*, **163**, 155, 1971.
34. D. J. Hollenbach, M. W. Werner, and E. E. Salpeter, *Ap. J.*, **163**, 165, 1971.
35. M. Jura, *Ap. J.*, **197**, 575, 1975.
36. D. G. York, *Ap. J.*, **204**, 750, 1976.
37. J. H. Black and A. Dalgarno, *Ap. J. (Lett.)*, **184**, L101, 1973.
38. W. D. Watson, *Ap. J. (Lett.)*, **182**, L73, 1973.

39. F. C. Fehsenfeld, D. B. Dunkin, E. E. Ferguson, and D. L. Albritton, *Ap. J. (Lett.)*, **183**, L25, 1973.
40. D. G. York, and J. B. Rogerson, *Ap. J.*, **203**, 378, 1976.
41. J. Barsuhn and C. M. Walmsley, *Astron. Astroph.*, **54**, 345, 1977.
42. A. Dalgarno, J. H. Black, and J. C. Weisheit, *Ap. Lett.*, **14**, 77, 1973.
43. W. D. Watson and E. E. Salpeter, *Ap. J.*, **175**, 659, 1972.
44. M. Oppenheimer and A. Dalgarno, *Ap. J.*, **200**, 419, 1975.
45. J. Black and A. Dalgarno, *Ap. Lett.*, **15**, 79, 1973.

6. Kinetic Temperature

The kinetic temperature of the gas in a steady state is determined by the condition that the total kinetic energy gained per cm^3 per second, which we denote by Γ, is equal to the corresponding energy lost per cm^3 per second, denoted by Λ. In general, both Γ and Λ will depend on the temperature T, as well as on the particle density; the value of T at which these two functions are equal will be an equilibrium temperature T_E.

More generally, the heating and cooling functions, Γ and Λ, are related to the change of T with time. We denote by subscripts ζ and η the interacting particles responsible for each contribution, $\Gamma_{\zeta\eta}$ or $\Lambda_{\zeta\eta}$, to the total Γ or Λ. The condition that $\Gamma-\Lambda$, the net thermal input of energy per cm^3 per second, equals the corresponding rate of increase of thermal energy, plus the work done by the gas, gives the result for a monatomic gas [1]:

$$n\frac{d}{dt}\left(\frac{3}{2}kT\right)-kT\frac{dn}{dt}=\sum_{\zeta,\eta}\left(\Gamma_{\zeta\eta}-\Lambda_{\zeta\eta}\right)=\Gamma-\Lambda, \tag{6-1}$$

where n is the total number of free particles per cm^3 in the interstellar gas. We assume that all components of the gas have the same kinetic temperature T. The left-hand side of equation (6-1) is $\rho T\,dS/dt$, where S is the entropy per gram of the interstellar medium, assumed to be a perfect monatomic gas.

Equation (6-1) neglects thermal conduction, which is negligible in normal H I and H II regions, with $T \leqslant 2\times10^{4}\,^{\circ}$K. For much higher T, and no magnetic field, the long mean free path gives a high conductivity K, and an additional term $\nabla\cdot(K\nabla T)$ must be added on to the right-hand side of equation (6-1); a magnetic field reduces markedly the conductivity transverse to $\mathbf{B}$. Thermal conduction plays an important role in the transition region between the hot coronal intercloud gas and the much cooler denser clouds [2], as well as in the early evolution of supernova remnants.

We may define a "cooling time" t_T by the relation

$$\frac{d}{dt}\left(\frac{3}{2}kT\right) = -\frac{3k(T-T_E)}{2t_T}. \tag{6-2}$$

Evidently for constant n, t_T equals the ratio of the excess of energy density (over its value in equilibrium) to the net cooling rate $\Lambda - \Gamma$. In the simple case where t_T and T_E are constant, $T - T_E$ approaches zero as $\exp(-t/t_T)$. Under certain unusual conditions, not often encountered, t_T can be negative in the neighborhood of T_E, in which case this equilibrium temperature is unstable [1]. In such a case, the gas will generally cool or heat toward a different equilibrium temperature which will, in general, be stable.

In order to compute T_E for interstellar conditions, we turn now to a derivation of formulae for Γ and Λ for some of the dominant processes. A primary mechanism for heating the interstellar gas is photoelectric ionization of neutral atoms. Let E_2 denote the kinetic energy of the ejected electron. Not all this energy can be counted as a gain, however, since in a steady state each photoionization must be offset by a corresponding capture of a free electron, whose kinetic energy we denote E_1. Let n_i denote the particle density of the ionized atoms, assumed to be all identical and all in the ground state. Since the number of captures to level j of the neutral atom per cm^3 per second is given by $n_e n_i \langle w\sigma_{cj}\rangle$, the final net gain associated with electron-ion recombinations, which we denote by Γ_{ei}, is given by

$$\Gamma_{ei} = n_e n_i \sum_j \left(\langle w\sigma_{cj}\rangle \bar{E}_2 - \langle w\sigma_{cj} E_1\rangle\right), \tag{6-3}$$

where $\langle\ \rangle$ again denotes an average over a Maxwellian distribution, whereas $\bar{E}_2$ denotes that an average is taken over all the ionizing photons. Under interstellar conditions, essentially all photoionizations take place from the ground level, hence $\bar{E}_2$ is independent of j; with use of equation (5-11), equation (6-3) may then be written

$$\Gamma_{ei} = n_e n_i \left\{ \alpha \bar{E}_2 - \tfrac{1}{2} m_e \sum_j \langle w^3 \sigma_{cj}\rangle \right\}, \tag{6-4}$$

where, as usual, we omit the superscript (1) from the total recombination coefficient α.

It will be noted that these equations for Γ_{ei} do not depend on the ionization probability β_{1f} or on the radiation density. As may be seen from equation (5-1), in a steady state the number of ionizing transitions per cm^3 per second must always equal the recapture rate $n_e n(X^{(r+1)})\alpha$. In non-steady states this simplification is not possible, and equation (6-3) must be modified to include the dependence on β_{1f}.

For energy loss in interstellar space, the primary mechanism is inelastic collisions between particles. In collisions between electrons and ions, the number of excitations from level j to level k per cm^3 per second will be $n_e n_{ij} \gamma_{jk}$, where n_{ij} is the number of ions in level j and γ_{jk} is the appropriate rate coefficient [see equation (4-5)]. The kinetic energy lost by the colliding electrons will be $E_k - E_j$, denoted by E_{jk}. The deexciting collisions will produce an offsetting energy gain, and if we sum over all transitions between all levels, the net rate of energy loss in electron–ion collisions, which we denote by Λ_{ei}, will be given by

$$\Lambda_{ei} = n_e \sum_{j<k} E_{jk} (n_{ij} \gamma_{jk} - n_{ik} \gamma_{kj}). \tag{6-5}$$

The assumption here that all photons escape is not valid for dense molecular clouds. Under many interstellar conditions which we shall consider, all the ions in question will be in the ground level, and the sum over j can be omitted, with n_{ij} for $j=1$ replaced by n_i, the total particle density of the ions under consideration.

In the following sections we apply these results to H I and H II regions. The other gain and loss processes considered, such as excitation of atoms, ions, or molecules by neutral H atoms, all satisfy equations similar to equations (6-4) and (6-5).

6.1 H II REGIONS

In a region where the hydrogen is nearly fully ionized, the energy gain Γ is produced primarily by the photoionization of H, with He providing a secondary contribution in some cases. The energy loss Λ results almost entirely from electron excitation of ionized atoms of the elements C, N, O, and Ne, since these have excited levels a few volts above the ground level; these levels are collisionally populated at temperatures of about 10^4°K, with radiation of energy resulting. The cross sections for such collisions exceed those for radiative capture by about 10^5, providing a very powerful cooling mechanism. Were it not for the relatively low abundance of all these heavier atoms with respect to H, the interstellar gas would cool to a very low temperature even in H II regions. In contrast, atomic hydrogen radiates only weakly in its equilibrium state. If the electron energies are high enough to excite the $n=2$ level of H at 10.2 eV energy, the hydrogen will generally be nearly fully ionized by collisions if not by radiation.

We discuss below the values which the heating and cooling functions Γ and Λ assume at different temperatures in H II regions, and the equilibrium temperatures which result.

a. Heating Function Γ

According to equation (6-4), the net gain Γ_{ei} is the average difference between the kinetic energy E_2 given to newly created photoelectrons per cm^3 per second and the corresponding kinetic energy E_1 lost by recombining electrons. The average of E_2 is given by

$$\bar{E}_2 = \frac{\int_{\nu_1}^{\infty} h(\nu-\nu_1) s_\nu U_\nu \, d\nu/\nu}{\int_{\nu_1}^{\infty} s_\nu U_\nu \, d\nu/\nu}, \tag{6-6}$$

integrated over all ν greater than the H ionization limit at ν_1. The exact determination of U_ν is complicated, since the absorption of the stellar ultraviolet radiation, of energy density $U_{s\nu}$, must be taken into account, and the contribution of the diffuse ultraviolet radiation field $U_{D\nu}$ included. Moreover, $U_{D\nu}$ affects the fraction of H atoms that are neutral and thus modifies the absorption of the direct stellar radiation.

Simple results are available in two situations. First, we may evaluate $\bar{E}_2$ relatively close to the exciting star. In this case, $U_{s\nu}$ is relatively large because of its inverse-square variation with distance, and $U_{D\nu}$ may be neglected in comparison. Second, we may evaluate $\bar{E}_2$ for the entire H II region. In both cases we shall assume that the stellar radiation field shortward of the Lyman limit can be represented as dilute blackbody radiation at the color temperature T_c, a parameter of the stellar atmosphere. In view of the relatively high stellar opacity at $\nu > \nu_1$, this assumption is reasonably realistic; T_c is generally somewhat less than the effective temperature, defined in terms of the total radiative flux from the star. In general, we define a dimensionless ratio ψ as

$$\psi = \frac{\bar{E}_2}{kT_c}. \tag{6-7}$$

The value of ψ for the unabsorbed stellar radiation field, which is relevant as r approaches zero, will be denoted by ψ_0, with $\langle\psi\rangle$ denoting the average of ψ over the ionized region as a whole.

To evaluate ψ_0, U_ν may be set equal to $4\pi B_\nu(T_c)/c$, where the Planck function is given in equation (3-4). Evaluation of the resultant integral as a series in ascending powers of $\exp(-h\nu/kT_c)$ gives the results in Table 6.1.

To determine $\langle\psi\rangle$ we do not use equation (6-6), which is valid in any one region. Instead we compute an average $\langle\bar{E}_2\rangle$ for the region as a whole; this average equals the total kinetic energy imparted to all photoelectrons per second, divided by $S_u(0)$, the total rate at which ultraviolet photons leave the star, which equals in turn the rate at which photoelectrons are produced throughout the ionized region. In computing the total energy gained, we need not consider the diffuse flux, since kinetic energy gained by one electron on absorption of a diffuse photon is exactly equal to the kinetic energy given up by a free electron somewhere else when this photon was produced. Thus the mean kinetic energy gain per photoelectron is simply given by averaging $h(\nu-\nu_1)$ over the stellar radiation field. Since each such photon is absorbed somewhere in the ionized region, we must omit the weighting factor s_ν which appears in equation (6-6); omission of s_ν and substitution of $U_{s\nu}$ for U_ν in equation (6-6) then gives $\langle\bar{E}_2\rangle$. With the use again of the Planck function for $U_{s\nu}$, and of the same series expansion as was used for ψ_0, we obtain the values of $\langle\psi\rangle$ which are also given in Table 6.1.

Table 6.1. Mean Photoelectron Energy E_2/kT_c

T_c (°K)	4,000	8,000	16,000	32,000	64,000
ψ_0	0.977	0.959	0.922	0.864	0.775
$\langle\psi\rangle$	1.051	1.101	1.199	1.380	1.655

The mean energy lost per recombining electron may be computed exactly. For the average value of $w^3\sigma_{cj}$, which occurs in equation (6-4), we write

$$\sum_{j=k}^{\infty}\langle w^3\sigma_{cj}\rangle=\frac{2A_r}{\pi^{1/2}}\left(\frac{2kT}{m_e}\right)^{3/2}\beta\chi_k(\beta), \tag{6-8}$$

where A_r and β are defined in equations (5-10) and (5-13). Table 6.2 gives values of $\chi_1(\beta)$ and $\chi_2(\beta)$ computed [3] with a series expansion for g_{fn}. The accuracy is about the same as that for ϕ_1 and ϕ_2 in Table 5.2.

When the diffuse radiation field is considered explicitly in the computation of $\bar{E}_2$, or is negligible at close distances from the central star, $\chi_1(\beta)$

Table 6.2. Energy Gain Functions χ_1 and χ_2

T (°K)	31.3	62.5	125	250	500	1000
χ_1	4.24	3.90	3.56	3.23	2.90	2.58
χ_2	3.46	3.12	2.78	2.45	2.12	1.80
T (°K)	2,000	4,000	8,000	16,000	32,000	64,000
χ_1	2.26	1.95	1.65	1.37	1.10	0.84
χ_2	1.49	1.20	0.92	0.67	0.46	0.30

must of course be used. Equation (6-4) then yields

$$\Gamma_{ep} = \frac{2.07\times10^{-11} n_e n_p}{T^{1/2}} \left\{ \bar{E}_2 \phi_1(\beta) - kT\chi_1(\beta) \right\} \frac{\text{erg}}{\text{cm}^3\,\text{s}}, \tag{6-9}$$

where we have used equations (5-12) and (5-14); Γ_{ep} denotes the value of Γ resulting from captures of electrons by protons. Close to the exciting star $\bar{E}_2$ may be determined from equation (6-7), with ψ set equal to ψ_0, given in Table 6.1. If $\langle\psi\rangle$ is to be used as an approximation, one should, for consistency, ignore $U_{D\nu}$ entirely and use the "on-the-spot" approximation, with all recaptures to the ground level ignored, and with ϕ_2 and χ_2 replacing ϕ_1 and χ_1. This approximation is an excellent one for considering the total volume of the ionized region (in the absence of dust absorption). However, for computing the detailed variation of T throughout the region, this approximation is not suitable (see the discussion in the next section).

b. Cooling Function Λ and Resultant T_E

The values of Λ_{ei} in H II regions may be found directly from equation (6-5) using the values of γ_{jk} obtained from Section 4.1, in particular from equations (4-11), (4-4), and the values of $\Omega(j,k)$ in Table 4.1. The ratio of the quantities n_{ij} and n_{ik} which appear in equation (6-5) is given approximately by equations (2-27) and (4-14) for a two-state atom; for more accurate results, collisional transitions among a number of levels may need to be considered.

The important radiating ions in H II regions all have low-lying excited levels similar to those of O II or O III shown in Fig. 4.1. These excited levels fall into two groups, metastable fine-structure levels in the ground spectroscopic term [S4.1a] (a ^3P term for O I, N II, O III, and Ne III), which have excitation energies less than 0.1 eV, and different spectroscopic terms, which are generally metastable and which have excitation energies more than an order of magnitude greater. Collisional excitation of the former levels, which produces radiation in the infrared, is quite insensitive to temperature for T greater than 1000°K. Excitation of the latter, however, increases quite sharply with increasing T, since only the electrons in the tail of the Maxwellian distribution have the 1.9 to 3.3 eV of energy required, and the number of these rises sharply with T. Hence these transitions to different spectroscopic terms provide a thermostatic mechanism that tends to keep the temperature in the neighborhood of 10,000°K.

Values of $\Lambda/n_e n_p$ for low-density H II regions ($n_e < 10^2\,\mathrm{cm}^{-3}$) are shown as a function of temperature in Fig. 6.1 [4], computed on the arbitrary assumption that O, Ne, and N, the only three radiating elements considered, are each 80 percent singly ionized and 20 percent doubly ionized. The abundances relative to H are essentially those in Table 1.1. At n_e significantly above $10^2\,\mathrm{cm}^{-3}$, collisional deexcitation reduces Λ below the

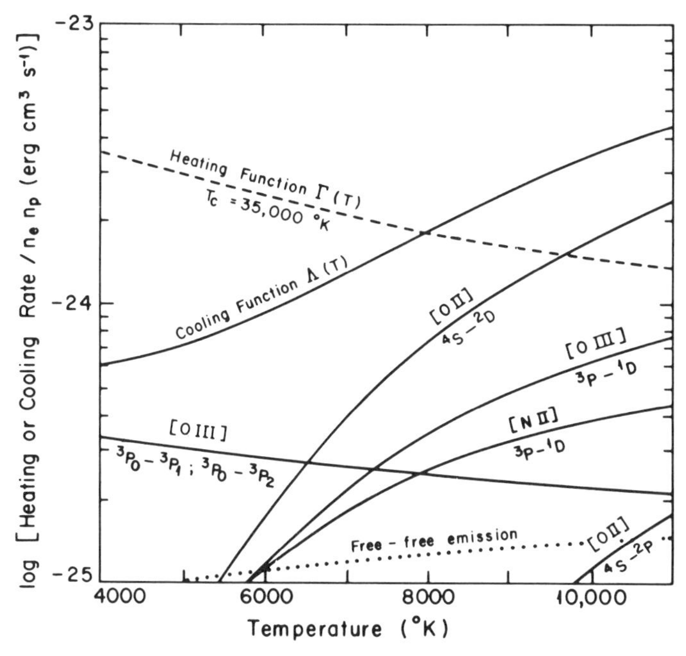

Figure 6.1 Heating and cooling functions in H II regions. Both $\Lambda/n_e n_p$ and $\Gamma/n_e n_p$ are shown as functions of the temperature T. In addition, the contributions to $\Lambda/n_e n_p$ from individual transitions in O and N ions are plotted for the low-density limit ($n_e < 10^2\ \mathrm{cm}^{-3}$) [4]. The dotted line shows $\varepsilon_{ff}/n_e n_p$. The heating function represents an average for the H II region as a whole (see text), for a central star of color temperature, T_c, of 35,000° at far ultraviolet wavelengths.

values indicated. The insensitivity to temperature of the O III, $^3P_0-^3P_{1,2}$, contribution results from the low excitation energy; the corresponding N II, $^3P_0-^3P_{1,2}$, contribution behaves similarly, but gives a Λ lower by about an order of magnitude because of the lower abundance of N. For the $^3P-^1D$ contribution, O III and N II contribute about the same; because of the assumed predominance of single ionization, the abundances of these two ions are comparable. The value of $\epsilon_{ff}/n_e n_p$, the contribution of free–free emission to the loss function, has been obtained from equation (3-56) and is plotted in Fig. 6.1; evidently ϵ_{ff} is unimportant for the conditions shown.

For comparison, approximate values of the average gain function Γ_{ep}, divided again by $n_e n_p$, are also shown in Fig. 6.1 for T_c equal to 35,000°. To portray mean conditions for the H II region as a whole, $\bar{E}_2$ is computed with ψ set equal to $\langle\psi\rangle$, and with ϕ_2 and χ_2 replacing ϕ_1 and χ_1 in equation (6-9). The intersection of the heating and cooling curves gives an equilibrium temperature, T_E, of 8000°K, in general agreement with the observations summarized in Sections 4.1 and 4.2. If $n_e < 10^2\,\text{cm}^{-3}$, Λ as well as Γ varies as $n_e n_i$, and T_E is nearly independent of the overall density.

The variation of T_E with distance r from the star in an idealized Strömgren sphere of uniform density can be found from detailed numerical integrations in which the diffuse flux is taken into account [5]. Ionization equilibrium and deviation of the stellar ultraviolet radiation from a blackbody curve have been included in the detailed numerical calculations. In general, T_E is found first to decrease as r increases; as $U_{s\nu}$ falls, because of inverse-square divergence, the diffuse flux becomes relatively more important, and recaptures to the ground level no longer produce significant heating. In a typical case, T decreases by some 1500°K as r/r_S increases from 0 to about 0.6. With a further increase in r, T_E increases again as the optical depth becomes significant, the lower frequencies become depleted relative to the higher ones, and $\bar{E}_2$ increases. At $r/r_S \approx 1$, the computed T_E is appreciably greater than its value at $r=0$, but it must decrease again in the transition region where the fractional ionization x becomes much less than unity. The presence of absorbing dust particles probably does not much alter these results, provided that r/r_i (where r_i is the actual radius of the ionized zone) replaces r/r_S [S5.1c]. However, the nonuniform character of actual H II regions makes it difficult to compare the observations with these idealized computations of temperature structure.

The rate at which the gas temperature changes, when T is not equal to T_E, may be determined from the general equation (6-1). If the equilibrium temperature is 8000°K [S4.1c], then for values of T in the neighborhood of T_E the value of the radiative cooling time t_T defined in equation (6-2) is given roughly by

$$t_T = \frac{2.0\times 10^4}{n_p} \text{ years.} \tag{6-10}$$

This result may be compared with the recombination time t_r for radiative electron capture, which equals $1/n_e\alpha$. If we use equation (5-14) for $\alpha^{(2)}$, we obtain

$$t_r = \frac{1.54\times 10^3 T^{1/2}}{Z^2 n_e \phi_2(\beta)} \text{ years.} \tag{6-11}$$

For T in the neighborhood of 10^4°K, t_r exceeds t_T by about an order of magnitude. Evidently if the radiating elements such as C and O have their cosmic abundances, the interstellar gas cools much more rapidly than it recombines.

Values of the cooling time for much higher temperature may be obtained from Fig. 6.2 [6], where values of Λ/n_H^2 are plotted as a function of T; cosmic composition is assumed and, for T greater than 10^4°K, collisional ionization [S5.2b]. If T_E is taken to be much less than T, then $t_T = 1.5\, kTn/\Lambda$, according to equations (6-1) and (6-2). If He is taken to be fully ionized, the total number of particles per nucleus, designated by n/n_H, equals 2.3; $t_T n_H$ is 2.2×10^3 and 5.8×10^6 years at $T = 10^5$ and 10^7°K, respectively. At $T = 10^6$°K, Λ is increasing so rapidly with decreasing T that t_T is no longer useful, since the time required for T to drop by a factor $1/e$ will be substantially less than the value obtained from equation (6.2). These results ignore cooling of the gas by collisions with dust grains; if the dust–gas ratio is the same as in cool clouds [S7.3b], such inelastic collisions will in fact dominate Λ for temperatures exceeding 10^6°K [7].

For comparison with these cooling times, $t_r n_e$ at 10^5 and 10^7°K, with $Z = 1$ and ϕ_1 replacing ϕ_2 in equation (6-11), is 4.6×10^5 and 6.6×10^7 years, respectively, again much longer than $t_T n_H$. For such highly stripped atoms as O VI, however, both Z^2 and ϕ_1 are increased, and $t_r n_e$ is less by some two orders of magnitude than the corresponding value for H; dielectronic recombination [S5.2] decreases t_r even further for some ions. Hence at the highest temperatures recombination of highly ionized atoms will be much more rapid than cooling.

6.2 H I REGIONS

When the hydrogen is neutral, there are no electron–proton recombinations, and the energy gain from photoionization now results from impurity atoms only and is much reduced. The value of Γ_{ei}/n_e is proportional to the ion density, which for a gas with some particular $n_{\rm H}$ is less by a factor 1/2000 in H I as compared with H II regions. In contrast, the corresponding loss-rate function per free electron, equal to Λ_{ei}/n_e, is less changed. Since the O, Ne, and N atoms are neutral, the loss rate at high temperatures shown in Fig. 6.1 is reduced to a lower value, determined by the lower excitation cross section for neutral atoms. However, the abundance of C^+ relative to hydrogen is much the same in many H I and H II regions, and at temperatures below 1000° the value of Λ_{ei}/n_e will be roughly the same in the two regions. Evidently the equilibrium temperature in H I regions tends to be relatively low.

A detailed calculation of T_E in H I regions requires the consideration of a number of processes, whose relative importance varies with temperature, density, chemical composition (i.e., the depletion), and other variables. It is not certain that all relevant processes have as yet been considered. In view of the many uncertainties, we consider only briefly some of the principal effects which are likely to be important. Since the known loss rates can be evaluated with only moderate uncertainty, these are considered first, with the uncertain rates of energy gain discussed subsequently.

a. Cooling Function Λ

Contributions to Λ will be produced by radiation from neutral atoms, ionized atoms, and molecules, each excited chiefly by electrons and H atoms. The rate coefficients for these processes have been determined by the methods described in Section 4.1. In particular, Λ_{ei} for excitation of C II and Si II can be obtained by substituting equation (4-11) into equation (6-5); the collisional deexcitation term in this latter equation is generally negligible for these ions in regions where n_e is less than 10 cm^{-3}. The corresponding values of Ω are listed in Table 4.1.

At temperatures of a few thousand degrees, electron excitation of the excited levels in neutral H can be an important source of energy loss. Since excitation of the $n=2$ level is mainly responsible for the radiation, the resulting loss rate, $\Lambda_{e\rm H}$, is nearly proportional to $\exp(-E_{12}/kT)$, where E_{12} is the excitation energy of the $n=2$ level from the ground state. At temperatures between 4000 and 12,000°K, we have

$$\Lambda_{e\rm H} = 7.3\times10^{-19} n_e n({\rm H\ I}) e^{-118{,}400^\circ/T}\ {\rm erg\,cm^{-3}\,s^{-1}}. \qquad (6\text{-}12)$$

Equation (6-12) differs by at most 3 percent from a detailed tabulation [6] based on quantum mechanical calculations. The Lα photons produced by electron excitation will mostly be absorbed by dust grains [S5.1c] with the energy reradiated in the infrared.

Collisional excitation of ions by neutral H atoms is frequently a dominant cooling process in H I regions. For cooling by excitation of C II, probably the most abundant of the heavy positive ions in H I regions, we obtain for T less than 100°K, using Table 4.2 and equation (4-4), with neglect of deexcitation in equation (6-5),

$$\begin{aligned}\Lambda_{\mathrm{H\,C\,II}} &= 7.9\times10^{-27} n_{\mathrm{H}}^2 d_{\mathrm{C}} e^{-92.0^\circ/T}\ \mathrm{erg\,cm^{-3}\,s^{-1}} \\ &= 2.5\times10^{-27} n_{\mathrm{H}}^2 d_{\mathrm{C}}\ \mathrm{erg\,cm^{-3}\,s^{-1}}\ \text{at } 80^\circ\mathrm{K},\end{aligned} \tag{6-13}$$

where we have assumed that all C and H nuclei are in C II and H I atoms. The quantity d_C is the depletion of carbon, defined as the ratio n_C/n_H divided by the cosmic value of this ratio, which equals 4.0×10^{-4} according to Table 1.1. The value of d_C shown for ζ Oph in Table 1.1 is 0.2, but this result is uncertain, and in less strongly reddened stars there is some indication that d_C is more nearly unity [S3.4c]; thus a range of possible values of d_C between 0.1 and 1 must be considered. For other radiating atoms, such as Fe II, C I, and O I, rate coefficients for collisional excitation by H atoms have been estimated [6].

Hydrogen molecules can be a source either of energy loss or of gain, depending on the circumstances. From the $J=0$ or 1 levels the rotational levels $J=2$ or 3, respectively, can be excited by H-atom impact, whereas the reverse deexcitation transitions produce an energy gain; the source of this gain is optical pumping, which can populate the upper rotational levels above their Boltzmann values at the kinetic temperature [S4.3b]. From equation (6-5) we have, using equation (4-4) for γ_{kj}/γ_{jk} and equation (2-27) for n_k/n_j,

$$\Lambda_{\mathrm{H\,H_2}} = n_{\mathrm{H}}^2 \sum_J \frac{n(J+2)}{n_{\mathrm{H}}} E_{J,J+2}\gamma_{J+2,J}\left\{\frac{b_J}{b_{J+2}} - 1\right\}, \tag{6-14}$$

where $n(J)$ is again the particle density of H_2 molecules in rotational level J, and $\gamma_{J+2,J}$ is the deexcitation rate coefficient. For observed diffuse clouds with strongly saturated H_2 lines, and with $2n(H_2)$ as much as 10 percent of n_H, the upper J levels tend to be overpopulated with respect to the lower levels; hence $b_J/b_{J+2}<1$ and $\Lambda_{\mathrm{HH_2}}$ is negative as a result of optical pumping. For reddened stars ($E_{B-V}\geqslant0.10$) with strong H_2 lines, a chief contribution to equation (6-14) comes from the $J=3$ to $J=1$ transition, since $b_3/b_1\gg1$. With $n(3)/n(\mathrm{H\ I})$ observed in the range from 10^{-6} to 10^{-4}, the corresponding gain is always less than $\Lambda_{\mathrm{HC\,II}}/25$, if d_C is not less than 0.2. Since b_2/b_0 is close to unity, the contribution from $J=2$ is uncertain, but is unlikely to be a significant heat source.

In the interiors of dense clouds, the opacity reduces the radiant energy density so sharply that optical pumping becomes unimportant, and Λ_{HH_2} becomes positive. However, the kinetic temperature of such clouds is normally much less than E_{02}/k, which equals 510°K, and the exponential factor in $n(2)$ [see equation (2-27)] reduces $n(2)/n_{\mathrm{H}}$ in equation (6-14) so sharply that Λ_{HH_2} is negligible. At low temperatures, HD radiates more strongly in the infrared than does H_2, since the HD transition from $J=1$ to 0 is permitted, and E_{01}/k is only 130°. Thus even if $n(\mathrm{HD})/n(\mathrm{H}_2)$ is as low as 1/20,000, HD will contribute more to Λ than will H_2 for $T<80°$ [6]. Cooling within dense clouds by line radiation from CO, CH, and CN molecules with even lower E_{jk} can be important [6], but the high opacity in the lines must be considered. Cooling and heating of the gas by collisions with dust grains, whose temperature is determined primarily by radiative processes [S9.1a], can be a dominant element in the thermal equilibrium of dense clouds.

A general picture of the overall loss function in H I as well as in H II regions is provided [6] in Fig. 6.2. For $T<10{,}000$°K, curves are shown for several different values of x, here defined as n_e/n_{H}; when x drops to 10^{-3}, collisional excitation of atoms and ions by neutral H atoms dominates Λ. Cooling by H_2 molecules is ignored in Fig. 6.2. The steep rise in Λ at about 10,000° results in part from the collisional excitation of H [see equation (6-12)] and in part from the marked increase of n_e as the hydrogen becomes ionized.

As in H II regions, we determine the cooling time t_T from Fig. 6.2 on the assumption that $T \gg T_E$; then Λ is correspondingly much greater than Γ in equation (6-1). With x taken at its minimum value of about 5×10^{-4}, corresponding to full ionization of C, Fe, and Si, we obtain for no depletion and with H_2 cooling again ignored

$$t_T \approx \frac{2.4\times10^5}{n_{\mathrm{H}}} \text{ years,} \tag{6-15}$$

with an accuracy of roughly 30 percent for n_{H} between 1 and 300 cm^{-3} and for T between 50 and 600°K. Thus the cooling time in H I gas is about 10 times that in the H II region with the same n_{H}, but with $T\approx10{,}000$°K. Since n_e is normally several orders of magnitude less than n_{H} in H I regions, the radiative recombination time t_r given in equation (6-11) is usually longer than t_T, even if the radiating atoms are depleted while the ionized ones are not.

b. Heating Function Γ

Among the energy gain processes, the only significant one that is known with reasonable accuracy is Γ_{ei}, the gain corresponding to electron cap-

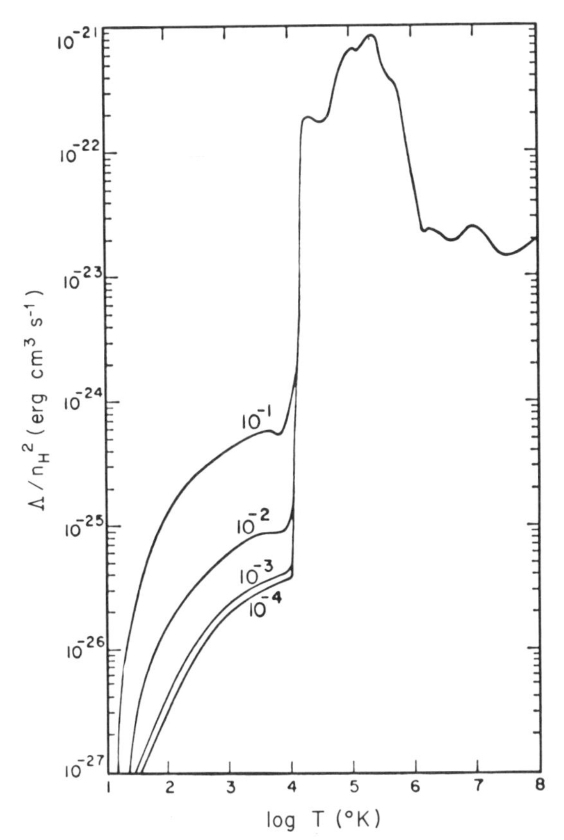

Figure 6.2 Cooling function for interstellar gas [6]. Values of $\Lambda(T)/n_{\mathrm{H}}^2$ are shown as functions of the temperature T. For $T < 10{,}000$°K, the different curves represent different values of $x \equiv n_e/n_{\mathrm{H}}$, while for $T > 10{,}000$°K, collisional ionization is assumed for all elements. Depletion and the possible presence of dust grains [7] or of H_2 are ignored.

tures by C II and other ions, with subsequent photoionization. This quantity is given by the equations obtained previously [S6.1a]; for C II the functions ϕ_2 and χ_2 must be used in equation (6-9), since the $n=1$ shell is occupied, and recaptures occur to the levels with $n \geqslant 2$. If only C II is considered and excitation of ions by neutral H atoms is ignored, equating Λ_{ei} to Γ_{ei} yields

$$T_E = 16\text{°K}, \tag{6-16}$$

independent of n_e, n_{H}, or the relative abundance of C II. To obtain this result we have assumed that $\bar{E}_2 \approx 2$ eV. Allowance for $\Lambda_{\mathrm{H}i}$ reduces T_E somewhat.

Evidently the temperatures [S11.1a] of about 80°K observed in diffuse clouds and of about 6000°K in H I regions of lower density require a much more powerful heating mechanism than can be provided by photoionization of C I. Many different processes have been proposed [8]. Among those analyzed that seem to have some promise of explaining the observed temperatures, we discuss here the following three: heating by cosmic-ray ionization of H, whose heating rate is designated by Γ_{HR}; heating by the formation of H_2 molecules on dust grains, designated by Γ_{Hd}; heating by photoelectric emission from grains, designated by Γ_{ed}.

The basic process involved in heating by cosmic rays is the ionization of neutral H [S5.1d]; if $n(\text{H I})$ is again the particle density of neutral H atoms per cm^3, the rate of production of free electrons by this process is $\zeta_H n(\text{H I})$ per cm^3 per second. Some of these electrons will be first-generation or primary electrons produced in direct encounters between a cosmic-ray particle and a neutral H atom; the mean kinetic energy of such electrons is about 35 eV. When allowance is made for energy lost by excitation of H and for the additional number of free electrons produced as secondaries, the mean value of E_h, the kinetic energy available for heating the gas, per free electron produced (primaries plus secondaries), is about 3.4 eV [9]. If we use for ζ_H the value of $7\times10^{-18}\,\text{s}^{-1}$ estimated [9] for cosmic rays reaching the Earth, then Γ_{HR} becomes

$$\Gamma_{HR} = n_H \zeta_H \langle E_h \rangle \approx 3.8\times10^{-29} n_H \,\text{erg}\,\text{cm}^{-3}\,\text{s}^{-1}. \tag{6-17}$$

We neglect ionization of He, which increases the overall heating rate somewhat.

If large numbers of cosmic rays are assumed at relatively low energies, between 1 and 100 MeV, ζ_H can be increased to as much as $10^{-15}\,\text{s}^{-1}$ and Γ_{HR}/n_H will be increased to about 5×10^{-27} erg s^{-1}, which for $n_H=20$ cm^{-3}, $d_C=0.1$ and $T=80$°K is just sufficient to offset the loss rate per H atom resulting from excitation of C II [equation (6-13)]. However, this mechanism would apparently give more HD (and also more OH) than is observed [S5.3b]. In addition, the propagation of cosmic rays through the interstellar gas, from a possible source to an interstellar cloud, poses theoretical difficulties [10]. As a result, serious doubt has been cast on this heating mechanism.

We have seen earlier [S5.3a] that H_2 molecules form at the rate $Rn_H n(\text{H I})\,\text{cm}^{-3}\,\text{s}^{-1}$ as a result of H-atom impacts on grains. Each such molecule formation releases a binding energy of 4.48 eV, which must go partly into heating the grain, partly into overcoming the energy of adsorption to the grain surface, partly into excitation energy of the newly formed molecules, and partly into the translational kinetic energy with which the

H_2 molecule leaves the grain. If we let z_H be the fraction of the binding energy which appears in this last form, and set R equal to $3\times10^{-17}\,\text{cm}^3\,\text{s}^{-1}$ [S5.3a], we obtain

$$\Gamma_{Hd}=2.2\times10^{-28}z_H n_H n(\text{H I})\,\text{erg}\,\text{cm}^{-3}\,\text{s}^{-1}. \tag{6-18}$$

Evidently Γ_{Hd} will balance $\Lambda_{\text{H C II}}$ if d_C is as low as 0.1, if at the same time $z_H=1$. There is serious question as to whether d_C can be as low as 0.1 in most clouds [S3.4c]. Similarly, the value of z_H is uncertain, with a rough theoretical computation giving a value of about 0.04 [11]. When an H_2 molecule is destroyed, the reaction products generally have kinetic energies which are available for heating the gas, effectively increasing z_H in equation (6-18). Thus dissociation by Lyman band absorption produces about 0.4 eV per H_2 molecule [12], corresponding to $z_H=0.09$. Much larger values of z_H can be obtained if most H_2 molecules are assumed to be destroyed by reaction (5-50), in which case dissociative recombination of the resulting CH_3^+ molecules gives an effective z_H of about unity; comparable contributions to z_H result from the ensuing reactions [13], which produce additional CH_2^+ molecules, helping to maintain reaction (5-50). However, detailed computations [14] indicate that even with z_H of about unity, Γ_{Hd} can account for an equilibrium temperature of about 80°K only with considerable straining of the parameters involved.

We turn now to an evaluation of Γ_{ed}, the heating rate associated with photoelectric emission from grains. In a steady state, the difference between the rates at which electrons and positive ions strike and stick to grains must equal the rate at which photoelectrons are emitted, with the balance between these two rates produced by the electric charge on the grain [S9.2b]. Here we express the heating rate in terms of the photoelectric emission. Again we let E_2 be the mean kinetic energy of the emitted electron for photons of frequency ν, and y_e, the photoelectric efficiency, also a function of ν. If we ignore the $\langle w\sigma_{cd}E_1\rangle$ term in equation (6-3), and replace the electron capture rate, $n_e n_d\langle w\sigma_{cd}\rangle$, by the photoelectron emission rate, we find

$$\Gamma_{ed}=n_H\Sigma_d c\int\frac{Q_a(\lambda)y_e U_\lambda E_2}{h\nu}d\lambda\equiv n_H\Sigma_d\bar{E}_2F_e, \tag{6-19}$$

where $Q_a(\lambda)$ is the absorption efficiency factor [S7.1], Σ_d is the total projected grain area per H nucleus [S7.3b], and $U_\lambda\,d\lambda$ is the energy density of radiation in the wavelength interval $d\lambda$ (see Table 5.5). Both y_e and $\bar{E}_2$ depend, in general, on the grain potential, but for electron densities exceeding about $10^{-2}\,\text{cm}^{-3}$ in diffuse H I clouds [S5.2a], this effect may be ignored. The quantity F_e defined by equation (6-19) is the photoelectron flux from the grain per projected grain area [S9.2b].

The most uncertain and consequently the most critical quantity in equation (6-19) is the photoelectric efficiency, or quantum yield y_e. For visible light, y_e is normally small, of order 10^{-4} except for specially prepared surfaces, but in the ultraviolet, at photon energies higher than 10 to 11 eV, values of 0.1 for y_e are plausible for most materials [15], especially for smaller grains [16]; the mean free path for slowing down a photoelectron can be about 100 Å in a semiconductor or insulator. It has been suggested [17] that for an appreciable fraction of the grains, the radius may be as small as 50 Å, giving y_e nearly unity.

In evaluating equation (6-19) we shall assume that y_e is negligible at wavelengths longer than 1100 Å (11.3 eV), and constant between 1100 and 912 Å; shortward of 912 Å, U_λ is assumed negligible because of absorption in the Lyman continuum. To evaluate the integral for the photoelectron flux F_e, we use the values of U_λ in Table 5.5, which we assume are reduced within a cloud by an absorption factor $\exp(-\tau_\nu)$, and obtain [17]

$$F_e = 2\times 10^7 y_e \overline{Q}_a e^{-\tau_\nu} \,\mathrm{cm}^{-2}\,\mathrm{s}^{-1}. \tag{6-20}$$

At these far ultraviolet wavelengths, $\overline{Q}_a$ is likely to be about unity if x exceeds unity, and to vary linearly with x for $x<1$ [see equation (7-7)]. If now we set $\overline{Q}_a = 1$ and $\Sigma_d = 1.1\times 10^{-21}\,\mathrm{cm}^2/\mathrm{H}$ atom, from equation (7-23), and replace $\overline{E}_2$ by an average value of 5 eV, appropriate for very small grains [17], we obtain

$$\Gamma_{ed} = 1.8\times 10^{-25} y_e n_{\mathrm{H}} e^{-\tau_\nu} \,\mathrm{erg\,cm}^{-3}\mathrm{s}^{-1}. \tag{6-21}$$

If τ_ν is small and $y_e = 1$, Γ_{ed} can balance $\Lambda_{\mathrm{H\ C\ II}}$ at 80°K in equation (6-13) even if d_C is 1 and n_H equals 100 cm^{-3}. The very small grains required for a high photoelectric efficiency and such a high $\overline{E}_2$ cannot account for all the observed $Q_e\Sigma_d$ (which is based [S7.3b] on the measured extinction at 1000 Å), but even with a considerable reduction, Γ_{ed} is apparently a very powerful source of heating for H I regions. If E_{B-V} through the cloud is 0.2 mag, τ_ν at 1000 Å half way through is 1.2 [S7.2b], reducing Γ_{ed} by a factor 0.3, and still permitting a computed T_E of 80°K for n_H about 30 cm^{-3}.

We consider next the question of how the gas can be heated to some 6000°K in regions of lower density. The value of n_{H} in these warm H I regions is uncertain, but is likely between 1 and 0.1 cm^{-3} [S11.1a]. From Fig. 6.2 it may be seen that Λ/n_{H}^2 at 6000°K is about an order of magnitude greater than at 80°K and is about $3\times 10^{-26}\,\mathrm{erg\,cm}^3\mathrm{s}^{-1}$. The heating rate Γ_{Hd} given in equation (6-18) is far below this required value; since this rate varies as n_H^2, as does Λ, its effectiveness does not increase as n_H drops. However, Γ_{HR} varies linearly with n_H, and cosmic rays become a relatively more effective heating source as n_H is decreased. This effectiveness is increased further by the increase of $\langle E_h \rangle$, the mean heating energy per electron produced; as the ratio $n_e/n(\text{H I})$ increases above about 10^{-2}, the primary electrons produced by the cosmic rays then give up most of their energy to heating the plasma of thermal electrons and ions rather than to exciting atomic H. Specific calculations [6] show that with ζ_H and n_H about equal to 10^{-15}s^{-1} and 0.2cm^{-3}, respectively, T_E is about 8000°K. For the reasons cited above it is doubtful whether so large a steady-state flux of such low-energy cosmic rays can pervade most of the interstellar medium.

Comparable heating of low-density gas can be achieved with photoelectric emission from grains. For several reasons the heating rate given in equation (6-21) is reduced when n_H is as low as 0.1 per cm^3. The rate of capture of free electrons is no longer sufficient to balance the photoemission rate given by this equation with $y_e \approx 1$. As pointed out in Section 9.2, positive charges accumulate on the grain under this condition, and the electric potential U rises, decreasing both y_e and E_2. In addition, at temperatures approaching 10,000°K, the kinetic energy which is lost when electrons strike the grains and which was ignored in equation (6-19) diminishes the net heating rate Γ_{ed} appreciably [equation (6-3)]. Nevertheless, some calculations indicate that when n_H is about 0.1 cm^{-3} and appreciable depletion of Fe and Si is assumed, the equilibrium temperature is between 2000 and 6000°K, with collisional excitation of atoms balanced in part by the photoelectric input of electron kinetic energy and in part by some X-ray and cosmic-ray heating (with $\zeta_H = 10^{-17} \text{s}^{-1}$).

A characteristic feature of some steady-state heating theories for warm H I gas is that often these theories predict the possible existence of two separate gas phases with the same pressure, but widely different densities and temperatures. As first pointed out for cosmic-ray heating [6] such a two-phase theory permits a simple explanation for the simultaneous existence of cold dense H I clouds and warm H I intercloud regions at the same pressure [S11.3]. Two-phase models may apparently be possible also with various different combinations of heating mechanisms.

The preceding discussion has all been based on the assumption that the temperature is determined by a steady-state thermal equilibrium. An entirely different possibility is that the gas is heated up occasionally by transient events, and then gradually cools and recombines. As we have seen in the discussion following equation (6-10), the cooling time is generally much shorter than the recombination time. Hence the fractional ionization of each element will depend on the detailed history, and must be computed at each time before the cooling rate at that time is known. Computations of cooling and recombination have been carried out [18] for a gas heated initially by a distant supernova explosion, assumed to be an intense source of soft (50-eV) X-rays. An alternative transient heating source involves dying stars [19], which are expected to pass through a phase of high ultraviolet luminosity during about 10^6 years, leaving ionized and heated regions behind them after the ultraviolet light source fades away. Such time-dependent effects must certainly be considered in a complete theory for the interstellar gas. However, it is not easy to find a crucial observational test of the complex models which have been computed.

Situations in which a gas is cooling may be unstable against the development of condensations [1, 20], since the cooling function Λ at constant pressure often increases with decreasing T. If a small region becomes colder than its surroundings, it will be compressed to correspondingly higher pressure and will cool more rapidly. However, the condensations formed in this way will be rather small; the requirement of constant pressure limits their size to the sound travel distance during the cooling time, which is at most a few parsecs before compression [20].

REFERENCES

1. G. B. Field, *Ap. J.*, **142**, 531, 1965.
2. L. L. Cowie and C. F. McKee, *Ap. J.*, **211**, 135, 1977.
3. M. J. Seaton, *M.N.R.A.S.*, **119**, 81, 1959.
4. D. E. Osterbrock, *Astrophysics of Gaseous Nebulae*, W. H.Freeman (San Francisco), 1974, Chapt. 3.
5. R. H. Rubin, *Ap. J.*, **153**, 761, 1968.
6. A. Dalgarno and R. A. McCray, *Ann. Rev. Astron. Astroph.*, **10**, 375, 1972.
7. J. P. Ostriker and J. Silk, *Ap. J. (Lett.)*, **184**, L113, 1973.
8. J. Silk, *P.A.S.P.*, **85**, 704, 1973.
9. L. Spitzer and M. G. Tomasko, *Ap. J.*, **152**, 971, 1968.
10. L. Spitzer and E. B. Jenkins, *Ann. Rev. Astron. Astroph.*, **13**, 133, 1975.
11. D. Hollenbach and E. E. Salpeter, *J. Chem. Phys.*, **53**, 79, 1970.
12. T. L. Stephens and A. Dalgarno, *Ap. J.*, **186**, 165, 1973.
13. A. Dalgarno and M. Oppenheimer, *Ap. J.*, **192**, 597, 1974.
14. A. E. Glassgold and W. D. Langer, *Ap. J.*, **193**, 73, 1974.
15. W. D. Watson, *Ap. J.*, **176**, 103, 1972.
16. W. D. Watson, *J. Opt. Soc. Am.*, **63**, 164, 1973.
17. M. Jura, *Ap. J.*, **204**, 12, 1976.
18. H. Gerola, M. Kafatos, and R. McCray, *Ap. J.*, **189**, 55, 1974.
19. J. Lyon, *Ap. J.*, **201**, 168, 1975.
20. J. Schwarz, R. McCray, and R. F. Stein, *Ap. J.*, **175**, 673, 1972.

7 Optical Properties of Grains

Dust grains scatter and absorb light in interstellar space. In addition, these small solid particles emit radiation at wavelengths much greater than those of the absorbed light. Knowledge of the optical properties of grains can be used to interpret the observed effects in terms of the nature and spatial distribution of these particles.

We discuss briefly some basic concepts of this subject. First we consider the "extinction," here defined as the sum of absorption and scattering; a collimated beam of light can be extinguished by either process. The rate of extinction is again given by equation (3-1). In the discussion of atomic processes the quantity κ_ν in this equation is generally called an absorption coefficient, even though the photon may be promptly reemitted, producing scattering; here κ_ν is called an "extinction coefficient." In measuring light from a single star one normally observes the flux $\mathcal{F}_\nu$, defined as the integral of I_ν over the solid angle of the stellar image. Since the light scattered or emitted by interstellar grains usually makes a negligible contribution to I_ν within a stellar image, the second term on the right-hand side of equation (3-3) may be ignored. If we express the optical thickness $\tau_{\nu r}$ in terms of s_ν, the extinction coefficient per particle, in accordance with equation (3-14), then equation (3-3) yields for A_λ, the interstellar extinction in magnitudes at the wavelength λ,

$$A_\lambda = -2.5 \log \frac{\mathcal{F}_\nu}{\mathcal{F}_\nu(0)} = 1.086 N_d Q_e \sigma_d, \tag{7-1}$$

where $\mathcal{F}_\nu(0)$ is the stellar flux at the Earth in the absence of extinction, N_d is the number of dust grains per cm^2 along the line of sight from the Earth to the star, and σ_d is the geometrical cross section of a single grain. We use the subscript d throughout to denote dust grains. The dimensionless quantity Q_e is the "extinction efficiency factor," defined in terms of the optical cross section s_ν by the relationship

$$Q_e = \frac{s_\nu}{\sigma_d}. \tag{7-2}$$

The extinction efficiency of the grains and the resulting extinction both depend on λ. These equations are based on the assumption that all the grains are identical. In the actual case equation (7-1) must be integrated over all the parameters characterizing the grains, including chemical composition, size, shape, and orientation.

Next we consider scattering of light by grains over an appreciable area of the sky, producing a diffuse source of illumination. In this case it is the intensity I_ν, rather than the flux $\mathcal{F}_\nu$, that is observed, and the emissivity for scattered light, which we denote by $j_{\nu S}$, makes the interesting contribution in equation (3-3). We define a "scattering efficiency factor" Q_s, so that a fraction Q_s/Q_e of the light extinguished is scattered rather than absorbed; this fraction is called the "albedo." We may write

$$j_{\nu S}(\boldsymbol{\kappa}) = n_d Q_s \sigma_d \int I_\nu(\boldsymbol{\kappa}') F(\boldsymbol{\kappa} - \boldsymbol{\kappa}') \, d\omega', \tag{7-3}$$

where $\boldsymbol{\kappa}'$ and $\boldsymbol{\kappa}$ are unit vectors indicating the directions of the incident and scattered photons, $d\omega'$ is an interval of solid angle about $\boldsymbol{\kappa}'$, and $F(\boldsymbol{\kappa} - \boldsymbol{\kappa}')$ is a phase function, which we may assume depends on the angle ϕ between $\boldsymbol{\kappa}$ and $\boldsymbol{\kappa}'$; the integral of F over all $d\omega'$ is unity.

Finally we consider the thermal emission from the grains. If we denote by Q_a the corresponding efficiency factor for pure absorption, evidently equal to $Q_e - Q_s$, then use of Kirchhoff's law, equation (3-5), gives for $j_{\nu E}$, the coefficient for thermal emission

$$j_{\nu E} = n_d Q_a \sigma_d B_\nu(T_s), \tag{7-4}$$

where T_s is the temperature of the solid material in the grain. In accordance with the discussion in Section 2.4, the validity of this equation results from the fact that energy is interchanged among the different modes of vibration of the solid grain much more rapidly than between these modes and the radiation field.

In the first section below, the extensive calculations of the optical efficiency factors Q_e, Q_s, and Q_a for spherical particles are summarized. Subsequent sections discuss the use made of these results in interpreting the observations of extinction, scattering, and thermal emission from interstellar grains.

7.1 OPTICAL EFFICIENCY FACTORS

The Mie theory of scattering and absorption by spheres with a complex index of refraction, m, has been applied numerically to a wide variety of cases [1]. Here we give a few typical results and discuss in particular the asymptotic behavior for spheres whose radius a is much larger or much smaller than the wavelength. It is customary to express the size of the sphere in terms of the dimensionless parameter x defined by

$$x = \frac{2\pi a}{\lambda}. \tag{7-5}$$

In Fig. 7.1 are shown values of Q_e for spheres with four different values for m, the index of refraction: (a) $m=\infty$, corresponding to an infinite dielectric constant; (b) $m=1.33$, corresponding to ice particles (at visual wavelengths); (c) $m=1.33-0.09i$, corresponding to ice with absorbing impurities, or "dirty ice"; (d) $m=1.27-1.37i$, corresponding to spheres of iron. The horizontal scale is x; as m approaches unity, the curves for Q_e shift to the right by an amount proportional to $1/(m-1)$. For small x, the scattering efficiency factor Q_s becomes very small; if $|mx|$ is much less than unity, we have the usual formula for Rayleigh scattering

$$Q_s = \frac{8}{3}x^4\left|\frac{m^2-1}{m^2+2}\right|^2, \tag{7-6}$$

while Q_a in this case is given by

$$Q_a = -4x\, Im\left(\frac{m^2-1}{m^2+2}\right), \tag{7-7}$$

where Im indicates that the imaginary part is to be taken. When m is purely real (spheres a and b in Fig. 7.1), Q_a vanishes, but for spheres c and d most of the extinction at low x results from absorption, giving rise to a linear variation of Q_e with x in accordance with equation (7-7). As a result, for the spheres with $m=1.33$, typical of ice, the introduction of a small imaginary part in m much increases Q_e for small x. It may be noted that for the ice sphere the large-scale oscillation of Q_s with changing x, which is due to interference between the transmitted and diffracted radiation, is much reduced when absorption within the sphere is assumed.

Figure 7.1 shows that as x increases, Q_e approaches an asymptotic value of about 2, both for absorbing and for dielectric spheres, while for absorbing spheres Q_a approaches a value somewhat less than unity as expected. Thus twice as much energy is removed from the beam as is

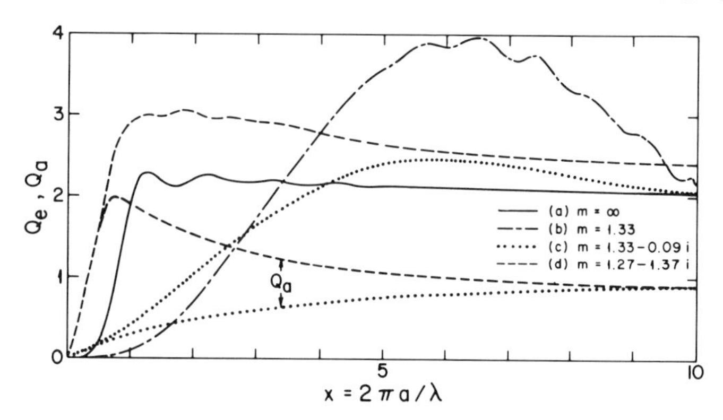

Figure 7.1 Extinction and absorption cross section for spheres. For each value of m, the index of refraction, the upper curves show Q_e, the ratio of extinction and geometrical cross sections, for spheres; x is the ratio of the circumference, $2\pi a$, to the wavelength, λ, of the incident radiation. For cases (c) and (d), with complex values of m, lower curves show Q_a, the corresponding ratio of absorption cross section to πa^2. These theoretical curves [1] are based on exact computations for cases (a), (b), and (d), but for case (c), on an approximate theory, valid for small $m-1$.

actually striking a large sphere. The reason for this behavior is that, in accordance with Babinet's principle, the diffraction pattern due to an obstacle is identical with that of an aperture of the same cross section, and thus contains the same energy as is striking the obstacle. The diffracted energy can be regarded as removed from the incident beam if two requirements are met. First, energy measurements must be made at such great distances behind the obstacle that the shadow has been "washed out," and the Fraunhofer pattern is applicable. Secondly, the diffraction angles must exceed both the initial angular spread of the incident beam and the angular resolution of the detection apparatus.

Evidently this diffracted light accounts for at least half of the energy scattered by a large sphere, with the rays passing through the sphere accounting for the other half if the material is nonabsorbing. We evaluate the contribution of this diffracted radiation to the phase function F, which occurs in equation (7-3) for the scattered light. If a spherical grain is considered, F is a function $F(\phi)$ of the angle ϕ between the incident light and the direction of the scattered or diffracted light; according to equation (7-3), $F(\phi)d\omega$ is the fraction of the scattered light energy which at great distances is traveling within a small cone, of solid angle $d\omega$, at an angle ϕ to the direction of the incident beam of light. The diffracted component of this radiation follows closely the Fraunhofer diffraction pattern for a circular aperture with a radius a equal to the radius of the sphere, and we have the familiar result,

$$F(\phi)=\frac{J_1^2(x\sin\phi)}{\pi\sin^2\phi}. \tag{7-8}$$

The normalization is exact only in the limit as x becomes large. In this equation $J_1(u)$ denotes a Bessel function of order 1, which approaches $u/2$ as u falls below unity. If we denote by ϕ_1 the value of ϕ at which $F(\phi)$ falls to half its central value, we obtain, if ϕ_1 is sufficiently small so that $\sin\phi_1$ may be replaced by ϕ_1,

$$\phi_1=\frac{1.617}{x}=0.257\frac{\lambda}{a}. \tag{7-9}$$

The total scattered light includes radiation which has been refracted through the sphere, and which is not included in equation (7-8). However exact computations [1] for a sphere with m equal to 1.33 show that equation (7-9) gives ϕ_1 rather accurately for the total scattered light when x is between 2 and 5.

The integral of Q_e over all λ can be obtained from the Kramers-Kronig relationship [2, 3], which for a spherical grain gives

$$\int_0^\infty Q_e\,d\lambda=4\pi^2a\left\{\frac{\varepsilon_0-1}{\varepsilon_0+2}\right\}\equiv4\pi^2aF_K, \tag{7-10}$$

where ε_0, the dielectric constant of the grain in the low-frequency limit, equals the square of m, the index of refraction; F_K is defined by equation (7-10). For spheroidal particles F_K is increased [3], but even for a two-to-one ratio of axes and ε_0 as great as 4, the increase is less than about 5 percent. If equation (7-1) is used to eliminate Q_e, and we let $\rho_s4\pi a^3N_d/3=\rho_dL$, where ρ_s is the density of solid material within the grain, and ρ_d is the mean density of dust along the line of sight, of length L, equation (7-10) becomes

$$\rho_d=\frac{1.01\times10^{-23}\rho_s}{F_K}\int_0^\infty\frac{A_\lambda\,d\lambda(\text{cm})}{L(\text{kpc})}. \tag{7-11}$$

The optical properties of nonspherical grains have also been considered [2, 4]. When grains are aligned, the extinction differs for the two directions of polarization (see Chapter 8) and in addition the extinction coefficients vary with the direction of the incident radiation. In the absence of alignment, the presence of grains with complex shapes apparently has no very large effect on the mean optical efficiency factors [4].

7.2 SELECTIVE EXTINCTION

While the absolute determination of A_λ, called the "general extinction," is observationally difficult, as shown in the next section, the determination of the change in A_λ with λ, called the "selective extinction," is reasonably straightforward. To make such measures, two stars whose spectral type and luminosity class are known to be identical from measurements of their spectral lines are compared, and the magnitude difference $\Delta m(\lambda)$ between them is measured at two wavelengths, λ_1 and λ_2. Measurements are usually made on stars of spectral type O and B, partly because they are far away and show substantial amounts of extinction, and partly because the shape of the intrinsic continuous spectrum for the hotter stars does not depend very sensitively on spectral type for wavelengths in the visible. Since the emitted spectra of the two stars are presumably identical, each value of $\Delta m(\lambda)$ is the sum of two terms, one varying as the logarithm of the distance ratio, one depending on the difference in A_λ between the two stars. If now we take the difference $\Delta m(\lambda_1)-\Delta m(\lambda_2)$, the terms depending on the distances cancel out, and we have

$$\Delta m(\lambda_1)-\Delta m(\lambda_2)=\Delta(A_{\lambda_1}-A_{\lambda_2}), \tag{7-12}$$

where Δ denotes the difference between the two stars. If for one of the two stars A_λ is thought to be zero, either because the star is very close, or because of agreement between the measured continuous spectrum and theoretical calculations, then equation (7-12) gives directly $A_{\lambda_1}-A_{\lambda_2}$ for the other star, a quantity which may be denoted by $E(\lambda_1,\lambda_2)$. In practice the shape of the intrinsic or unreddened spectrum is usually determined separately for a group of relatively close stars. Since there may be some uncertainty in this determination, a small zero-point error in $E(\lambda_1,\lambda_2)$ is usually a possibility. If λ_1 and λ_2 are taken to be 4350 and 5550 Å, about equal to the mean wavelengths for the blue (B) and visual (V) photometric bands [5], respectively, then $E(\lambda_1,\lambda_2)$ becomes the standard color excess E_{B-V}.

Measures of $E(\lambda_1,\lambda_2)$ have been used in two ways. With λ_1 and λ_2 fixed, observations of many stars indicate how the color excess, and presumably the number of grains in the line of sight, vary with direction and distance. With λ_2 fixed, measurements have been made with variable λ_1 to determine the variation of selective extinction with wavelength. We discuss each of these applications in turn.

a. Spatial Distribution of Grains

Photoelectric measures of some 1300 stars of known spectral type [6] have been extensively analyzed for their statistical properties. (Color excesses used for these data must be multiplied by 2.06 [7] to convert them to values of E_{B-V}.) Since absolute magnitudes can be estimated from the spectra, and the total extinction A_V may be set equal to $3E_{B-V}$ [S7.3a], an approximate distance is known for each star. These data indicate that at a fixed distance E_{B-V} varies greatly from star to star, and a uniform reddening coefficient in magnitudes per kpc cannot be used to predict E_{B-V} at a distance L.

The statistical distribution of color excesses may be used to give information on the irregularity of the dust distribution. For example, the data can be fitted with a hypothetical model in which the grains are concentrated in identical obscuring clouds, each producing a color excess E_0; such clouds may be assumed to have a random distribution, with a line of sight in the galactic disc intersecting k such clouds per kiloparsec. The probability $p(n)$ that a line of sight along a distance r will intersect n clouds is given by the usual Poisson distribution,

$$p(n) = \frac{(kr)^n e^{-kr}}{n!}. \tag{7-13}$$

The average color excess at the distance r is given by

$$\langle E_{B-V} \rangle = krE_0, \tag{7-14}$$

where in general $\langle\ \rangle$ denotes here an average over stars all at the same distance. For the average value of $(E_{B-V})^2$ one obtains, making use of equation (7-13),

$$\langle (E_{B-V})^2 \rangle = (kr + k^2r^2)E_0^2. \tag{7-15}$$

If now we eliminate kr by the use of equation (7-14), we find

$$E_0 = \frac{\langle (E_{B-V})^2 \rangle}{\langle E_{B-V} \rangle} - \langle E_{B-V} \rangle. \tag{7-16}$$

Once E_0 is known, k may be obtained from equation (7-14).

These equations have been used [8] to compute E_0 and k for some 500 stars within 1000 pc of the Sun and within 100 pc of the galactic plane, yielding

$$kE_0 = \frac{\langle E_{B-V} \rangle}{L} = 0.61 \text{ mag/kpc}. \tag{7-17}$$

This determination of the mean selective extinction is useful for determining approximately the total amount of dust in the galactic disk within

1000 pc from the Sun, but should not be used to predict the value of E_{B-V} for any one object. The value of k found from the analysis was 4.3 kpc^{-1}, corresponding to E_0 equal to about 0.14 mag.

A more complete analysis [8] of the data takes into account the detailed skewness of the distribution (measured by the value of $\langle (E_{B-V})^3 \rangle$), and also the relative number of unreddened stars. Fitting the data requires two types of clouds; the resultant values of E_0 and k for each of the two cloud types, here called "standard clouds" and "large clouds," respectively, are given in Table 7.1. The errors given are average deviations obtained by comparison of the two distance groups analyzed—250 to 500 pc and 500 to 1000 pc. A similar analysis [9] which assumes a spherical cloud shape gives about the same results as Table 7.1.

Table 7.1. Statistical Properties of Dust Clouds

Type of Cloud	Standard Cloud	Large Cloud
Mean E_{B-V} per cloud, E_0	0.061 ± 0.006	0.29 ± 0.06
Number of clouds per kpc, k	$6.2 \quad \pm 0.3$	$0.8 \quad \pm 0.2$
Selective extinction per kpc, kE_0	$0.38 \quad \pm 0.05$	0.23 ± 0.01

While the numerical values in Table 7.1 would presumably differ in different regions, in view of the large-scale structure of the interstellar medium, the small-scale patchy distribution of dust seems well established. Measured color excesses of A0 stars within 250 pc of the Sun [10] indicate similar patchiness, with no selective extinction for most stars, and with about 10 percent of the stars (concentrated in limited areas, particularly in Taurus and Scorpius-Ophiuchus) showing color excesses of 0.056 mag or more. The total mean extinction per kpc averaged over all these A0 stars is about one-fifth of the value given by equation (7-17), in agreement with other indications that the density near the Sun of dust and gas [S3.4b] is substantially less than its mean value at greater distances.

The constancy of the dust-to-gas ratio is shown by the relatively close correlation between E_{B-V} and N_H obtained from $L\alpha$ and H_2 measures [11] (see Fig. 1.1). The line of best fit shown in that figure gives a value of 5.4×10^{21} mag^{-1} cm^{-2} for N_H/E_{B-V}. While this ratio includes hydrogen in H_2 as well as in neutral H, an additional correction of roughly 10 percent [12] must be made for ionized H in the H II zones surrounding the early-type stars being observed. With this correction we obtain

$$N_H = 5.9 \times 10^{21} \; E_{B-V} \; mag^{-1} \; cm^{-2}. \tag{7-18}$$

About this same result is obtained [13] if N_H is determined from X-ray absorption or if $N(\mathrm{H\ I})$ is obtained from 21-cm emission in the line of sight to globular clusters.

From equation (7-18) we find that for the "standard cloud" in Table 7.1, N_{H} [which here equals $N(\mathrm{H\ I})$] is 4×10^{20} cm^{-2}, in rough agreement with the mean value of 3×10^{20} cm^{-2} for the mean H I cloud seen in absorption in extragalactic sources, with $|b| > 6°$ [S3.4a].

The value of 4 per kpc for k obtained from 21-cm absorption components in the galactic plane [S3.4a] is somewhat less than the value of 6.2 per kpc given for standard clouds in Table 7.1; the former value is averaged over a radius of more than 2 kpc, and includes interarm regions, while the latter value refers to closer stars and is more heavily weighted by the spiral arms in which the observed O and B stars are concentrated. The "large cloud" in Table 7.1 is not evident in the 21-cm absorption data, possibly because much of the hydrogen is molecular in these more opaque clouds. In any case this discussion clearly indicates that a cloud of dust is also a cloud of gas.

The mean value of E_{B-V}/L given in equation (7-17), and indicated also in Table 7.1, has a number of important applications. This average may be combined with a mean value of 0.059 found for $E_{B-V} \sin|b|$ from 38 globular clusters at $|b| > 10°$ [14] to give an effective thickness $2H$ of the dust layer of about 200 pc, somewhat less than the 250 pc found for H I but probably not significantly so in view of the small number of clusters in this average. Also, equations (7-17) and (7-18) may be combined to yield

$$\langle n_{\mathrm{H}} \rangle = 1.2 \text{ cm}^{-3}. \tag{7-19}$$

Other applications are discussed in Section 7.3.

To determine the mean radii of the clouds with different E_0, measures of the spatial correlation of E_{B-V} are required. The limited data available indicate [15] radii of about 35 pc for the large clouds in Table 7.1, with radii about an order of magnitude smaller for the standard clouds.

While the evidence for a patchy distribution of gas and grains together is overwhelming, the real situation is certainly more complex than the idealized model underlying Table 7.1. As pointed out below [S7.3c], many different types of clouds may be distinguished. In addition, the distribution of clouds is probably far from random, with the smaller clouds doubtless correlated with the larger ones in complex ways.

b. Variation of Extinction with Wavelength

To combine measures of selective extinction from different stars, each with different numbers of absorbing grains in the line of sight, it is customary to normalize $E(\lambda_1, \lambda_2)$, dividing by E_{B-V}. Also λ_2 is generally taken to be

5550 Å, the reference wavelength for the V band; the ratio $E(\lambda, V)/E_{B-V}$ evidently equals 0 or 1 for λ equal to 5550 or 4350 Å, respectively. The variation of this normalized selective extinction with $1/\lambda$, measured over the entire accessible spectrum from 1000 Å to 2 μ, is shown in Fig. 7.2 [2, 16]. Most stars show about the same normalized extinction curves, but for a few objects the curves are different, particularly in the infrared and far ultraviolet. The points for $\theta^1+\theta^2$Ori [17], with perhaps the most unusual extinction curve, are indicated in Fig. 7.2.

By virtue of equations (7-1) and (7-12), the normalized extinction curves are directly comparable, apart from scale factors and zero points, with the curves for Q_e in Fig. 7.1. The theoretical curves in Fig. 7.2 have been obtained [16] with a three-component model, taking into account the variation with λ of the index of refraction m of each component. Small graphite grains ($\bar{a}=2.5\times 10^{-6}$ cm) have been included to fit the observed peak in the extinction curve in the ultraviolet. Grains of SiC ($\bar{a} = 7.5\times 10^{-6}$ cm) have been added to fit the visual and infrared regions, whereas smaller magnesium and aluminum silicate grains ($\bar{a}=4.5\times 10^{-6}$ cm) yield

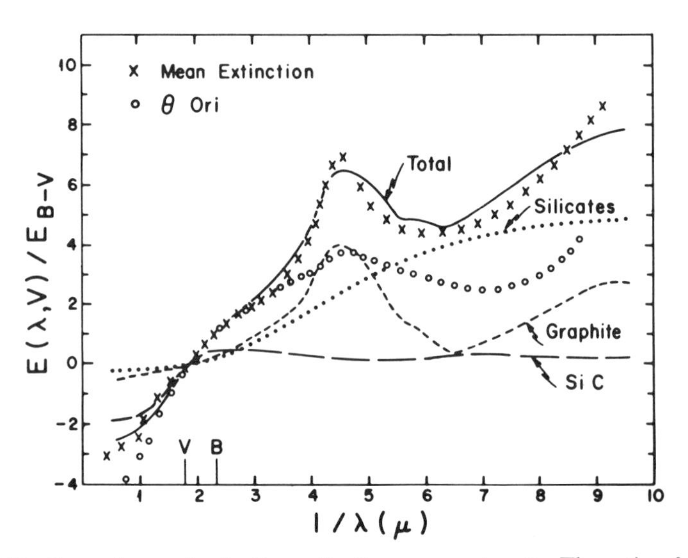

Figure 7.2 Dependence of selective extinction on wavelength. The ratio of $E(\lambda, V)$ to E_{B-V} is plotted against the reciprocal wavelength in microns. The crosses give the mean observed extinction for normal stars [2]; in the ultraviolet these are based on 14 observed stars, excluding 3 abnormal ones; the circles give observed values for $\theta^1+\theta^2$ Ori, showing abnormal extinction. The other curves are computed theoretically [16] for grains of three different types (see text), with the sum of the three shown by the solid line.

the increasing extinction in the ultraviolet, although not so steep as observed, and also explain the infrared feature observed at 9.7μ, usually in absorption but occasionally in emission [18].

These three types of grains may be referred to as "UV-peak grains," "visual grains," and "far-UV grains,", respectively. While these three regions of the selective extinction curve might conceivably be fitted by particles of one type, the fact that these three regions seem to vary from star to star somewhat independently in a few cases suggests that in fact at least three different types of grain are involved. Thus in θ Ori both the UV-peak grains and the far-UV grains are less abundant than normal, whereas the visual grains may have a radius larger than normal. In σ Sco, on the other hand, the normalized far ultraviolet extinction [17] is about the same as in θ Ori, but the UV peak is more nearly normal. As shown in Section 8.2, the relative abundance of the far-UV particles, as indicated by the relative strength of the normalized extinction at the shortest wavelengths, does not seem well correlated with the size of the visual grains, as deduced from the observations of polarization and selective extinction. One might expect that the distribution function of interstellar grains over various sizes and types would show variations between different regions of space.

While the theoretical fit to the mean observational curve shown in Fig. 7.2 is one of the best of the many that have been proposed [2], it is not quite satisfactory. The data at wavelengths as short as 1000 Å, obtained [19] after the theoretical fit, show a continuing steep rise of selective extinction with decreasing wavelength, suggesting that grains with a radius smaller than 2×10^{-6} cm, giving $x<1.3$ at 1000 Å, make a dominant contribution. In addition, the model contains several times more silicon than would be available for the usual cosmic composition (see Table 1.1). Consistency with cosmic composition is obtained if the visual grains are assumed to consist mainly of ice [20]. In any case the fits obtained with grain models are not unique. Even the UV peak at 2200 Å can be explained by certain silicates or by irradiated quartz as well as by graphite [21]. The Mie theory [S7.1] predicts that the theoretical shape and position of this peak tends to be highly sensitive to grain size and shape; this sensitivity, which is particularly marked for silicates, is difficult to reconcile [21] with the observed constancy of the UV peak (relative to E_{B-V}) in most stars other than θ Ori.

From the standpoint of grain evolution it seems likely that at least some of the grains have a complex structure with mantles of more volatile compounds overlying more refractory nuclei [S9.4a]. As shown by curves (a) and (b) in Fig. 7.1, the $Q_e(x)$ curves shift to the right, toward larger x, as $m-1$ becomes small. Hence mantles of ice or of other light-weight volatile substances of low m must have outer radii appreciably larger than $\lambda/2\pi$ to produce much extinction. For example, a good fit with the selective extinction data is obtained if the visual grains are assumed to be cylinders with silicate cores of radius 8×10^{-6} cm surrounded by ice mantles with an outer radius of about 1.4×10^{-5} cm [20], about twice the value assumed for the SiC visual grains assumed in Fig. 7.2. If no core is present, the radius required for the ice cylinder is about 2×10^{-5} cm [S8.2a]. However, there is some doubt whether ice can be a major constituent of interstellar grains, since the mass of ice relative to that of silicates within grains, as measured by the strengths of the absorption bands at 3.1 and 9.7μ, respectively, in molecular clouds, is between 2 and 20 percent [22], as compared with the roughly equal masses of ice and silicates in the model core-mantle grains described above. Possibly chemical changes produced by UV radiation might explain the weakness of this absorption [S9.4a].

7.3 GENERAL EXTINCTION

a. Ratio of General-to-Selective Extinction

The general or total extinction is measured by A_V, the value of A_λ at the V wavelength, 5550 Å. The quantity R_V is defined as

$$R_V = \frac{A_V}{E_{B-V}}. \tag{7-20}$$

In Fig. 7.2 $-R_V$ equals $E(\infty, V)/E_{B-V}$, that is, the normalized selective extinction at infinite wavelength, where A_λ is assumed zero. A value of about 3 seems indicated by the data in this figure, extending out to 2μ. An extension of the measured extinction out to 11.4μ gives a value of 3.4 for R_V [23]. Evidently dielectric grains with radii between 2 and 10μ which could produce an appreciable difference of A_λ between wavelengths of 2 and 10μ, are not sufficiently abundant to produce appreciable interstellar extinction.

An independent value of R_V can be obtained if A_V can be determined separately for stars of known E_{B-V}, without regard to the selective extinction curve at long wavelength. However, precise measures of A_V are difficult, since they require objects of known true brightness at a known distance, or several objects whose relative true brightness and relative distances are known. Different attempts [2] to determine A_V directly and thus R_V have used open clusters and luminous stars, with the distances determined by the cluster diameter in the first case and the radial velocity resulting from galactic rotation in the second case. These methods are not very precise, but give an average of about 3.2 for R_V. This same value has been obtained from the brightest E and S0 galaxies in obscured clusters at low b, with distances determined from red shifts [24]. Since the uncertainty in this mean is probably comparable with 0.2, we shall use here the generally accepted value of 3.0 for the mean value of R_V.

Variations in R_V undoubtedly exist between different regions, as indicated by the differences of the normalized extinction curve in the infrared between different stars. Thus for θ Ori, a value of R_V at least as great as 6 seems indicated by the infrared data shown in Fig. 7.2. Apparently in the center of the Orion nebula the radius of the visual grains may be as much as twice its normal value. Infrared color excesses, which are needed for a determination of R_V from $E(\lambda, V)$, are somewhat uncertain because of possible infrared emission by circumstellar dust grains, which at 20μ is unusually strong from the Orion nebula regions surrounding the Trapezium stars that constitute θ^1Ori [25]. However, the mean correlation between λ_{max}, the wavelength of maximum polarization, and infrared color indices [S8.2b] provides good evidence for a correlation between R_V and λ_{max} and indicates the range over which R_V varies.

b. Mean Density and Surface Area of Grains

A definite value of R_V determines the zero point of the extinction scale in Fig. 7.2, and makes it possible to determine the total average projected surface area and mass density per unit volume of interstellar space. We consider first the surface area, since this is more easily determined.

We denote by $\langle n_d \sigma_d \rangle$ the mean projected surface area of the grains per cm^3. To evaluate this quantity we make use of the measured selective extinction at 1000 Å [19], the shortest wavelength at which data are available; at this shortest λ, Q_e should have its relatively greatest value. For stars with the mean extinction curve (Fig. 7.2) we have

$$\frac{E(1000\text{ Å}, V)}{E_{B-V}} = 10. \tag{7-21}$$

If we set $R_V = 3$, then A_λ at 1000 Å is $13E_{B-V}$; if also we assume that $Q_e = 2$ at 1000 Å, we obtain from equations (7-17) and (7-1)

$$\langle n_d \sigma_d \rangle = 1.2 \times 10^{-21} \text{ cm}^{-1}. \tag{7-22}$$

At 1000 Å, where the extinction is still rising rapidly [19], Q_e may be well below its maximum possible value, and equation (7-22) is a conservatively low value for the projected geometrical area of all the grains per cm^3. If we divide this result by the mean value of n_H found in equation (7-19), we obtain for Σ_d, defined as the mean projected area of dust grains per H atom,

$$\Sigma_d \equiv \frac{\langle n_d \sigma_d \rangle}{\langle n_H \rangle} = 1.0 \times 10^{-21} \text{ cm}^2. \tag{7-23}$$

The total density ρ_d of matter within dust grains per unit volume of interstellar space can be determined directly from equation (7-11). From the selective extinction curve shown in Fig. 7.2 one finds [2],* with $R_V = 3$ and spherical grains assumed,

$$\langle \rho_d \rangle = 1.3 \times 10^{-27} \rho_s \left\langle \frac{A_V}{L} \right\rangle \frac{\varepsilon_0 + 2}{\varepsilon_0 - 1}, \tag{7-24}$$

where L is again in kpc. For the model grains whose extinction is portrayed in Fig. 7.2, ρ_s and ε_0 are about equal to 3 gm/cm^3 and 4, respectively, giving a mean density of 1.4×10^{-26} gm/cm^3, if we set $\langle A_V/L \rangle = 1.8$ mag kpc^{-1} from equation (7-17), with R_V again equal to 3. Allowance for infrared extinction peaks at 10 and 20μ, not included in Fig. 7.2, plus a small contribution from extinction at wavelengths as short as 1000 Å [19], increase the integral of $A_\lambda \, d\lambda$ by a factor 1.3 [26], and for the same values of ρ_s and ε_0 we obtain

$$\langle \rho_d \rangle = 1.8 \times 10^{-26} \frac{\text{gm}}{\text{cm}^3}. \tag{7-25}$$

For grains (or mantles) of H_2O both ρ_s and ε_0 are less, with $\langle \rho_d \rangle$ not greatly changed.

Equations (7-25) and (7-19) give for the dust-to-gas ratio, if a H–He ratio of 10 by number is assumed,

$$\frac{\rho_d}{\rho_{gas}} = 0.6 \times 10^{-2}. \tag{7-26}$$

Equations (7-25) and (7-26) are strictly lower limits, since large grains could contain much mass without detectable extinction. We use these equations to denote the density of the grains which are seen at optical wavelengths, from 10μ to 1000 Å.

*Equation (6) of ref. (2), which is essentially equation (7-24) here, is based on the assumption that $F_K = 0.3$, corresponding to $\varepsilon_0 = 2.3$ [equation (7-10)].

For the cosmic composition in Table 1.1, the fraction of mass in all elements heavier than He and Ne is 1.4×10^{-2} and 0.27×10^{-2}, respectively. Unless extinction in the far ultraviolet or far infrared raises appreciably the integral in equation (7-11), thereby increasing ρ_d/ρ_{gas}, the grains can contain only about one-third of the C, N, and O atoms in addition to most of the heavier elements. This conclusion is in agreement with the depletions found towards unreddened stars, where abundances of O, N, and possibly C seem nearly normal [S3.4c]. Perhaps the marked depletion of C, N, and O apparently characterizing the line of sight to ζ Oph is atypical.

c. Visible Nebulae and Representative Clouds

Even a casual glance at Milky Way photographs shows that bright and dark features with a wide variety of sizes and shapes can be discerned. A tabulation of some 1800 dark nebulae [27] shows that the mean projected area per cloud is about 0.4 square degrees for clouds whose opacity A_V at visible wavelengths is 3 mag; this mean area varies about as A_V^{-3}. A few extended clouds are seen with projected areas exceeding 10 square degrees, but these mostly have low values of A_V, about 1 or 2 mag.

Table 7.2 gives a somewhat arbitrary list of various representative clouds. The globules [28, 29] are seen visually against bright nebulae and their CO emission has been measured; some globules may have much greater opacities and correspondingly greater masses. Information on the small infrared clouds adjacent to H II regions has been obtained from the infrared continuous emission produced by heated grains and the many molecular emission lines observed [30]. Detailed properties of the dark clouds are obtained from CO emission data [31], which indicate a mean value of 5×10^{17} cm^{-2} for N(CO) in clouds with A_V about equal to 4 mag, provided that the $^{12}C/^{13}C$ isotope ratio is assumed to equal the terrestrial

Table 7.2. Representative Interstellar Clouds

Cloud Type	Globule	IR/H II Cloud	Dark Cloud	Diffuse Cloud	Cloud Complex
A_V (mag)	4	30	4	0.2	4
N_H (cm^{-2})	8×10^{21}	6×10^{22}	8×10^{21}	4×10^{20}	8×10^{21}
n_H (cm^{-3})	7×10^{3}	4×10^{4}	2×10^{3}	20	2×10^{2}
R (pc)	0.3	0.4	1	5	10
T (°K)	10	50	10	80	10
$M/M_\odot$	30	400	300	400	3×10^{4}

value of 89. The indicated value of N_H is then obtained from A_V, with use of equation (7-18) and with $R_V = 3$; for the cosmic C/H ratio in Table 1.1, these results correspond to one-sixth of the carbon atoms in CO molecules. The diffuse or standard cloud (see Table 7.1) produces too low a value of A_V to be seen directly in visible light; the values of A_V and N_H, taken from Table 7.1 and equation (7-18), agree roughly with the values of N_H obtained from 21-cm absorption measures [S3.4a].

The properties listed for the cloud complex are determined from CO emission [30], and thus refer to those large clouds which have a substantially greater A_V than the value of 0.9 mag found from Table 7.1; molecules will be dominant constituents only if they are shielded from the stellar ultraviolet light which can dissociate them. Typically a cloud complex contains some much denser regions such as the IR/H II cloud. It should be emphasized that certainly for cloud complexes and probably for most of the other clouds as well (except perhaps for some globules), the assumption of spherical symmetry, used in computing n_H and M from N_H and r, is an idealization which probably gives reasonable average results but is certainly not valid in detail.

7.4 SCATTERING

Measurement of the light scattered from dust grains can also give information on the properties of these grains. Geometrical uncertainties tend to complicate the interpretation; in contrast to the relatively simple equation (7-1) relating the observed extinction to the column density of grains, the scattered intensity obtained from equation (7-3) depends in detail on the angles at the grain between the incident and scattered light, as well as on the phase function F. Despite these difficulties, significant results have been obtained in two situations which are discussed below, the scattering of galactic light by the layer of dust in the galaxy, producing the so-called "diffuse galactic light" and the scattering by dust close to bright early-type stars.

a. Diffuse Galactic Light

Attempts to measure the diffuse galactic light must be corrected for the light from many faint stars. The residual intensity, which is usually comparable to that of the starlight subtracted, is then accounted for by starlight scattered from grains. Since starlight is peaked in a direction parallel to the galactic plane, most of the diffuse galactic light can be accounted for by scattering through a rather small angle, as might be expected from equations (7-8) and (7-9) for a appreciably greater than $\lambda/4$. Alternatively, this diffuse light can be accounted for by nearly isotropic scattering if the albedo Q_s/Q_e is assumed to be nearly unity. In practice, the effects of high albedo and of a sharply forward-throwing phase function are difficult to separate [2], but the observations at least rule out the possibility that the grains are nearly isotropic scatterers with low albedo, such as would be expected for absorbing particles small compared to the wavelength. A wide class of particle models, including the one assumed for Fig. 7.2, are roughly consistent with the diffuse galactic light at visible wavelengths. The ultraviolet measurements show [32] that the albedo drops to about 0.35 in the UV extinction peak at 2200 Å, as compared with values between 0.6 and 0.7 at 1500 Å and at wavelengths from 3000 to 4500 Å. This result is consistent with increased absorption by small grains in this UV-peak wavelength band.

b. Scattered Light in H II Regions

Measurement of the starlight scattered by dust within H II regions can be used to give an approximate determination of the dust–gas ratio. As in the previous theoretical discussions of H II regions, we neglect density gradients in general and clumpiness in particular, assuming uniform ρ_d. The scattered intensity along a line of sight, passing the star at a projected distance a, is determined by integrating $j_\nu ds$ along this line, where j_ν is given in equation (7-3). The intensity of starlight, $I_\nu(\kappa')$ in this equation, varies as $1/r^2$, where r is the distance of any point from the star. Along an interval of length comparable with a, r will not differ much from a, hence the scattered intensity will vary as $1/a$. By contrast, the hydrogen emission lines have a more nearly uniform intensity across an idealized H II region of uniform density, since j_ν in this case is everywhere the same, apart from a weak variation with temperature.

A rough numerical analysis, based on the ratio of surface brightnesses in Hβ and the dust-scattered stellar continuum in four H II regions, gives dust–gas ratios of about 1/200 [33], roughly consistent with equation (7-26). However, uncertainties in the assumed optical properties of the grains make this result uncertain by at least half an order of magnitude.

Different regions show large scatter in their values of ρ_d/ρ_{gas}. In particular, the scattered light from the inner region of the Orion nebula (M42, NGC 1976) yields a dust–gas ratio less than found for other H II regions by about an order of magnitude. As pointed out above, the radius of the visual grains may be greater than normal along the line of sight to θ Ori, and both the UV-peak grains and the far-UV grains are relatively much less abundant. While the ρ_d/ρ_{gas} values are uncertain, and variation of clumpiness of n_e between H II regions may explain some of the observed differences, it seems that, at least at the center of the Orion nebula, the distribution of dust grains differs markedly from that in average H I clouds.

7.5 INFRARED EMISSION

Infrared radiation from H II regions has been observed over a wide spectral range, about 3 to 300μ [18]. The spectrum is mostly smooth, and has a shape corresponding to temperatures in the general neighborhood of 50 to 150°K. Temperatures in this range might be expected for grains within 0.1 pc from an early-type star [S9.1b], and the observed radiation at the shorter infrared wavelengths, 3 to 10μ, is generally attributed to emission from grains within H II regions. At a temperature of 150°K the blackbody spectrum, which should provide a crude first approximation to grain emission if the relatively slow variation of Q_a in equation (7-4) is ignored, has its peak at about 20μ. At the longer wavelengths, especially at 100μ and longward, radiation from the cooler grains in H I regions, at temperatures of 20 to 30°K, would be expected to contribute.

As pointed out in Section 7.2b, absorption features which may be attributed to ice and silicate in interstellar grains have been observed in the most highly reddened objects at 3.1 and 9.7μ. These absorption features may be attributed to a region of cooler grains in front of a region where warmer grains produce strong emission, or alternatively to cool grains, with high κ_ν at 3.1 and 9.7μ, interspersed with warmer grains of a different material for which κ_ν is featureless [34].

Detailed interpretations of the dust emission at intermediate infrared wavelengths, roughly from 10 to 20μ, require some assumptions on the variation of grain temperature with distance from the central star as well as on the optical properties of the grains themselves, including the index of refraction and the range of values of the radius a to be assumed; for essentially all the grains present a is believed to be significantly less than $\lambda/2\pi$, which exceeds 10^{-4} cm for λ greater than 6μ, hence the small-x limit for Q_s and Q_a should be used. Detailed theoretical models have been constructed to explain the 20-μ emission [35], but uncertainties in the distribution of dust grain temperatures T_s throughout the emitting region and in the absorption efficiency factor Q_a make it difficult to draw definite conclusions. Thus the data can be fitted with models in which the dust–gas ratio by mass is about 10^{-3} [36], one-sixth of the value found from equation (7-26) for interstellar clouds generally, or with models in which this ratio is reduced to 10^{-4} [35] or 10^{-5} [37] within the HII region.

In principle, a somewhat simpler analysis is required to interpret the wide-band infrared emission from the vicinity of H II regions. The fluxes from such regions measured between 40 and 350μ with a balloon-borne telescope contain about 70 percent of the total infrared radiation emanating from H II regions and from the H I clouds surrounding them. The distances of these regions can be determined approximately, attributing the radial velocities of the observed recombination lines (or of the 21-cm absorption lines produced by intervening H I clouds) to galactic rotation. Thus values of L_{IR}, the infrared luminosity, can be determined and have been plotted in Fig. 7.3 [18], expressed in terms of the solar bolometric luminosity. Similarly, from the measured flux F_ν in the radio continuum the total radio luminosity L_ν from the ionized gas per unit frequency

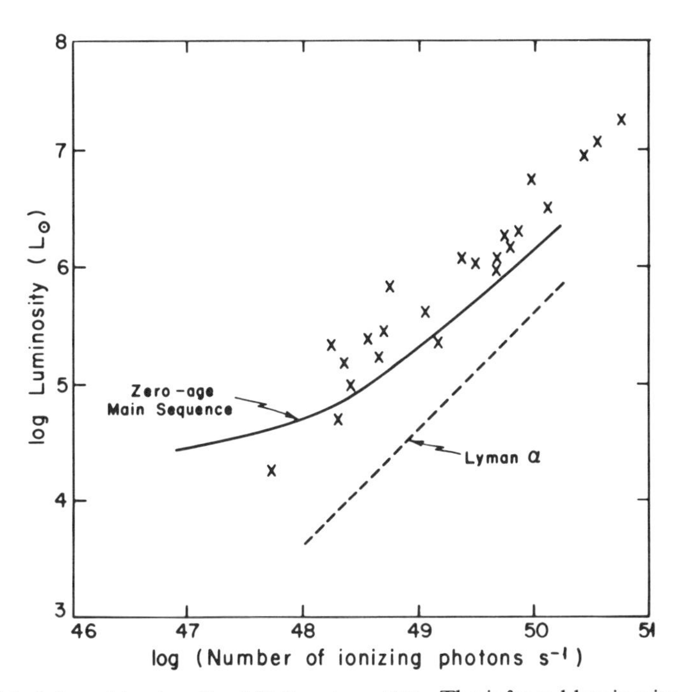

Figure 7.3 Infrared luminosity of H II regions [18]. The infrared luminosity of H II regions, obtained from measured fluxes between about 40 and 350 μ, is plotted against the total number of ionizations per second, computed from the free–free radio luminosity. The solid curve represents the total luminosity of zero-age main sequence stars of spectral types from B0 to O4, plotted against the number of ionizing photons radiated per second shortward of the Lyman limit. For the dashed curve the ordinates represent the total luminosity in Lyman α if each ionization produces one such photon.

interval can be computed. The total number of ionizations per second in the H II region (equal to the corresponding recombination rate) can then be determined from L_ν, since both quantities vary as $n_e n_p r_i^3$, and has been used as the abscissa in Fig. 7.3.

Also shown in the figure are two theoretical curves, both obtained from the theoretical spectra of early-type main sequence stars. The upper one is a plot of the total stellar luminosity L against N_u, the number of photons shortward of the Lyman limit radiated by the star per second. This curve is an appropriate one for an H II region if two assumptions are made: (a) there is no dust absorption in the H II region ($f = y_i^3 = 1$ [S5.1c]); (b) virtually all the stellar energy radiated is converted into infrared radiation by dust absorption in the surrounding H I clouds ($g = L_{IR}/L = 1$). The stellar data assumed are those for zero-age main sequence stars of spectral type ranging from B0 to O4. The lower curve is a plot of the infrared luminosity resulting from absorption of Lα photons only; that is, $L_{IR} = N_u h\nu_{12}$, where ν_{12} is the frequency of Lα equal to three-quarters of ν_1, the frequency of the Lyman limit, and one Lα photon is assumed to result from each absorbed photon in the Lyman continuum [S9.1b]. Evidently the upper curve gives a reasonable fit to the observed points, with some small but systematic discrepancy.

The general agreement shown in Fig. 7.3 permits two important conclusions. First, the excess of the plotted points over the curve for infrared emission solely of the Lα energy indicates that an appreciable fraction of the total stellar luminosity, including wavelengths longer as well as shorter than the Lyman limit, must be absorbed either in the H II regions or in surrounding H I clouds. Hence appreciable optical depths for dust absorption in such regions must be the rule rather than the exception. Second, the rough agreement between the plotted points and the upper curve indicates that the number of electron-ion recombinations does not differ greatly from the computed stellar N_u available for ionizing the H atoms, and suggests that even in the denser, more compact H II regions the optical thickness $\tau_{\nu r}$ for stellar ultraviolet radiation shortward of the Lyman limit is not very large.

More quantitatively, the small deviations between the upper theoretical curve and the plotted points in Fig. 7.3 may be used to determine the amount of dust present [38]. If only a fraction f of the ultraviolet stellar photons are absorbed by H atoms rather than by dust grains within the H II region, the theoretical curve is shifted to the left by a factor f. If only a fraction g of the total radiant energy is absorbed and reemitted by grains, the curve is shifted downward by a factor g. The data are not sufficient to determine f and g separately, but have been analyzed first for the case $g = 1$, with complete ultraviolet absorption outside the H II region, and second for the case in which ultraviolet absorption is restricted to the H II region only; if $\tau_{\nu r}$ were independent of ν, this case would correspond to $f=1-g$. In the first case, the value of $\tau_{\nu r}$ in the H II zone, at wavelengths just shortward of the Lyman limit, is found to range from 0 to 2.9, averaging 1.27, whereas in the second case, the mean $\tau_{\nu r}$ is 1.66, not much greater than in the first case. More complete data were used than those shown in Fig. 7.3. If for each region, the average $\tau_{\nu r}$ for these two cases is taken, and is attributed to particles with $a\rho_s/Q_a$ equal to 10^{-5} gm cm^{-2}, the resultant dust-to-gas ratio ρ_d/ρ_{gas} ranges from 0.002 up to 0.04, averaging about 0.01. Allowance for clumpy distribution of the gas would decrease ρ_{gas}, whereas ρ_d would be decreased if within each H II region the grains were concentrated near the exciting star. Again, the results are not conclusive but give no compelling reason to believe that ρ_d/ρ_{gas} is systematically different between H II regions and H I clouds.

REFERENCES

1. H. C. van de Hulst, *Light Scattering by Small Particles*, Wiley (New York), 1957.
2. P. A. Aannestad and E. M. Purcell, *Ann. Rev. Astron. Astroph.*, **11**, 309, 1973.
3. E. M. Purcell, *Ap. J.*, **158**, 433, 1969.
4. J. M. Greenberg and S. S. Hong, *The Dusty Universe*, G. B. Field and A. W. Cameron, Editors, N. Watson Academic Publ. (New York), 1975, p. 131.
5. S. Sharpless, *Stars and Stellar Systems*, Vol. **3**, University of Chicago Press (Chicago), 1963, p. 225.
6. J. M. Stebbins, C. M. Huffer, and A. E. Whitford, *Ap. J.*, **91**, 20, 1940; **92**, 193, 1940.
7. H. L. Johnson, *Stars and Stellar Systems*, Vol. **3**, University of Chicago Press (Chicago), 1963, p. 218.
8. G. Münch, *Ap. J.*, **116**, 575, 1952.
9. E. Schatzman, *Ann. d'Astroph.*, **13**, 367, 1950.
10. B. Strömgren, *Quart. J. Roy. Astr. Soc.*, **13**, 153, 1972.
11. R. C. Bohlin, *Ap. J.*, **200**, 402, 1975.
12. E. B. Jenkins, *The Structure and Content of the Galaxy and Galactic Cosmic Rays*, C. E. Fichtel and F. W. Stecker, Editors, NASA-CP-002, 1977, p. 215.
13. L. Spitzer and E. B. Jenkins, *Ann. Rev. Astron. Astroph.*, **13**, 133, 1975.
14. G. R. Knapp and F. J. Kerr, *Astron. Astroph.*, **35**, 361, 1974.
15. H. Scheffler, *Z. Astrophys.*, **65**, 60, 1967.
16. D. P. Gilra, *Nature*, **229**, 237, 1971.
17. R. C. Bless and B. D. Savage, *Ap. J.* **171**, 293, 1972.
18. C. G. Wynn-Williams and E. E. Becklin, *Publ. Astr. Soc. Pacific*, **86**, 5, 1974.
19. D. G. York, J. F. Drake, E. B. Jenkins, D. C. Morton, J. B. Rogerson, and L. Spitzer, *Ap. J. (Lett.)*, **182**, L1, 1973.
20. J. M. Greenberg and S. S. Hong, *Galactic Radio Astronomy*, IAU Symp. No. 60, F. J. Kerr and S. C. Simonson, Editors, D. Reidel Publ. Co. (Dordrecht, Holland), 1974, p. 155.
21. B. D. Savage, *Ap. J.*, **199**, 92, 1975.
22. K. M. Merrill, R. W. Russell, and B. T. Soiffer, *Ap. J.*, **207**, 763, 1976.

23. J. A. Hackwell and R. D. Gehrz, *Ap. J.*, **194**, 49, 1974.
24. A. Sandage, *Publ. Astr. Soc. Pacific*, **87**, 853, 1975.
25. E. P. Ney, D. W. Strecker, and R. D. Gehrz, *Ap. J.*, **180**, 809, 1973.
26. L. J. Caroff, V. Petrosian, E. E. Salpeter, R. V. Wagoner, and M. W. Werner, *M.N.R.A.S.*, **164**, 295, 1973.
27. B. T. Lynds, *Ap. J. Supp.*, **7**, 1, 1962.
28. B. J. Bok, C. S. Cordwell and R. H. Cromwell, *Dark Nebulae, Globules and Protostars*, B. T. Lynds, Editor, University of Arizona Press, 1971, p. 33.
29. P. Thaddeus, *Star Formation*, IAU Symp. No. 75, T. de Jong and A. Maeder, Editors, D. Reidel Publ. Co. (Dordrecht, Holland), 1977, p. 75.
30. B. Zuckerman and P. Palmer, *Ann. Rev. Astron. Astroph.*, **12**, 279, 1974.
31. R. L. Dickman, *Ap. J.*, **202**, 50, 1975.
32. C. F. Lillie and A. N. Witt, *Ap. J.*, **208**, 64, 1976.
33. D. E. Osterbrock, *Astrophysics of Gaseous Nebulae*, W. H. Freeman (San Francisco), 1974, Section 7.3.
34. C. Sarazin, *Ap. J.*, **220**, 165, 1978.
35. T. de Jong, F. P. Israel, and A. G. G. M. Trilens, *Lecture Notes in Phys.*, J. Springer (Berlin), **42**, 1975, p. 123.
36. N. Panagia, *Astron. Astroph.*, **42**, 139, 1975.
37. E. L. Wright, *Ap. J.*, **185**, 569, 1973.
38. N. Panagia, *Ap. J.*, **192**, 221, 1974.

8. Polarization and Grain Alignment

Measurement of the linear polarization of starlight requires a determination of the intensity I_μ in some wavelength band as a function of the direction $\boldsymbol{\mu}$ of the electric vector in the radiation. If I_μ varies from a maximum value I_{max} to a minimum value I_{min} as the direction of the electric vector passed by the polarimeter is rotated about the direction $\boldsymbol{\kappa}$ of propagation, the polarization P is defined by

$$P = \frac{I_{\text{max}} - I_{\text{min}}}{I}, \tag{8-1}$$

where the total intensity I of the beam is given by

$$I = I_{\text{max}} + I_{\text{min}}. \tag{8-2}$$

Frequently $100P$, the percentage polarization, is used. Alternatively, the polarization may be expressed in magnitudes and denoted by p; if P is small, $p = 2.17P$. The circular polarization, denoted by V/I, is also defined by an equation similar to (8-1), but with I_r and I_l, the intensity for right-handed and left-handed circular polarization, replacing I_{max} and I_{min}. Sometimes the polarized light is characterized by the four Stokes parameters, I, V, Q, and U, where these last two quantities are defined in equations (8-7) and (8-8).

Plane polarization has been observed in many distant stars [1], with values of P ranging up to a maximum of about 0.07. The direction of the electric vector for which I_μ equals I_{max} tends to be parallel to the galactic plane. The plane defined by this direction of the electric vector and the direction of propagation $\boldsymbol{\kappa}$ is called the "plane of vibration." The circular polarization V/I produced by interstellar grains is much weaker, about 10^{-4}, and has been observed in only a few stars.

Here we discuss first the optical properties of elongated grains, since it is presumably such nonspherical particles that are responsible for the polarization. Subsequent sections treat the observed polarization and the possible methods which have been proposed for aligning the grains.

8.1 OPTICAL PROPERTIES OF NONSPHERICAL PARTICLES

For a nonspherical particle, the efficiency factors defined in Section 7.1 will depend on the orientation of the particle with respect both to $\boldsymbol{\kappa}$ and the direction of the electric vector **E**. Because of the enormous variety of possible shapes for interstellar grains, and because of computational complexities, such efficiency factors have been determined mostly for infinite cylinders [2, 3] (with all cross sections taken per unit length) and for spheroids [3, 4], both of which are axially symmetrical. We denote by Q_{eE} or Q_{eH} the values of Q_e when the plane defined by $\boldsymbol{\kappa}$ and the axis of symmetry contain the electric or the magnetic vector, respectively, of the radiation. Values of these two efficiency factors for infinite cylinders are shown [3] in Fig. 8.1 for m equal to 1.33 in the simple case where the wave front is parallel to the cylinder axis; Q_{eE} and Q_{eH} now refer to waves for which the electric vector is parallel and perpendicular, respectively, to the cylinder axis. Figure 8.1 shows that Q_{eE} tends to exceed Q_{eH} for wavelengths appreciably longer than the cylinder radius ($x<2\pi$); under these conditions radiation is more strongly scattered when **E** is parallel to the cylinder axis, permitting currents in the grain to flow without building up an opposing electrostatic field. For $x<1$, perfectly reflecting cylinders ($m=\infty$) behave as small antennae, and Q_{eE}/Q_{eH} approaches infinity as x goes to zero. For dielectric cylinders, the addition of some absorption (a small imaginary component of m) changes Q_{eE} and Q_{eH} in much the same way as Q_e is changed for a sphere [compare curves (b) and (c) in Fig. 7.1]; however, $Q_{eE}-Q_{eH}$ is not much altered.

As $(m-1)$ increases, the curves of Q_{eE} and Q_{eH} shift to smaller x, exactly as in Fig. 7.1. Thus for an infinite cylinder with $m=2^{1/2}(1-i)$, Q_{eE} and Q_{eH} are close to their maximum value of about 2 for $x \geqslant 1$, whereas $10(Q_{eE}-Q_{eH})$ reaches its maximum value of about 7 at $x \approx 0.3$ [2].

For spheroids the optical properties have been measured in the laboratory at microwave frequencies with model spheroids scaled to give the correct $2\pi a/\lambda$, where a is the semiminor axis. For a prolate spheroid with its axis of symmetry perpendicular to $\boldsymbol{\kappa}$ and with its semimajor axis b equal

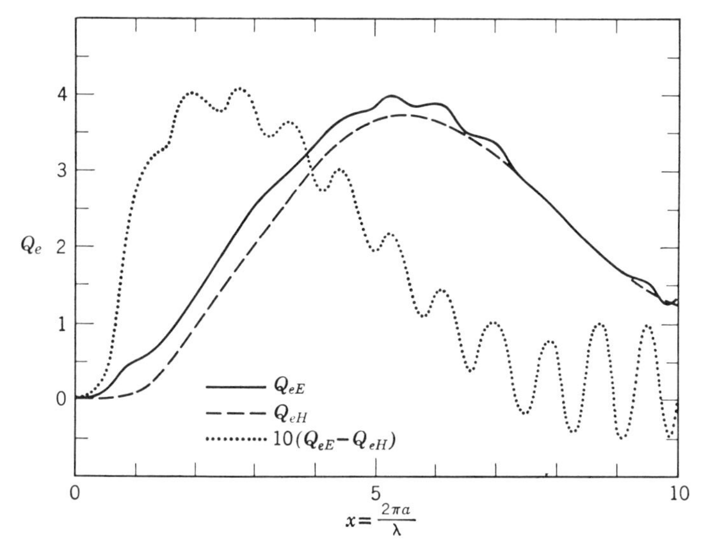

Figure 8.1 Extinction cross sections for cylinders. The theoretical curves [3] show Q_{eE} and Q_{eH}, the extinction efficiency factors for cylinders whose axes are parallel to the electric or magnetic vectors, **E** or **H**, respectively, in the incident radiation. The direction of propagation is taken to be perpendicular to the cylinder axis; a is the cylinder radius, and m, the index of refraction within the cylinder, is assumed to be 1.33.

to $2a$, the measured values [3] of Q_{eE} and Q_{eH} are not very different from the theoretical values for an infinite cylinder of the same index of refraction, with a radius equal to a. In particular, the values of $Q_{eE} - Q_{eH}$ for such spheroids are generally at least half or more of the corresponding values for the cylinders. Comparable results have also been obtained from the exact theory [4], based on solutions of the wave equation in spheroidal coordinates.

The polarizing effectiveness of cylindrical particles with m, the index of refraction, equal to 1.33 is summarized [3] in Table 8.1. We define the quantity r_p as the ratio of $Q_{eE} - Q_{eH}$ to $\overline{Q}_e$, where $\overline{Q}_e$ is defined as $(Q_{eE} + Q_{eH})/2$. As may be seen from equation (8-5) below, $0.46r_p$ is the

Table 8.1. Polarizing Effectiveness of Infinite Cylinder

x	1	2	3	4	5
$\overline{Q}_e$ $(m = 1.33)$	0.285	1.11	2.20	3.18	3.76
$0.46r_p$	0.48	0.17	0.078	0.039	0.023

ratio of polarization to extinction in the ideal case where cylindrical grains are all aligned with their axes parallel to each other and perpendicular to the line of sight. For metallic absorbing cylinders with $m=2^{1/2}(1-i)$, the range of values of r_p is roughly similar, but the corresponding values of x are about one-fifth as great.

8.2 OBSERVED POLARIZATION

Extensive measures of P have been made for many stars at a few wavelengths [1] and over most of the accessible spectrum for a few stars [5]. The relationship between P and the observed extinction and the variation of P with wavelength λ can be interpreted in terms of the intrinsic properties of the particles and the extent of their alignment. The variation over the sky both of P and of the direction of polarization gives information on the magnetic field. We discuss here each of these aspects of the observations, with a brief treatment of circular polarization at the end of this section.

a. Dependence on Color Excess

In general, unreddened stars show no polarization to within the observational error. For stars with a large color excess, the values of P show a wide distribution, ranging somewhat uniformly from zero up to a maximum value [1, 3] given by

$$\left(\frac{P}{E_{B-V}}\right)_{\max} = 0.090 \text{ mag}^{-1}; \tag{8-3}$$

here P is measured at visual wavelengths, where the dependence on wavelength is generally very small [S8.2b]. The mean value of P/E_{B-V} for observed O and B stars is about half this upper limit.

This correlation between P and E_{B-V} provides strong evidence that extinction by grains is responsible for the observed polarization, and we proceed to express P in terms of the grain properties. If we denote by $Q_{e\max}$ and $Q_{e\min}$ the maximum and minimum mean values of Q_e as the electric vector is rotated, combination of equations (8-1) and (3-3), with $j_\nu=0$, followed by use of equations (3-2), (7-2), and (3-14), gives

$$P = \tfrac{1}{2} N_d \sigma_d (Q_{e\max} - Q_{e\min}), \tag{8-4}$$

provided that $P \ll 1$ so that only the linear term in $(Q_{e\max} - Q_{e\min})$ need be retained in expanding the exponent. If we use equation (7-1) to eliminate

$N_d \sigma_d$, we find the relation

$$\frac{P}{A_\lambda} = 0.46 \frac{(Q_{e\,\max} - Q_{e\,\min})}{\bar{Q}_e}, \tag{8-5}$$

where $\bar{Q}_e$ is again the mean of $Q_{e\,\max}$ and $Q_{e\,\min}$.

If the grains are identical cylinders or highly elongated spheroids, aligned exactly parallel to each other, and perpendicular to the line of sight, $Q_{e\,\max}$ and $Q_{e\,\min}$ will equal Q_{eE} and Q_{eH}, respectively, and P/A_λ will equal $0.46 r_p$, whose values are given in Table 8.1 above. If we set A_V/E_{B-V} equal to 3, then equation (8-3) for the peak observed polarization gives $(P/A_V)_{\max} = 0.030$. Comparison between the observed and theoretical wavelength dependence of polarization shows [S8.2b] that $x = 2.5$ at 5500 Å if the grains are taken to be cylinders with $m = 1.33$; for this value of x, Table 8.1 gives $0.46 r_p = 0.12$, about four times the observed peak value, and eight times the average value of 0.015 for P/A_V. If a greater m is assumed, x is reduced, but the resultant polarizing effectiveness at the V wavelength is not greatly changed. In fact, not all grains will be sufficiently elongated to produce so much polarization nor will the alignment be perfect. These two effects, perhaps with others also, reduce the actual peak value of P/A_V well below its ideal theoretical value.

b. Dependence on Wavelength

Over the wavelength range from 4000 to 6000 Å, the linear polarization, which we denote by $P(\lambda)$, is remarkably flat for most stars, but falls off at shorter or longer wavelengths. Extensive measures of $P(\lambda)$ over a wide spectral range for many stars indicate that the wavelength of maximum polarization, denoted by $\lambda_{\max}$, differs for different stars, with observed values ranging mostly between 4500 and 8000 Å [5], averaging about 5500 Å. However, if the polarization is expressed as a function of $\lambda/\lambda_{\max}$, and divided by $P(\lambda_{\max})$, then the normalized polarization curves for all stars are identical.

The normalized points obtained in this way are plotted in Fig. 8.2 [5]. The open circles in the middle of the spectral range represent averages over groups of 60 stars each, whereas the solid circles represent data for single stars. The curve represents a theoretical calculation of $P(\lambda)$ for partially aligned cylinders with $m = 4/3$, about the value for ice grains, with an assumed distribution of radii. Evidently, the general shape of the $P(\lambda)$ curve is very similar to that of $Q_{eE} - Q_{eH}$ in Fig. 8.1; the distribution of particle sizes broadens the theoretical curve and smoothes out the wiggles

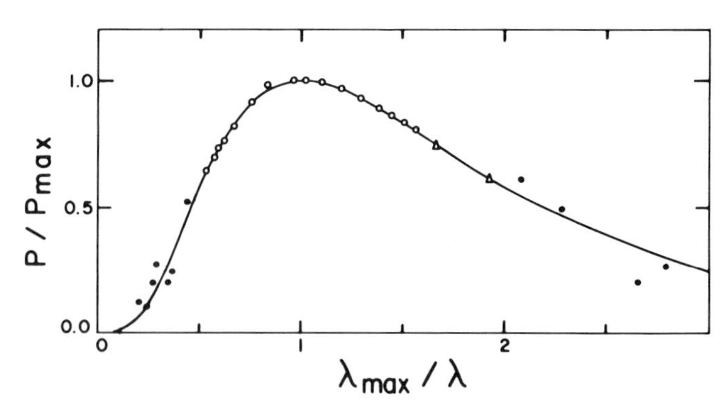

Figure 8.2 Normalized wavelength dependence of polarization. The ratio of polarization P observed in each star, to P_{max}, the maximum value in that star, is plotted against λ_{max}/λ, where λ_{max} is the wavelength at which P_{max} is observed. The open circles and triangles represent averages of 60 and 25 values, respectively, while the filled circles represent individual stars [5]. The solid curve gives a theoretical P/P_{max} for extinction by partially aligned cylinders with $m = 1.33$.

which result from interference effects. Thus the maximum value of $Q_{eE} - Q_{eH}$ at $x \approx 2.5$ yields an effective mean cylinder radius of about 2×10^{-5} cm for $\lambda_{max} = 5500$ Å. The fit shown in Fig. 8.2, although satisfactory, is not unique, and there is some question as to whether ice grains produce much of the observed selective extinction [S7.2b]. Particles of smaller a must be present to explain the ultraviolet observations, and the presence of some silicates is indicated by the 9.7-μ absorption as well as by theories of grain formation and development. A similar close fit can be obtained [6] with cylinders of about half the radius found above, but with m increased to 1.6, a typical value for silicates in the visible, giving about the same value of $(m-1)x$ at each wavelength.

The difference of λ_{max} between different stars suggests a corresponding difference in the sizes of the dust grains between different clouds. Corresponding small differences in the selective extinction curves observed for different clouds appear to be present. More specifically, stars in which λ_{max} exceeds the average value have relatively more extinction in the infrared. Thus for stars with $\lambda_{max} = 7000$ Å, the mean $E(\lambda, V)/E_{B-V}$ decreases more steeply than normal as λ increases above 1μ, approaching about -4 with increasing λ, indicating an excess of larger radii among the visual grains. Evidently R_V, defined in equation (7-20), varies with λ_{max}; detailed measures give [7]

$$R_V = \frac{5.5\lambda_{max}(\text{Å})}{10{,}000\ \text{Å}}, \tag{8-6}$$

yielding a variation in R_V from 2.5 to 3.9 as λ_{max} varies from 4500 to 7000 Å. A mean value of 5500 Å for λ_{max} corresponds to the usually adopted value of 3 for R_V; consideration of the further variation of $E(\lambda, V)$ out to 11.4μ would increase all these values of R_V slightly [S7.3a].

This correlation between selective extinction and the wavelength of maximum polarization cannot readily be explained by stellar peculiarities and appears to provide firm evidence for a difference in grain properties between different interstellar clouds.

Similar changes seem to be present between the outer and inner, presumably denser, regions of a single large cloud. In particular, within the extended obscuring cloud surrounding ρ Oph, systematic changes of λ_{max} and of the infrared extinction have been observed with increasing extinction [8]. Thus for moderately reddened stars, for which E_{B-V} is in the range 0.2 to 0.4 mag, and which are presumably near the boundary of this cloud, λ_{max} averages 6200 Å. For stars with E_{B-V} in the range from 1.0 to 1.3 mag, which are presumably behind denser inner regions of the cloud, λ_{max} rises to 8000 Å. The infrared extinction also changes with changing E_{B-V} and λ_{max} in a manner consistent with equation (8-6). These data may be interpreted as observational evidence for larger particle sizes in the denser regions of the cloud.

The observed change with wavelength of the direction of polarization also provides evidence for a change of grain properties with position. About 13 percent of the stars examined show smooth variations in the direction of the electric vector with wavelength over the spectral range from 3300 to 9800 Å, with a total angular change ranging between 3° and 27°, averaging about 10° [9]. While it is difficult to derive quantitative results about the interstellar medium from these data, it seems clear that regional variations are required both in the direction of alignment and in the wavelength dependence of ΔQ_e, the change in Q_e with changing inclination between **B** and the plane of vibration; ΔQ_e depends both on $Q_{eE} - Q_{eH}$ for individual dust grains and on the degree of alignment.

While the variation of λ_{max} from one region to another is well correlated with the shape of the normalized extinction curve from visual wavelengths into the infrared, the correlation with extinction in the far ultraviolet seems much weaker. We compare σ Sco (a star embedded in the ρ Oph cloud discussed above), whose normalized extinction curve in the far ultraviolet is about that of θ Ori (see Fig. 7.2), with ζ Oph, whose normalized extinction at 1100 Å is about 20 percent above the average. The values of λ_{max} for these two stars are 5600 Å for the former and 5900 Å for the latter [7]; this difference is relatively small and is in the opposite direction one might expect if the relative lack of far UV grains [S7.2b] in σ Sco were accompanied by a larger diameter for the visual grains. This result contrasts with that for θ Ori [S7.2b], where the absence of either the UV peak grains or the far UV grains is apparently accompanied by an appreciably larger radius for the visual grains.

c. Dependence on Galactic Longitude

While λ_{max} apparently shows systematic variations over the sky [7], we shall consider here only the variation of the direction of polarization; that is, the orientation of the plane of vibration. As shown in Section 8.3a, the interstellar grains are likely to precess around the magnetic field **B** regardless of the specific process responsible for alignment. Hence the direction of polarization gives information on the direction of **B**. This general argument does not indicate whether **B** is parallel to the plane of vibration (which is mostly parallel to the galactic plane) or perpendicular to it. However, the Faraday rotation measures leaves little question but that **B** is parallel to the galactic plane; hence the plane of vibration must be parallel to **B**.

In order to average data from different stars, the Stokes parameters Q and U are introduced, defined by the relations

$$\frac{Q}{I} = P\cos 2\left(\theta_P - \frac{\pi}{2}\right), \tag{8-7}$$

$$\frac{U}{I} = P\sin 2\left(\theta_P - \frac{\pi}{2}\right), \tag{8-8}$$

where I is the total intensity, defined in equation (8-2), and θ_P is the angle between the plane of vibration projected on the plane of the sky and the great circle from the star to the north galactic pole. Evidently Q is positive if this plane of vibration is parallel to the galactic plane ($\theta_P = \pi/2$), and negative if perpendicular to this plane ($\theta_P = 0$). When the line of sight to a star passes through several clouds, each producing a polarization P_j with a position angle θ_{Pj}, it may be shown [10] that the resulting values of Q/I and U/I are simply the sums of Q_j/I and U_j/I over all clouds, provided the polarization is small, as is generally the case.

Mean values of the Stokes parameter q, in magnitudes ($q = 2.17Q/I$), are plotted in Fig. 8.3 as a function of galactic longitude l; each point represents an average for some 25 stars, all further away than 600 pc and with $|b|$ less than 3° [11]. The double-sine-wave curve drawn in the figure corresponds to what would be expected if the grains were prolate spheroids, oriented with their major axes perpendicular to a magnetic field in the direction $b = 0°$, $l = 50°$; evidently the mean Q falls to a relatively low value between the maxima. For the closer stars the observed minima in Q are more nearly at about 60° and 240°. The direction of the local

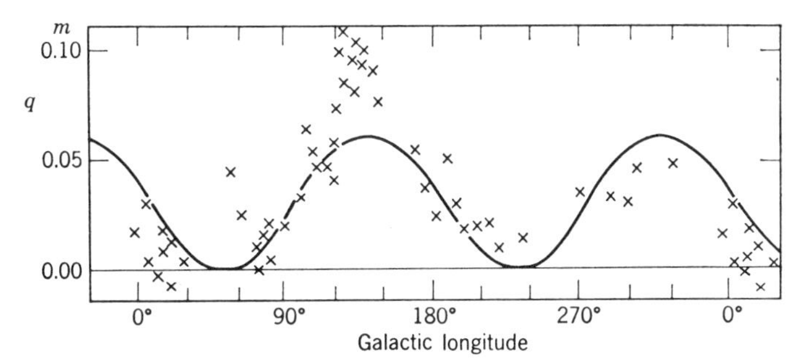

Figure 8.3 Dependence of polarization on galactic longitude [11]. Mean values of the Stokes parameter q (the difference in magnitudes of stellar intensity for **E** parallel and perpendicular to the galactic plane) are plotted for stars in different intervals of galactic longitude l. Each point represents an average for a group of about 25 stars, all at galactic latitudes less than 3° and at distances greater than 600 pc. The double sine wave is arbitrarily drawn with minima at $l = 50°$ and 230°.

Orion arm is roughly 70° [S1.5]. The magnetic field directions determined from these optical polarization data and from the Faraday rotation measurements in pulsars [S3.6a] are somewhat discordant, with a field indicated in the direction of l equal to 50° and 94°, respectively. While the observational uncertainty is appreciable, the magnetic field configuration is certainly more complicated than the simple uniform **B** assumed in analyzing the data.

The optical polarization observations give quantitative data on the nonuniformities in the magnetic field. The mean value of U/I is nearly zero for stars at low galactic latitude, as would be expected for a mean magnetic field parallel to the galactic plane. However, the rms dispersion of U/I, denoted by σ_U, is appreciable as a result of the different directions of **B** in different regions. The magnitude of σ_U may be used to determine the rms fluctuations of the position angle θ_{Pj} from one region to another, provided a number of statistical assumptions are made. An idealized model will be assumed in which the polarization is produced by a random distribution of clouds, each producing the same P_j, but with a distribution of θ_{Pj} about some mean value. We consider a group of stars in the galactic plane, all at the same distance, viewed in a direction perpendicular to **B**. If **B** is parallel to the galactic plane, the mean θ_{Pj} is $\pi/2$ radians, and we define

$$\alpha^2 = \left\langle \left(\theta_{Pj} - \frac{\pi}{2} \right)^2 \right\rangle, \tag{8-9}$$

where the brackets denote a mean value for all the stars considered.

Evidently, α is the dispersion of θ_{Pj} in radians. On the assumption that the values of θ_{Pj} in different clouds along the line of sight are uncorrelated, equation (8-8) yields

$$\sigma_U^2 = P_j^2 \sum_j \left\langle \sin^2 2\left(\theta_{Pj} - \frac{\pi}{2}\right)\right\rangle = \frac{4P^2\alpha^2}{n}, \tag{8-10}$$

where n is the average number of clouds along the line of sight, equal to P/P_j. To obtain equation (8-10), use has been made of equation (8-9), together with the assumption that $\theta_{Pj} - \pi/2$ is small compared to unity.

To determine n we use the dispersion of Q/I, which we denote by σ_Q. For $\theta_{Pj} - \pi/2$ small, the cosine in equation (8-7) is unity, and Q/I equals the sum of the P_j. Since the rms dispersion in the number of clouds along the line of sight is $n^{1/2}$, we have

$$\sigma_Q^2 = nP_j^2 = \frac{P^2}{n}. \tag{8-11}$$

Equations (8-10) and (8-11) now yield

$$\frac{\sigma_U}{\sigma_Q} = 2\alpha. \tag{8-12}$$

The dispersions σ_U and σ_Q have been determined for the stars in the double cluster h and χ Per, with an observed ratio of 1.0 for σ_U/σ_Q, giving a value of about 0.5 for α, or roughly 30°. The above analysis is rather approximate for such large α, but a more detailed treatment [12] yields about the same result. The corresponding ratio of the fluctuating magnetic field in one direction to the mean field is about 0.6. For stars more than about a parsec apart, the fluctuations of Q/I and U/I show little correlation, indicating that the scale size for these fluctuations in magnetic field is less than a parsec. The value of n found from equation (8-11) is about 100 for this double cluster, consistent with a small scale for the magnetic field fluctuations—about an order of magnitude smaller than is obtained from the variation of σ_Q with distance (see below).

These numerical values depend, of course, on the statistical model used. Other assumptions would give somewhat different results. In particular, if the number n of statistically independent regions is set equal to 10 in equation (8-10), a value more typical for the interstellar medium generally, α is reduced to about 0.2 for the double cluster in Perseus, corresponding to about 10°. Similar values are obtained in this way from other regions [11], and it seems very likely that the ratio of the fluctuating magnetic field in one direction to the mean field is at least 0.2 and may be as great as 0.6.

Undoubtedly fluctuations with a wide variety of scales are present in **B** as well as in n_H, and different types of measurements will give different results for the scale size. For example, analysis of the dispersion in Q/I among all stars at a fixed distance r shows [13] a square-root variation with r, in accordance with equation (8-11), if r exceeds about 150 pc; for lesser r, σ_Q varies roughly linearly with r. These data can be fitted with a model in which fluctuations of the interstellar medium have a mean scale size of 150 pc [13]. This result presumably gives the cell size for a change in θ_P of at least unity, so that the polarizations produced by different cells are no longer additive.

On a small scale the direction of the magnetic field appears correlated with the narrow filaments or streamers seen in the distribution of dust. Thus for stars seen through filamentary nebulae, such as the Pleiades, there tends to be good agreement [11] between the direction of the filaments and the orientation of the plane of vibration, which is assumed parallel to **B**.

d. Circular Polarization

The possibility of interstellar circular polarization was first derived theoretically [2]. The extinction produced by interstellar grains is equivalent to a small imaginary part in the effective index of refraction m_e of interstellar space. According to optical theory, taking into account the phase and amplitude of the scattered waves, the solid particles also produce a comparable change in the real part of m_e and consequently in the phase velocity. If the particles are aligned and elongated, m_e depends on the direction of **E** relative to the direction of alignment. Circular polarization will then develop if radiation which is already polarized (perhaps by absorption in a previous cloud) passes through a region where the grains are aligned in a direction which is neither parallel nor perpendicular to the plane of vibration. In this case, the components of the polarized beam with **E** parallel and perpendicular, respectively, to the direction of alignment (projected on the plane of the sky) will show different wave velocities, resulting in a relative phase shift which is typically a fraction of one cycle over the entire line of sight to a star and which produces slight circular polarization.

Circular polarizations of 0.8 to 2.9×10^{-4} have been measured in a dozen stars [14, 15], with errors of about 0.3×10^{-4}; the amplitude of the circular polarization V/I is determined from the observed intensity of left-handed and right-handed components [see text following equation (8-1)]. Analysis of some of these data indicates that a relatively large change of alignment must be present along the line of sight; on the assumption that the direction of alignment rotates uniformly with distance from the Sun, the total angle of rotation is found to be about 100° [15].

The change of circular polarization with wavelength depends on the optical properties of the grains and thus yields information on the grain material. The reversal of sign of this polarization at about 6000 Å excludes metallic particles [15], suggesting once again [S7.2b] that the visual grains are composed of some dielectric material, with a rather small imaginary part of m at visual wavelengths. Either ice or SiC would satisfy this requirement. Quantitative agreement with these circular polarization measures has been shown theoretically for oblate grains of magnetite (Fe_3O_4) [16].

8.3 ALIGNMENT

The correlations between P and E_{B-V} and between λ_{max} and R_V point clearly to the conclusion that the interstellar grains are responsible for the observed polarization. Two requirements must be satisfied if dust grains are to polarize light. In the first place, the grains must be optically anisotropic in the sense that for some orientations Q_{eE} and Q_{eH} [S8.1] are different. In the second place, the grains must be aligned at least partially. Grains of irregular shape will be anisotropic, although appreciable elongation or flattening would seem to be necessary to explain the observed polarization. In this section we discuss the physical mechanisms that might produce alignment.

One basic property of the grains that affects all alignment processes is their rapid spin. It is clear that random collisions with atoms and molecules in the gas will give the grains rotational as well as translational kinetic energy. If such collisions are elastic and no other nonconservative torques are present, the mean square value of ω, the angular velocity about a principal axis, with a moment of inertia equal to I, is given by

$$\tfrac{1}{2}I\langle\omega^2\rangle=\tfrac{1}{2}kT, \tag{8-13}$$

where T is the kinetic temperature of the gas. For spherical grains of radius a, I equals $2Ma^2/5$, and for grains of internal density of 1 gm cm^{-3} at a gas temperature of 80°K, we find

$$\langle\omega^2\rangle^{1/2}=\frac{8.1\times10^{-8}}{a^{5/2}}\,\mathrm{s}^{-1}. \tag{8-14}$$

Thus if a equals 3×10^{-5} cm, the rms ω is 1.6×10^4 s^{-1}, increasing to 2.6×10^5 s^{-1} for $a=10^{-5}$ cm. As we shall see below, other effects probably increase ω to an even greater value.

In general, a grain will not be spherical, and its moments of inertia will be different around different principal axes. In the absence of external torques, the angular momentum **J** of a freely rotating body is fixed in space, whereas the axis of rotation ω revolves about **J**, a process called nutation. To follow the motion of such a particle when various torques are present we adopt a very simplified treatment of the problem, describing some of the principal torques which are present and the effects which they might be expected to produce.

a. Conservative Torques

Since a magnetic field **B** is known to exist in interstellar space [S3.6c], and since dust grains will have a small magnetic moment **M**, the interaction between **B** and **M** will produce a torque, $\mathbf{M}\times\mathbf{B}$. This torque is to a first approximation conservative in that it produces no steady change in the rotational energy, and we denote it by $\mathbf{L}_c$. In general $\mathbf{L}_c$ is very small, giving a total change of potential energy with changing orientation which is much smaller than $\frac{1}{2}kT$. Hence L_c will produce a negligible change of rotational energy during one period of nutation about **J**, and cannot by itself produce any appreciable alignment of the grains with respect to **B**. Instead, $\mathbf{L}_c$ will produce a slow precession of **J** about **B**.

The analysis of this process is simplest for the magnetic moment produced by the electrical charge of the rotating grains [17]. In general, a grain will have a mean charge [S9.2] equal to $Z_d e$, where e is the proton charge in esu. The value of Z_d is uncertain, but in an H I cloud with $n_{\mathrm{H}}=20\ \mathrm{cm}^{-3}$, and a value of at most 0.05 for n_e, a value of at least 20 for Z_d seems reasonable if $a=3\times10^{-5}$ cm. The mean time for a neutral grain to capture an electron is at most a few days, in a typical H I cloud, and over the period of years required for precession or alignment, only the mean value of Z_d is relevant. The magnetic moment of a charge q (in emu) revolving in a circle of radius a with a frequency ν is $\pi a^2 q\nu$; hence for a rotating dust grain, the magnetic moment is given by

$$M=\frac{Z_d e\langle z^2\rangle\omega}{2c}, \tag{8-15}$$

where $\langle z^2\rangle$ is the mean square distance of the surface charge from the axis of rotation, and c is the velocity of light. The torque $\mathbf{L}_c$ is given by

$$\mathbf{L}_c=\mathbf{M}\times\mathbf{B}. \tag{8-16}$$

For a spherical particle **M** is parallel to **J**, and $\mathbf{L}_c$ is perpendicular both to **J** and **B**. As in the usual theory of gyroscopes, the resultant rate of change of **J** perpendicular to **J** produces a precession of **J** about **B**, with an angular rate Ω_c given by

$$\Omega_c = \frac{MB \sin\theta}{J}, \tag{8-17}$$

where θ is the angle between **J** and **B**. For a sphere of radius a, $\langle z^2 \rangle$ equals $2a^2/3$, and equation (8-17) yields, with the substitution of $I\omega$ for J and replacement of I by $(2a^2/5)\times(4\pi a^3 \rho_s/3)$,

$$\Omega_c = \frac{5Z_d eB \sin\theta}{8\pi c \rho_s a^3} = \frac{2.5\times 10^{-19} Z_d \sin\theta}{a^3} \text{ year}^{-1}, \tag{8-18}$$

where ρ_s, the density of solid material in the grain, has again been set equal to unity, with B, to 2.5×10^{-6} G [S11.1a]. For a, Z_d, and $\sin\theta$ equal to 3×10^{-5} cm, 20, and 0.5, respectively, $1/\Omega_c$ is about equal to 1.1×10^4 years. As we shall see below, this is about one-tenth the interval t_m required for the grains to change their rotational energy or their orientation by interaction with the gas in a typical diffuse cloud. For $\rho_s = 3$ gm cm^{-3} and $a = 10^{-5}$ cm, $1/\Omega_c$ is relatively even shorter.

With grains of more complex shape, **M** is not parallel to **J**. However, as $\boldsymbol{\omega}$ revolves about **J**, the only component of **M** that does not average out is the component parallel to **J** and the results are essentially unchanged. We conclude that for all grains, except possibly for those few whose radius a much exceeds 3×10^{-5} cm, the precession about **B** is probably considerably more rapid than any alignment process. Magnetic moments produced in other ways tend to reinforce this conclusion.

This result has the important corollary that regardless of how the grains are aligned, the direction of their average orientation in diffuse clouds tends to be controlled by the magnetic field, since the angular distribution of grain orientation will be rotationally symmetric about **B**. It follows immediately that the measured orientation of the plane of vibration gives information on the direction of **B**. This analysis does not by itself indicate whether the grains are oriented so that the absorption is least with the plane of vibration parallel or perpendicular to the interstellar **B**. However, the Faraday rotation [S3.6c] shows clearly that **B** must be roughly parallel to the galactic plane, on the average; since the observed plane of vibration also tends to be parallel to the galactic plane, there seems little doubt that the grains are aligned with their greatest dimension perpendicular to **B**, giving the greatest absorption for **E** perpendicular to the $\boldsymbol{\kappa}$,**B** plane, where $\boldsymbol{\kappa}$ is again a unit vector in the direction of wave propagation. For the less absorbed plane-polarized component, **E** will then be parallel to the $\boldsymbol{\kappa}$,**B** plane; that is, the plane of vibration will be parallel to **B**.

b. Accelerating Collisional Torques.

As pointed out above, elastic collisions of gas atoms with grains tend to give these solid particles rotational energies of $\frac{1}{2}kT$ around each of three axes. However, these collisions are, in fact, far from elastic. In particular, about one-third or more of the H atoms which collide with grains will stick together to form H_2 molecules, and when they leave they will carry off some translational energy; as pointed out in Section 6.2b, the amount of this energy is uncertain, but it is likely to lie in the range from 0.2 to 2 eV, corresponding to z_H between about 0.04 and 0.4 and thus to an ejection velocity between 4 and 14 km s^{-1}. These departing molecules will each give the grain an appreciable increment $\mathbf{\Delta J}$ of angular momentum. If the surface of the grain were uniform, $\langle \mathbf{\Delta J} \rangle$, the mean value of $\mathbf{\Delta J}$ for these outgoing molecules, would vanish, and $\langle (\Delta J)^2 \rangle$ would be increased by one or two orders of magnitude over its value for elastic collisions. The net result [18] would be to increase the rotational energy of the grains in equilibrium by the same factor.

Actually the surface of a grain is probably not uniform, and conversion of H atoms into H_2 molecules is most likely to occur at certain spots, which we shall call "active sites." Depending on the location of these sites, $\langle \mathbf{\Delta J} \rangle$ will not vanish, and a steady acceleration of the grains will ensue [19], as though jets were located at various fixed sites on the grain surface as on a pinwheel. With suitable locations for these active sites, the grains may be accelerated preferentially around any one of the three principal axes, although stability considerations lead to a grain rotation about the axis either of minimum or maximum I [19]. The angular velocity of each grain will increase until the acceleration is balanced by frictional torques, such as are produced by atoms that do not stick to the grain. In equilibrium the rotational velocity can be an appreciable fraction of the ejection velocity of the H_2 molecules, with a resultant value of ω perhaps as great as 10^9 s^{-1}. In any case, ω should much exceed the rms thermal value of 10^4 to 10^5 s^{-1} obtained from equation (8-14). This systematic acceleration to very high ω is sometimes referred to as "spin-up," and may be caused also by nonuniformities in photoelectric emission or in sticking probability [19] as well as by nonuniformities in the formation of H_2 molecules.

The time required to approach an equilibrium value of ω, when spin-up is present, should be about equal to the time interval t_m required for a grain to collide with a total mass of gas equal to its own mass. For a sphere of radius a we have

$$t_m = \frac{4\rho_s a}{3\times 1.2 n_H m_H \langle w_H \rangle} = 1.6\times 10^{11} \frac{a}{n_H} \text{ years,} \tag{8-19}$$

where the factor 1.2 allows for a 10 percent ratio of He to H atoms (and for a He velocity half that of H). The numerical value has been obtained with $\rho_s = 1$ gm cm^{-3}, and $\langle w_H \rangle = 1.31 \times 10^5$ cm s^{-1} from equation (2-19) for $T = 80°$K. For n_H we may take the typical value of 20 cm^{-3} for a diffuse cloud [S11.1a]. Then for a grain radius a of 3×10^{-5} cm, $t_m = 2.4 \times 10^5$ years, appreciably longer than the precession time $1/\Omega_c$ found from equation (8-18). For $a = 10^{-5}$ cm and $\rho_s = 3$ gm cm^{-3}, t_m is unchanged.

Evidently this effect changes the picture substantially from one of rotation in thermal equilibrium. It would seem that the grains must be rotating with energies enormously greater than $\frac{1}{2}kT$. However, whether they are rotating about the axis of greatest or of least moment of inertia is not yet clear from dynamical arguments [19]. Since, as we have seen [S8.3a], the direction of **J** must be aligned with **B**, on the average, and according to measures of P and of **B** the longest grain dimensions are perpendicular to **B**, it follows that the grains must preferentially be rotating about their axes of maximum moment of inertia.

Corresponding to the systematic angular acceleration there will also be a linear acceleration, of which only the component parallel to **J** will produce a systematic effect. Since Ω_c, the rate of precession of **J** about **B**, much exceeds $1/t_m$, the component of acceleration perpendicular to **B** tends to average out, and there results a motion of the grains through the gas parallel to **B**; this translational velocity, like the corresponding rotational velocity, will be less than the ejection velocity of the H_2 molecules, and will increase the number of low-speed collisions between grains [S9.4b].

The preceding discussion assumes that the active sites on the grain surface remain fixed with time. In fact, both the theory of grain growth [S9.4a] and observations on the increase of λ_{max} with distance into an extended cloud [S8.2b] indicate that within a cloud each dust grain will grow with time. Hence the old active sites will be covered up and will be replaced by new ones, whose location may or may not be correlated with that of the old sites. One would expect that $\langle \Delta \mathbf{J} \rangle$ would be entirely altered if the grain radius increased by about the thickness of a monomolecular layer. In 2.4×10^5 years, the value of t_m above, the radius a of the icy mantle on a spherical grain will increase by about 5×10^{-8} cm, according to equation (9-30), if $T = 80°$K, $n_H = 20$ cm^{-3}, and all heavy atoms stick ($\xi_a = 1$). Since this increase of a exceeds the thickness of a monomolecular layer, the systematic acceleration of a grain either in linear or angular velocity may change entirely over times less than t_m. As the acceleration changes, reversing sign occasionally, the rotational velocity ω will occasionally decrease to a low value, when a grain will presumably lose all memory of its former orientation. Numerical computations show [20] that the mean duration of high-ω spin-up about equals t_m if the mean life of active sites is less than t_m. In the following discussion we shall consider two alternatives, with the duration of systematic spin-up taken to be either equal to t_m ("short-lived spin-up") or essentially infinite ("long-lived spin-up").

c. Retarding Magnetic Torque

The most generally accepted mechanism for aligning **J**, the rotational angular momentum of the grains, with **B** involves the retarding torque associated with paramagnetic relaxation [1, 21]. When a paramagnetic substance is subject to an oscillating or rotating magnetic field, the magnetic susceptibility is complex, and it may be written as

$$\chi = \chi' + i\chi''. \tag{8-20}$$

The imaginary part, $i\chi''$, represents the absorption of energy from the changing magnetization in the material. When the paramagnetic material is rotating in a fixed field, the χ'' term drags the internal magnetization $\chi\mathbf{B}$ along with the material and away from **B**. For a wide range of conditions, the value of χ'' in typical paramagnetic substances is given in cgs units by the equation

$$\chi'' = 2.5 \times 10^{-12} \frac{\omega}{T_s}, \tag{8-21}$$

where T_s is the temperature of the solid material in the grains, and ω is the angular rotational frequency. For angular frequencies approaching $10^9\ \mathrm{s}^{-1}$, χ'' approaches the static susceptibility (the value of χ' at zero frequency), typically about 4×10^{-4} if T_s is 15°K; equation (8-21) is no longer valid for such high ω. Because of spin-up [S8.3b], the value of ω reached by grains may be comparable with $10^9\mathrm{s}^{-1}$.

The imaginary part of χ produces an internal magnetization of the grain which is perpendicular to **B** and which equals $\chi''\boldsymbol{\omega}\times\mathbf{B}/\omega$. The corresponding magnetic moment **M** is the product of this induced field and the grain volume V. The retarding torque $\mathbf{L}_r$, given by $\mathbf{M}\times\mathbf{B}$ [see equation (8-16)], is perpendicular to **B**; we have

$$L_r = V\chi'' B^2 \sin\theta. \tag{8-22}$$

where θ is the angle between **B** and ω.

For a spherical grain the dynamical effect of L_r is relatively simple; the angular momentum $I\omega_\perp$ about an axis perpendicular to **B** decreases exponentially. The magnetic retardation time t_r, the time constant for this decrease, may be written as

$$t_r = \frac{I\omega_\perp}{L_r}, \tag{8-23}$$

which equals

$$t_r = 1.6\times 10^{11}\frac{a^2\rho_s T_s}{B^2}\,s = 1.2\times 10^{16}\, a^2 \text{ years}, \tag{8-24}$$

where equations (8-21) and (8-22) have been used with I set equal to 0.4 $a^2V\rho_s$. The numerical value in years has again been computed with $\rho_s = 1$ gm cm^{-3}, $T_S = 15$°K, and $B = 2.5 \times 10^{-6}$ G. Thus $t_r = 1\times 10^7$ years for $a = 3\times 10^{-5}$cm. If $a = 10^{-5}$ cm and $\rho_S = 3$ gm cm^{-3}, t_r is decreased to about 3×10^6 years. In either case the value of t_r resulting from normal paramagnetic relaxation in typical H I clouds is nearly two orders of magnitude greater than t_m obtained from equation (8-19).

For a spheroidal grain the situation is more complex, but one would expect physically that the component of **J** perpendicular to **B** would be damped out during the time interval t_r, while the component parallel to **B** would be unaffected, since rotation parallel to **B** does not change the magnetization within the grain. Thus a spheroidal grain, like the sphere, will end up with **J** nearly parallel to **B**. If, as seems possible, the spin-up process produces rotation of a grain about the principal axis of greatest I [19], the observed alignment of grains can then be accounted for at least qualitatively. If spin-up failed to align the grain rotation with respect to the principal axes, the magnetic relaxation process would itself tend to produce such alignment, leaving the particles spinning about their axes of maximum I. This result follows from the fact that L_r is independent of I, according to equations (8-21) and (8-22). Hence as the grain nutates, varying the angle between each principal axis and **B**, one might expect the rotational velocity about axes of relatively lower I to damp out earlier. A detailed dynamical analysis [21] shows that this expectation is confirmed, and if L_r is the only decelerating torque acting, the particles tend toward the state in which each grain is rotating about its axis of maximum I, and this axis is aligned parallel to **B**.

We now consider the conditions under which paramagnetic relaxation can account quantitatively for the observed polarization of starlight. If short-lived spin-up is assumed [S8.3b], the basic problem is that according to equations (8-19) and (8-24) t_m and the assumed spin-up duration are several orders of magnitude less than t_r, with relatively weak alignment resulting; before paramagnetic relaxation will have any effect, the spin-up will usually reverse sign, and the grain will then spin up about some axis in a different direction. The simplest solution of this difficulty is to assume long-lived spin-up. This assumption leads to good alignment of the grains with respect to **B** provided that during times as long as t_r some other torques do not cause the grains to precess in different directions.

Another possibility is to retain the assumption of short-lived spin-up, but to assume that t_m/t_r exceeds the value found from equation (8-24) because of changes in the various parameters. We first consider the least increase needed of t_m/t_r on the assumption that the mean duration of systematic spin-up equals t_m. On this basis the effective alignment of grains should vary as t_m/t_r when this ratio is small. More precise computations [20] indicate that the effective alignment factor for polarization (defined as the ratio of the actual polarization to the ideal value for complete alignment) equals about 0.3 t_m/t_r. We have seen that the mean value of P/A_V (at $\lambda=\lambda_{\max}$) is about one eighth the theoretical maximum for aligned cylinders. It follows that t_m/t_r must equal about 0.4 on the average, with a higher value required if the grains are not much elongated. To increase t_m/t_r to 0.4 requires that $B^2\chi''/(\omega_\perp n_{\mathrm{H}}\langle w_{\mathrm{H}}\rangle a)$ be increased by at least an order of magnitude. If only B is increased, we find $B>8\times10^{-6}$ G. So large a B in H I clouds is excluded by the upper limit of less than 3×10^{-6} G obtained from the Zeeman effect measures [S3.4a]. Simultaneous decreases in the other relevant parameters, to the lowest values consistent with observations, might well give the value of t_m/t_r required, but seem somewhat implausible.

Much larger decreases in t_r result if the Fe atoms within dust grains are assumed to occur in clumps [18]. If these clumps contain about 100 atoms each, the material will be "superparamagnetic," whereas if an appreciable fraction of the grain volume is composed of metallic Fe, or of Fe compounds such as magnetite [S8.2d], the material will be ferromagnetic. In either case, χ'' can exceed by as much as 10^5 the value given by equation (8-21). Then t_m/t_r exceeds unity with appreciable orientation likely, even for fields as low as 10^{-7} G. While the required arrangement of the Fe atoms may seem somewhat special, there is some reason to expect such segregation of different refractory elements during the initial formation of grain nuclei [S9.4a]. The various nonmagnetic orientation schemes which have been proposed do not seem sufficiently effective [1]. We conclude that the grain alignment required to explain the interstellar polarization can be explained by magnetic relaxation without straining all the interstellar parameters if either long-lived spin-up is assumed or χ'' is much enhanced by a clumpy distribution of Fe atoms within grains.

REFERENCES

1. P. Aannestad and E. M. Purcell, *Ann. Rev. Astron. Astroph.*, **11**, 309, 1973.
2. H. C. van de Hulst, *Light Scattering by Small Particles*, Wiley (New York), 1957.
3. J. M. Greenberg, *Stars and Stellar Systems*, Vol. **7**, University of Chicago Press (Chicago), 1968, p. 221.
4. S. Asano and G. Yamamoto, *Appl. Op.*, **14**, 29, 1975.
5. G. V. Coyne, T. Gehrels, and K. Serkowski, *A. J.*, **79**, 581, 1974.
6. T. Gehrels, *A. J.*, **79**, 590, 1974.
7. K. Serkowski, D. S. Mathewson, and V. L. Ford, *Ap. J.*, **196**, 261, 1975.
8. L. Carrasco, S. E. Strom, and K. M. Strom, *Ap. J.*, **182**, 95, 1973.
9. G. V. Coyne, *A. J.*, **79**, 565, 1974.
10. S. Chandrasekhar, *Radiative Transfer*, Clarendon Press (Oxford), 1950, Section 15.
11. J. S. Hall and K. Serkowski, *Stars and Stellar Systems*, Vol. **3**, University of Chicago Press (Chicago), 1963, p. 293.
12. K. Serkowski, *Advan. Astron. Astroph.*, **1**, 289, 1962.
13. J. R. Jokipii, I. Lerche, and R. A. Schommer, *Ap. J. (Lett.)*, **157**, L119, 1969.
14. J. C. Kemp and R. D. Wolstencroft, *Ap. J. (Lett.)*, **176**, L115, 1972.
15. P. G. Martin, *Ap. J.*, **187**, 461, 1974.
16. P. R. Shapiro, *Ap. J.*, **201**, 151, 1975.
17. P. G. Martin, *M.N.R.A.S.*, **153**, 279, 1971.
18. R. V. Jones and L. Spitzer, *Ap. J.*, **147**, 943, 1967.
19. E. M. Purcell, *The Dusty Universe*, G. B. Field and A. G. W. Cameron, Editors, N. Watson Academic Publ. (New York), 1975, p. 155.
20. E. M. Purcell, *Ap. J.*, **231**, 404, 1979.
21. L. Davis and J. L. Greenstein, *Ap. J.*, **114**, 206, 1951.

9. Physical Properties of Grains

While little is known about the detailed internal structure of interstellar dust grains, certain physical properties of the grains are determined by relatively simple processes. The rotation and orientation of the grains has already been discussed. Here we treat the temperature T_s of the solid material in the grains, the surface charge $Z_d e$, and the motion of the grains under the joint influence of radiation pressure and collisions with the gas. A final section discusses very briefly the processes likely to affect the evolution of the grains, including their origin, growth, and disruption.

9.1 TEMPERATURE OF THE SOLID MATERIAL

Within the solid material composing the interstellar grains, the exchange of vibrational energy between atomic nuclei is generally much more rapid than the exchange of energy with the external radiation field. Just as in the interstellar gas, this condition assures the existence of an equilibrium velocity distribution at the kinetic temperature T_s, where the subscript s (as in ρ_s) denotes properties of the solid material composing the dust grains. In a steady state the temperature of a grain is determined by the condition that the rate of gain of internal energy equals the rate of loss, just as for the gas (see Chapter 6). For a solid particle, the energy gain normally results from absorption of photons and from collisions with the gas, whereas the loss results from infrared radiation.

In principle, the rates of gain and loss can be computed in a straightforward manner, although in practice inadequate information on the properties of the grains makes the results uncertain. We let G denote the rate of energy gain per second per unit projected area of dust, with G_r and G_c representing the values of G for absorption of radiation and for collisions with atoms and molecules of the gas. Then we have

$$G_r = c\int_0^\infty Q_a(\lambda)U_\lambda\,d\lambda, \tag{9-1}$$

where $Q_a(\lambda)$ is the absorption efficiency factor [S7.1] (with the dependence on λ indicated specifically), $U_\lambda\,d\lambda$ is again the energy density of radiation within the wavelength range $d\lambda$, and c is the velocity of light. For G_c we have

$$G_c = \sum_{\mathrm{X}} n_{\mathrm{X}}\left[\left\langle \frac{\sigma_{de}}{\sigma_d}\frac{m_{\mathrm{X}}w_{\mathrm{X}}^3}{2}\right\rangle + \left\langle \frac{\sigma_{de}}{\sigma_d}w_{\mathrm{X}}\right\rangle\left(E_{c\mathrm{X}} - \bar{E}_{2\mathrm{X}}\right)\right]. \tag{9-2}$$

In this equation n_{X} and w_{X} are the particle density and random velocity of component X of the gas, and the brackets denote an average over a Maxwellian distribution. The quantity σ_{de}/σ_d is the ratio of the effective collision cross section, including the effect of electrostatic forces, to the geometrical cross section of the dust grain. As in equation (6-3), $\bar{E}_{2\mathrm{X}}$ is the mean energy of the particles which escape, following a collision between particles of type X and the grain, whereas $E_{c\mathrm{X}}$ is the mean chemical energy given up to grain heating by each such colliding particle. For collisions with atomic H, the term $E_{c\mathrm{H}}$, representing part of the 4.48 eV released in formation of a H_2 molecule, generally outweighs the other two terms.

The energy radiated by the grain is essentially the same as in thermodynamic equilibrium at the temperature T_s. Hence from the principle of detailed balancing in such equilibrium [S2.4] we find that $L_r(T_s)$, the total energy lost by radiation per second per unit projected area of dust, is given by equation (9-1), with $4\pi B_\lambda(T_s)/c$ substituted for U_λ, giving

$$L_r(T_s) = 4\pi\int_0^\infty Q_a(\lambda)B_\lambda(T_s)\,d\lambda. \tag{9-3}$$

The Planck function per unit wavelength in this expression evidently equals cB_ν/λ^2, where B_ν is the corresponding intensity per unit frequency [see equation (3-4)].

a. H I Regions

We apply this analysis now to H I regions, first showing that collisional heating is generally negligible in such regions. For neutral atoms colliding with grains, $\sigma_{de}=\sigma_d$ in equation (9-2). Since E_{cX} vanishes for He, and other atoms are relatively scarce, only H need be considered in this equation. As a first approximation in H I clouds we may set U_λ equal to its typical value for starlight in interstellar space, for which the total energy density U is about 7×10^{-13} ergs cm^{-3}. This does not include the 2.7°K blackbody radiation, whose energy density is 4×10^{-13} erg cm^{-3} and which in all but the densest, coldest clouds is not significantly absorbed by dust particles. This value of U also ignores the extinction of radiation, which becomes important in the more opaque clouds, discussed at the end of this section. If now we divide equation (9-1) by equation (9-2), retaining only the E_{cH} term in the latter and utilizing equation (2-19), we find

$$\frac{G_r}{G_c}=\frac{c\bar{Q}_a U}{2n(\text{H I})E_{cH}}\left(\frac{\pi m_{\text{H}}}{2kT}\right)^{1/2}, \tag{9-4}$$

where the quantity $\bar{Q}_a$ is an average of $Q_a(\lambda)$ over U_λ. The value of E_{cH} is somewhat uncertain. The H_2 dissociation energy is 4.48 eV, but the escaping molecules will carry off some of this as vibrational and rotational excitation energy and also as translational kinetic energy. If one-third of the colliding atoms form molecules, and each such molecule gives 1.5 eV of energy to the grains, E_{cH} is 0.5 eV, some two orders of magnitude greater than the thermal energy given up. If we let $T=80$°K, and $U=7\times10^{-13}$ erg cm^{-3}, we then obtain

$$\frac{G_r}{G_c}=2.0\times10^5\frac{\bar{Q}_a}{n(\text{H I})}. \tag{9-5}$$

Since $\bar{Q}_a$ is at least 0.1, even for grains with a as small as 3×10^{-6} cm [S7.2b], collisional heating of the grains in H I regions is evidently unimportant unless the clouds are very opaque or very dense, in which case most of the hydrogen is likely to be molecular rather than atomic.

If we ignore collisions, the dust temperature T_s is then determined entirely by radiative processes, with the total absorption of energy, mostly in the near infrared, the visible, and the near ultraviolet, balanced by emission in the far infrared. Hence the steady-state equation of equilibrium is that $G_r=L_r$; from equations (9-1) and (9-3) we find

$$c\int_0^\infty Q_a(\lambda)U_\lambda\,d\lambda=4\pi\int_0^\infty Q_a(\lambda)B_\lambda(T_s)\,d\lambda. \tag{9-6}$$

Solutions of equation (9-6) depend, of course, entirely on how $Q_a(\lambda)$ varies with λ. If this efficiency factor were independent of λ, equation (9-6) would yield

$$T_s = \left(\frac{U}{a}\right)^{1/4} = 3.5°\mathrm{K}, \tag{9-7}$$

where a is the radiation density constant, and the total energy density U includes the blackbody radiation as well as starlight. In fact, $Q_a(\lambda)$ must decrease with increasing λ, as is evident from equation (7-10). If m is constant with λ, equation (7-7) indicates that $Q_a(\lambda)$ varies as $1/\lambda$, which seems to be a good first approximation for many substances over visual and infrared wavelengths. If in accord with equation (4-13) we represent U_λ as a Planck function at 10^4°K, diluted by a factor W of 10^{-14} (consistent with the energy density of starlight cited above), then equation (9-6) yields

$$T_s = 10^{4°}\,\mathrm{K}W^{1/5} = 16°\mathrm{K}. \tag{9-8}$$

More generally, the energy density U_λ may be expressed as a sum of Planck functions, each at a temperature T_j, and each multiplied by $4\pi W_j/c$, where W_j is the dilution factor at that temperature. The quantity G_r may then be expressed in terms of $L_r(T)$, given in equation (9-3); equation (9-6) then becomes

$$\Sigma_j W_j L_r(T_j) = L_r(T_s), \tag{9-9}$$

which may be solved numerically when the function $L_r(T)$ has been determined both for the low values of T characteristic of T_s and the high values characterizing U_λ.

Computations of $L_r(T)$ for several grain radii and materials [1] have provided detailed determinations of T_s. The temperatures found with only one term on the left-hand side of equation (9-9) ($T_j = 10{,}000$°K and $W_j = 10^{-14}$) differ by less than 10 percent from those computed with more exact fits to the observed U_λ, and for ice grains of radius 10^{-5}cm are about 14°K, about the same as obtained from equation (9-8). For graphite grains of this size, the temperature is appreciably higher, about 35° [1, 2], since for pure graphite $Q_a(\lambda)$ is relatively large for visible light and small in the far infrared. Depending on the variation of m and Q_a with λ, appreciable variation of T_s is obtained, although 15° seems to be the most likely value for actual grains.

For all materials considered, T_s decreases with increasing radius. This effect results from the fact that for x greater than about unity, Q_a levels off and no longer increases with x. Thus the ratio of Q_a in the far infrared, where $x \ll 1$, to that in the visible, where x is about unity if $a=0.1\,\mu$, increases with increasing radius. For a equal to 100 μ, for example, x would exceed unity even in the far infrared, and equation (9-7) would be a better approximation than equation (9-8). For ice grains, T_s decreases from 14 to 8°K as a increases from 0.1 to 10μ, with an even steeper decrease for graphite particles.

As a result of the steep dependence of $L_r(T_s)$ on T_s, these temperatures are not very sensitive to the assumptions made. Increasing $Q_a(\lambda)$ in the infrared to about the maximum amount permitted by equation (7-10) lowers the temperatures of ice and graphite grains, with $a=10^{-5}$ cm, to values between 8 and 11°K [1, 3]. Allowance for absorption of the radiation penetrating a relatively opaque cloud changes T_s rather slowly, since starlight in the near infrared is absorbed much less than in the visible. Thus T_s for a grain with an icy mantle 1.5×10^{-5} cm in radius, surrounding a graphite core 0.5×10^{-5} cm in radius, is about 15°K at the surface of a cloud, and falls to 8°K [3, 2] only at an optical depth τ_V of 10 in the visible, corresponding to E_{B-V} of 3 mag; an increase of τ_V to 100 is required to reduce T_s to 4°K.

b. H II Regions

When the hydrogen is ionized, the presence of strong Lα radiation can provide the dominant mechanism for heating the grains. As we have seen [S5.1c], all Lα photons produced in an H II region are likely to be absorbed by the grains within that region. On the usual idealized model of constant density, the Lα photons are produced at a uniform rate and also are absorbed uniformly. Hence the rate of absorption by grains per unit volume must equal the corresponding rate of production. We let $F_{\mathrm{L}\alpha}$ be the flux of such photons crossing the surface of a grain per unit projected area per second. Then the steady-state condition for Lα production and absorption becomes

$$\sigma_d n_d Q_a F_{\mathrm{L}\alpha} = z_{\mathrm{L}\alpha} n_e n_p \alpha^{(2)}. \tag{9-10}$$

The quantity $z_{\mathrm{L}\alpha}$ is the fraction of recombinations in levels $n \geqslant 2$ that lead to emission of a Lα photon; the fraction $1 - z_{\mathrm{L}\alpha}$ leads to population of the $2s$ level, which is deexcited with simultaneous emission of two photons [4]. Values of $z_{\mathrm{L}\alpha}$ for n_e much less than $1.5 \times 10^4\,\mathrm{cm}^{-3}$ are given [5] in Table 9.1. At higher densities, collisions chiefly with protons produce transitions from $2s$ to $2p$ and $z_{\mathrm{L}\alpha}$ increases.

To evaluate $F_{\mathrm{L}\alpha}$ from equation (9-10) at $T=8000$°K and at electron densities of $10^3\,\mathrm{cm}^{-3}$ or less we make use of equation (7-23) and Table 9.1,

Table 9.1. Fraction $z_{L\alpha}$ of Recombinations Producing $L\alpha$ Photons

T_e (°K)	5,000°	10,000°	15,000°	20,000°
$z_{L\alpha}$	0.70	0.68	0.66	0.64

and assume nearly complete H ionization, with $n_e \approx n_p \approx n_H$, obtaining

$$F_{L\alpha} = \frac{0.69 n_H \alpha^{(2)}}{\Sigma_d Q_a} = 2.1 \times 10^8 \frac{n_H}{Q_a} \frac{\text{photons}}{\text{cm}^2\,\text{s}}. \tag{9-11}$$

The recombination coefficient $\alpha^{(2)}$ has been set equal to $3.1 \times 10^{-13}\,\text{cm}^3\text{s}^{-1}$ from equation (5-14) and Table 5.2. The corresponding energy gain, which we denote by $G_{L\alpha}$, is given by

$$G_{L\alpha} = Q_a F_{L\alpha} E_{L\alpha} = \frac{z_{L\alpha} n_H \alpha^{(2)} E_{L\alpha}}{\Sigma_d}, \tag{9-12}$$

where we have used equation (9-10) to eliminate $F_{L\alpha}$, and $E_{L\alpha}$ is the energy of a $L\alpha$ photon. If the dust-gas ratio is reduced below the average or the relatively small grains are missing, Σ_d will be less than the value given in equation (7-23) and $G_{L\alpha}$ will be increased.

For comparison, the collisional energy gain G_c is less by at least two orders of magnitude. Most impacts of protons with grains will presumably lead to H neutralization. To maximize G_c we assume that all the energy of each electron–proton recombination on the grain surface goes into grain heating; that is, E_{cp} in equation (9-2) is set equal to 13.6 eV. On the other hand, most electron–proton recombinations in the gas yield a heating energy $E_{L\alpha}$ of 10.2 eV, about the same as the assumed value of E_{cp}. Thus $G_c/G_{L\alpha}$, the ratio of the energy gains from these two processes, about equals the ratio of $n_d \sigma_d \langle w_p \rangle$ to $z_{L\alpha} n_e \alpha^{(2)}$. With n_e set equal to n_H, with the same value of $\alpha^{(2)}$ used as in equation (9-11), with $\langle w_p \rangle$ set equal to $1.3 \times 10^6\,\text{cm}\,\text{s}^{-1}$ at 8000°K from equation (2-19), and with use again of equation (7-23), we see that this ratio is about 0.006. Consideration of electrostatic effects on E_{cp} and on the effective grain cross section [S9.2] does not change the order of magnitude of this computed ratio. We shall therefore neglect G_c in H II as well as in H I regions.

The equilibrium grain temperatures resulting from the balance of $L\alpha$ absorption and infrared emission are significantly higher than the values of T_s found for H I regions. From the computed values [1] of $L_r(T_s)$ we find that for $n_H = 10\,\text{cm}^{-3}$, T_s equals 20 and 40°K for grains of ice ($a=$

10^{-5} cm) and graphite ($a = 5 \times 10^{-6}$ cm), respectively. At higher densities $G_{L\alpha}$ is increased correspondingly, and at $n_H = 10^3 \text{cm}^{-3}$, T_s increases to 45 and 80°K for the ice and graphite grains, respectively.

The value of T_s found for a graphite grain depends on the assumption that no other grains are present. For grains all of the same type, Q_a at 1216 Å cancels out when equation (9-11) is substituted into equation (9-12). As a result, the difference of T_s between the different grains quoted above results entirely from the assumed lower infrared emissivity of graphite grains. If grains of different types are present in the same region, $F_{L\alpha}$ must be the same for all grains, depending on the effective absorption by all grains at Lα, whereas $G_{L\alpha}$ will be greatest for those grains with a relatively large Q_a (1216 Å). If Q_a (1216 Å) is believed to be substantially greater for graphite grains of the assumed radius than for the ice grains, it follows that within a given H II region the values of T_s found above for graphite particles relative to those for ice grains would be increased.

Very close to the central star of an H II region the stellar radiation field becomes very strong, and G_r outweighs $G_{L\alpha}$ for sufficiently small r. If $G_{L\alpha}$ is ignored, then T_s for the same ice grain considered above, exposed to a dilute radiation field with a color temperature of 30,000°K, increases from 35°K at $W = 10^{-14}$ to 85° and 200°K at $W = 10^{-12}$ and 10^{-10}, respectively, where W is the dilution factor, equal to the ratio of the solid angle subtended by the star to the total solid angle, 4π. When the stellar distance r much exceeds the stellar radius R_s, W is given by

$$W = \frac{R_s^2}{4r^2}. \tag{9-13}$$

The dilution factors found above correspond to distances of 1, 0.1, and 0.01 pc if $R_s = 9R_\odot$. Evidently, close to a star, stellar radiation is likely to be important for heating and possibly evaporating the grains.

Another way of viewing this same result is to look at the total optical thickness τ_{di} of dust in the H II region [S5.1c]. When τ_{di} is less than unity for the Lyman continuum, absorption of stellar radiation by grains is relatively small at most wavelengths, and it is only the Lα radiation that is mostly absorbed in its many passages back and forth across the ionized region. Under this condition $G_{L\alpha}$ is the dominant source of grain heating. When τ_{di} is unity or more, grains will absorb an appreciable fraction of all stellar energy shortward of 912 Å, and some longer wavelength photons as well, and $G_{L\alpha}$ becomes less important. For example, around an O7 star, $\tau_{di} \approx 1$ for $n_H = 50 \text{cm}^{-3}$; in more detail, we find from Table 5.3 that r_S and τ_{Sd} equal 4.2 pc and 1.33, respectively, whereas from Table 5.4, $y_i = 0.77$, giving r_i and τ_{id} equal to 3.2 pc and 1.0, respectively. Evidently, for this condition stellar radiation begins to be comparable with Lα radiation as a heating source even at the boundary of the H II region. With n_H increased to $5\times10^4\,cm^{-3}$, τ_{di} increases to 4 and r_i falls to 0.01 pc, with stellar heating important throughout. Thus in compact H II regions, Lα heating can be ignored.

9.2 ELECTRIC CHARGE

In a steady state the electric charge on a grain, like other physical quantities, must be constant on the average. The chief physical processes which tend to modify the charge on a grain and which must therefore cancel out on the average are photoelectric emission and collisions with electrons and positive ions. Because of uncertainties in the various physical constants affecting these processes, we do not give the complete steady-state equation [6], but consider the two limiting cases where photoelectric emission is unimportant or is dominant.

a. Electron and Ion Collisions

If photoelectric emission is ignored, the electric charge on a grain is determined by the requirement that the number of positive ions captured by the grain per second equals the corresponding capture rate for electrons. Since the electrons are moving much more rapidly, and thus collide with a neutral grain much more often than do the positive ions, the grain must be negatively charged so that the electrons will be repelled, the positive ions attracted, and a balance produced.

To compute the charge in equilibrium we consider the effect of the electric potential U of the grain on the effective collision cross section πp_c^2 for an ion of charge Z_ie; we take $Z_i=+1$ for protons and -1 for electrons. The quantity p_c is the critical collision parameter (distance of closest approach in the absence of forces) for which the ion orbit is just tangent to the surface of the grain, which we take to be spherical. From conservation of energy and angular momentum we obtain

$$\pi p_c^2=\sigma_d\left\{1-\frac{2Z_ieU}{m_iw_i^2}\right\}; \tag{9-14}$$

U is in esu, w_i is the ion velocity, and σ_d is again the geometrical cross section. To compute the mean collision cross section, equation (9-14) must be integrated over a Maxwellian distribution of velocities. If U is negative, then for a positive ion the integral extends over all w_i from zero to infinity.

For an electron, however, the lower limit on w_i is then the value for which p_c vanishes and for which the electron kinetic energy is only just sufficient to overcome the grain potential. If we assume that only one type of positive ion is present, then $n_e = Z_i n_i$, and the condition that $\langle \pi p_c^2 w \rangle$ be equal for electrons and positive ions yields

$$e^{eU/kT} = \left(\frac{m_e}{m_i}\right)^{1/2}\left(1 - \frac{Z_i eU}{kT}\right). \tag{9-15}$$

This result is based on the assumption that the probability of sticking to a grain, rather than bouncing off again, is the same for electrons and positive ions. This assumption is plausible but unproven.

Equation (9-15) may be applied to a region where the hydrogen is ionized. In this case, $Z_i = 1$, $(m_e/m_i)^{1/2} = 1/43$, and equation (9-15) has the solution

$$\frac{eU}{kT} = \frac{Z_d e^2}{akT} = -2.51, \tag{9-16}$$

where Z_d is the number of elementary charges, in units of the proton charge, on the grain surface. The negative electric charge increases the rate of proton collisions with a grain by a factor 3.51, decreasing the rate of electron collision by a factor 1/12.3.

As we shall see below, this simple result is probably not applicable to much of the interstellar gas, since photoelectron emission is likely to be appreciable even from relatively large grains. However, equation (9-16) may be applicable in a collisionally ionized gas [7]. If $T = 10^6$ °K, then $kT = 86$ eV and the grain potential is -0.72 esu; if we denote by U(V) the grain potential in volts, equal to 300 U, then $U(\mathrm{V}) = -220$ V. The corresponding charge is 1.5×10^4 electrons if $a = 10^{-5}$ cm.

b. Photoelectric Emission

When the rate of photoelectric emission from a grain exceeds the rate of electron capture by a neutral grain, the grain potential becomes positive, decreasing the rate of photoemission and increasing the collision rate with electrons. Capture of positive ions by the grain is negligible in this case of strong photoelectric emission, and the equation of charge equilibrium may be written

$$\xi_e n_e \left(\frac{8kT}{\pi m_e}\right)^{1/2}\left(1 + \frac{eU}{kT}\right) = F_e, \tag{9-17}$$

where F_e, the flux of photoelectrons per unit projected area of dust grains, is related to grain parameters and the radiation field by equation (6-19), and ξ_e is the sticking probability for electrons striking the grain. In equation (9-17) other factors multiplying n_e are the mean electron velocity $\langle w_e \rangle$ [see equation (2-19)] and the factor by which the mean cross section is increased by electrostatic forces (see above). We assume below that $\xi_e = 1$.

As pointed out earlier [S6.2b] the value of F_e depends critically on the photoelectric efficiency y_e, which for radiation shortward of 1100 Å is very much greater than at visible wavelengths, with a value of 0.1 expected for many substances. For grains with radii less than 10^{-6} cm, y_e may rise to nearly unity. The value of y_e generally vanishes for photon energies less than some threshold value E_e, which increases linearly with increasing eU, the negative energy of an electron on the grain surface. As pointed out below, the decrease of y_e and thus of F_e with increasing eU can be important in the intercloud medium. The value of y_e for $U=0$ is referred to here as the "intrinsic" y_e.

The value of F_e in H I regions is given in equation (6-20). For comparison, $n_e\langle w_e \rangle$, the corresponding electron collision rate per unit area of neutral grains, is about $5.6\times 10^6 n_e$ cm^{-2}s^{-1} at 80°K. Evidently, if n_e is much less than $4y_e\overline{Q}_a \exp(-\tau_v)$ cm^{-3}, which is likely in H I diffuse clouds, photoelectric emission will be strong and eU/kT obtained from equation (9-17) will be positive and large. Detailed computations [8], with a somewhat different numerical constant in equation (6-20), show that for $T=$ 80°K, U(V) decreases from 0.6 V for $n_e = 5\times 10^{-3}$cm^{-3} to about 0.07 V for $n_e = 5\times 10^{-2}$cm^{-3}. These values of U are sufficiently small so that they do not affect much either the photoelectric efficiency y_e or the mean energy $\overline{E}_2$ per emitted photoelectron [S6.2b]. For $a = 10^{-5}$cm, the corresponding range of Z_d values is from 42 to 5, with three times this value if a is increased threefold to 3×10^{-5}cm.

The rapid rise of selective extinction down to wavelengths as short as 1000 Å seems to require that an appreciable fraction of the grains have radii 2×10^{-6}cm or less [S7.2b]. Some of these may be sufficiently small that photoelectrons escape with no appreciable loss, giving $y_e \approx 1$ for a neutral grain. To satisfy equation (9-17), if $n_e/T^{1/2}$ is relatively low, the grain potential U must increase to a level that reduces y_e. We assume here that the work function of the grain material is 4 eV. The threshold voltage for photoelectric emission then becomes $4+U$(V); a value of 9.6 V for U(V) would reduce y_e to zero for all photons of energy less than 13.6 eV, corresponding to the Lyman limit at 912 Å. Hence the greatest possible value of the left-hand side of equation (9-17) is obtained with $U=9.6/300$ esu, giving $eU/kT = 1.4\times 10^3$ if $T=80°$. For very low $n_e/T^{1/2}$ and high intrinsic y_e, $U(\mathrm{V})$ must approach sufficiently close to 9.6 V to reduce F_e below this upper limit. Computations for a low-density H I gas, with $n_e = 3\times 10^{-3}\,\mathrm{cm}^{-3}$ and $T=8000°\mathrm{K}$, show that $U(\mathrm{V}) \approx 8.6$ V under these conditions [9], giving a maximum energy E_2 of the emitted electrons equal to 1 eV. For so large a value for U, the heating rate Γ_{ed} for the gas [S6.2b] actually decreases with increasing intrinsic y_e and increasing U, since the decrease of $\bar{E}_2$ more than offsets the small increase of F_e.

Within H II regions, the Lα radiation can similarly produce strong photoelectric emission. In this case we have

$$F_e = Q_a y_e F_{\mathrm{L}\alpha}, \tag{9-18}$$

where $F_{\mathrm{L}\alpha}$ is given in equation (9-11). It is readily seen that for y_e about equal to 0.3, F_e equals $n_e\langle w_e\rangle$, the number of electrons incident on a neutral grain per cm^2 per second. Thus the resultant charge on the grain will depend on the balance of these effects, and may be positive or negative. If F_e were zero, then according to equation (9-16), $U(\mathrm{V})$ would equal -1.7 V, corresponding to Z_d equal to 120 for $a = 10^{-5}\,\mathrm{cm}$; again we have let $T = 8000°\mathrm{K}$. If F_e and $n_e\langle w_e\rangle$ are equal to within about 10 percent, a coincidence which seems somewhat improbable, the potential is reduced by an order of magnitude to about ± 0.2 V. Evidently, the grain potential in H II regions resulting from $F_{\mathrm{L}\alpha}$ may well be -1 or -2 V, but positive values are also possible, and a potential near zero cannot be excluded. Sufficiently close to the central star the photoemission produced by intense starlight may give large positive U [10].

9.3 RADIATIVE ACCELERATION

The force exerted on grains by radiation pressure is often much greater than the local gravitational force, and in the absence of other forces an interstellar particle can acquire appreciable velocities. However, a grain tends to be coupled to the gas, both through interaction with the magnetic field and through collisions, especially with positive ions. As a result, the selective effect on grains resulting from radiation pressure is often much reduced. We discuss here the magnitude of these effects in a few typical cases.

If a unidirectional radiation field, with a flux $\mathcal{F}$ per cm^2 per second, is incident on a grain, the radiative force F_r is given by

$$F_r = \frac{\mathcal{F}\sigma_d Q_p}{c}, \tag{9-19}$$

where c is again the velocity of light and Q_p is the efficiency factor for radiation pressure, averaged over all frequency. For pure absorption, $Q_p = Q_a$, but for a scattering grain, Q_p may be substantially less than Q_s; if the radiation is scattered predominantly in the forward direction, the transfer of momentum to the grain is quite small. For example, if $m = 1.33$ and $x = 4$, $Q_p = 0.7$ [11] as compared to the value of 2.7 for Q_s shown in Fig. 7.1.

a. Gyration around Magnetic Field

The first constraint on the response of a grain to the radiative force results from the magnetic field. Because of their electric charges, grains will tend to move in helical paths around the magnetic lines of force. In general, for a particle with mass and charge equal to m and Ze, respectively, the angular frequency ω_B of this helical motion, called the "gyration frequency" or "cyclotron frequency," is

$$\omega_B = \frac{|Z|eB}{mc}. \tag{9-20}$$

The radius a_B of the helix is called the "radius of gyration" and is given by

$$a_B = \frac{w_\perp}{\omega_B}, \tag{9-21}$$

where $w_\perp$ is the velocity component perpendicular to $\mathbf{B}$. For a grain of radius 3×10^{-5} cm and of density 1 gm cm^{-3}, we obtain in a field of 2×10^{-6} G

$$\frac{1}{\omega_B} = \frac{4\pi a^3 \rho_s c}{3|Z_d|eB} = \frac{1.1 \times 10^5}{|Z_d|} \text{ years.} \tag{9-22}$$

Thus if Z_d is about 20, a plausible value for a grain of this radius in H I regions, the "gyration time" $1/\omega_B$ is only some 6000 years, much less than most dynamical times of interest, which are often at least 10^6 years. In H II regions a larger value of $|Z_d|$ is likely and $1/\omega_B$ is even shorter. In addition, for smaller grains $1/\omega_B$ decreases as a^2, since a/Z_d is constant for constant grain potential U.

In some regions the mean value of $|Z_d|$ may be very small, as can occur in an H I region if F_e vanishes or in an H II region if F_e is nearly equal to $n_e\langle w_e \rangle$. However, a detailed analysis [12] of the relative number of dust grains with each value of Z_d indicates that an appreciable fraction of the grains may have a charge of ± 1 in such cases. Even with $|Z_d| = 1$, $1/\omega_B$ is still shorter than the nominal dynamical time of 10^6 years. We conclude that in most situations dust grains, like the positive ions, cannot move across the magnetic field.

b. Dynamical Friction with Gas

If motions of the grains parallel to **B** are considered, another constraint results from collisions of the gas atoms and molecules with the grains. The retardation of a test particle moving through the gas is called dynamical friction [S2.1], and can be described in terms of the slowing-down time t_s, defined in equation (2-1). To give an example of this constraint we consider the equation of motion of a grain situated at a distance r from a star of luminosity L. From equation (9-19) we obtain, replacing $\mathcal{F}$ by $L/(4\pi r^2)$ and σ_d by πa^2,

$$\frac{dw}{dt}=\frac{La^2Q_p}{4cr^2m_d}-\frac{w}{t_s}. \tag{9-23}$$

We consider the solution of this equation both with and without the dynamical friction term, $-w/t_s$.

If the second term in equation (9-23) is ignored, and the grain is assumed to be initially at a distance r_0 from the star, w will increase to an asymptotic value w_a, given from usual potential theory as

$$w_a^2=\frac{3LQ_p}{8\pi cr_0a\rho_s}, \tag{9-24}$$

where m_d, the mass of the dust grain, has been replaced by $4\pi a^3\rho_s/3$. For example, if a grain 3×10^{-5} cm in radius and with ρ_s equal to unity is initially 1 pc from a star of luminosity $10^5\ L_\odot$, and if $Q_p=0.7$, then $w_a=34$ km s^{-1}.

If now we assume that the drag term is relatively large, then at each r, w will reach a steady value w_D, called the "drift velocity," at which the outward force equals the drag. At the radius r_0 this drift velocity is given by

$$\frac{w_D}{w_a}=\frac{w_at_s}{2r_0}. \tag{9-25}$$

This equation is valid only if w_at_s/r_0 is small compared to unity.

To evaluate this result in an H II region, we compute t_s from equation (2-7), with protons and He^+ ions replacing electrons as field particles, and the grains replacing positive ions as test particles. To maximize the grain charge, we ignore photoelectric emission and determine Z_d from equation (9-16), obtaining

$$t_s = 0.17 \frac{a\rho_s}{n_H (kTm_H)^{1/2} \ln \Lambda}, \qquad (9\text{-}26)$$

where n_{He}/n_H has been set equal to 0.10. If $n_H = 10\,\text{cm}^{-3}$ and $T = 8000°$K, then t_s is about 500 years for the same grain parameters used above; if we set r_0 and w_a equal to 1 pc and 34 km s^{-1}, respectively, $w_a t_s/2r_0$ obtained from equation (9-25) is about 10^{-2}. Hence w_D is only about 0.3 km s^{-1}, and during 10^6 years the particles will move only 0.3 pc with respect to the interstellar gas. As we have seen [S9.3a], this relative motion is restricted to the direction parallel to **B**. From equations (9-24) through (9-26), it follows that in this limit of small $w_a t_s/r_0$, w_D is independent both of ρ_s and a.

As pointed out earlier [S9.2b], $|Z_d|$ may be much reduced by photoelectric emission, especially near the central star. If $a = 3\times10^{-5}$ cm and $Z_d = -40$, giving the relatively low value of -0.2 V for U(V), t_s is increased by 10^2 to 5×10^4 years, and $w_a t_s/2r_0$ equals unity. In this case the coupling with the gas is only just beginning to be important. If U is negative at large distance r from the central star and positive for small r, then U and Z_d for grains of a particular type will vanish at some intermediate r [10], increasing t_s to that for neutral atoms [see equation (9-27)]. We may conclude that grains and gas are probably well coupled most of the time in H II regions, but there may be significant exceptions where the coupling is weak because the grain charge is low.

In an H I region, such as would be expected around a late-type supergiant star, Z_d is probably not much greater than 40 in any case, and coupling between the dust and the positive ions is relatively weak as well as uncertain. Collisions with neutral H and He now produce a minimum drag, and the corresponding t_s is obtained from equation (2-14). With the usual formulae for m_d and σ_d, and with $n_{He}/n_H = 0.10$ as before, we obtain

$$t_s = 0.70 \frac{a\rho_s}{n_H (kTm_H)^{1/2}}. \qquad (9\text{-}27)$$

If T is equal to 80°K, but a, ρ_s, and n_H have the same values as before, $t_s = 5\times10^5$ years. The dynamical friction produced by collisions of C^+ and Si^+ with the charged grains depends both on Z_d and on the depletion of C and Si; with plausible assumptions t_s resulting from this interaction can range from about 5000 to about 5×10^5 years. If the overall t_s is as long as 5×10^5 years, $w_a t_s/2r_0 \approx 10$, and the drag near the luminous star considered above may be ignored, but with lower t_s the coupling becomes significant.

Evidently the conclusions indicated are not very definite. While grains and gas in H II regions are probably coupled together dynamically much of the time, in H I regions the grains and gas may or may not be uncoupled, depending on the density, temperature, and composition of the gas as well as on the physical properties of the grains. Thus in the intercloud medium between the H I clouds, the few grains present may well reach appreciable velocities [13].

With larger directed forces on grains than we have considered here, the coupling between grains and gas may be unimportant in general. For example, dust grains within a few parsecs of a supernova, for which L is some 10^{10} $L_{\odot}$, will experience a radiative force much greater than the collisional drag. A sufficiently small particle may reach a velocity exceeding 10^4 km s^{-1}, although the total energy input into such high-speed grains would appear to be small [14].

9.4 EVOLUTION OF GRAINS

To trace the life cycle of a grain with any assurance would require much more detailed information than is now available, especially on the detailed properties of the grains themselves. However, an examination of the relevant physical processes does set certain constraints on possible evolutionary theories. Some of the more important processes are discussed in the present section.

a. Formation and Growth

In the diffuse H I clouds which have been chiefly studied in the previous chapters, and which are characterized by values of n_H between 10 and 100 cm^{-3}, the rate of formation of most molecules is expected to be very slow, and, in fact, the relative abundance of all molecules except for H_2 is low. Hence it is difficult to see how the condensation nuclei required for the birth of grains can be produced in sufficient numbers within such clouds. Higher density regions seem required, and either the dense clouds observed in molecular emission lines [S7.3c] or the outer regions of stellar atmospheres would seem likely possibilities. It is often assumed that the observed dust grains have mostly formed in the atmospheres of late-type giant stars, where the temperatures are sufficiently low to permit condensation of graphite or silicates. Infrared measures confirm the presence of heated grains surrounding such stars, but some estimates of the total mass loss rate are only about one-tenth of the value required to account for the interstellar grains [15].

Regardless of whether the grains were born in stellar atmospheres or in dense interstellar clouds, the depletion found for different elements seems consistent with a relatively simple model of grain formation in a gas of decreasing temperature and density [16]. On this model, the elements with the highest condensation temperature T_c should condense first, with other elements condensing later. If the density decrease is sufficiently rapid, the fraction of atoms of each element condensing into grains will be less for the low-T_c elements than for those of high T_c. Figure 9.1 shows a plot of the depletion observed [17] in the interstellar cloud in the line of sight to ζ Oph [S3.4c]; the values of the depletion given in Table 1.1 are plotted against computed values of T_c [16]. In the computations, the gas with a temperature T is assumed to have cosmic composition and to be at a pressure of about 1000 dyne cm^{-2}. The abundances of the different molecular species are determined by the appropriate relations for thermodynamic equilibrium. The value of T at which half the atoms of a given element have condensed out in one form or another is defined as T_c. At different pressures, the values of T_c are different, but the general ap-

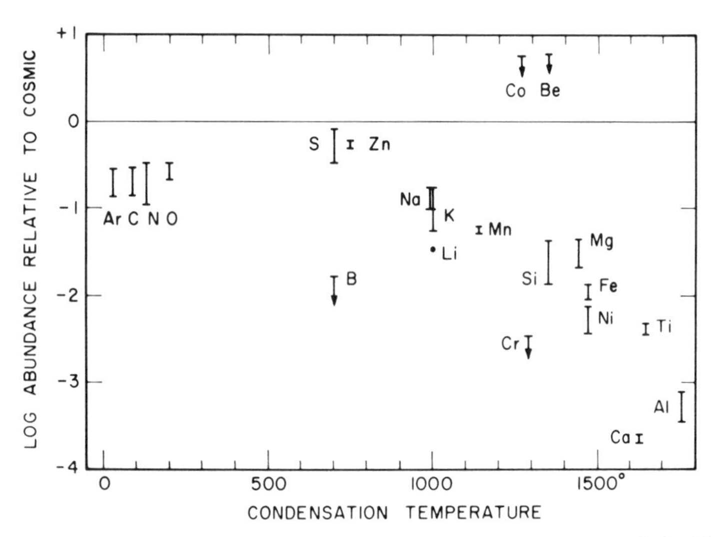

Figure 9.1 Dependence of depletion on condensation temperature [17]. The observed values of the depletion for each element observed in ζ Oph (Table 1.1) are plotted against values of the condensation temperature T_c (°K) [16], defined as the temperature at which half the atoms of an element have condensed out in one form or another in thermodynamic equilibrium. Each vertical bar represents an estimated possible error in the depletion resulting from uncertainties in the various curves of growth assumed.

pearance of the plot in Fig. 9.1 is not much changed. There seems no question but that the depletions tend to be large for the elements of highest T_c.

One might expect that small grains composed of the more refractory materials (characterized by higher melting temperatures) might form first from condensation in a relatively dense cloud, and that mantles of CH_4, NH_3, and H_2O would form at a later time, perhaps within the denser observed diffuse clouds. The depletion of C, N, and O shown in Fig. 9.1 is somewhat uncertain, but is consistent with the apparent observed increase in grain radius with E_{B-V} in extended cloud complexes [S8.2b]. Also consistent with this point of view is the relatively slight depletion observed for N and O in more transparent clouds [S3.4c].

While the theory of initial grain condensation in a dense warm gas yields some impressive agreement with the observations, the model raises certain difficulties. The very large depletion of elements such as Ca and Ti seems easily understood if all the atoms of these elements have passed through this condensation phase. However, it is generally assumed that the atoms of these heavy elements were formed in supernovae rather than in late-type giants; it seems unlikely that almost all the heavy elements in interstellar space have passed either through stars or through dense high-temperature interstellar clouds at least once since being ejected from supernovae. In addition, the agreement shown in Fig. 9.1 is about as good if the depletion is plotted against first ionization potential, provided that the three alkali metals, Li, Na, and K, are arbitrarily excluded [18].

If one assumes that the condensation nuclei have formed somehow, the rate at which the grains grow by accretion in observed diffuse clouds can be computed on the basis of a relatively simple model. For grains with more than a certain number of atoms, one might expect that the sticking probability ξ_a for most atoms would be not far below unity. Physical adsorption to the surface, which requires no activation energy, should be possible, and an atom colliding with the grain is likely to give up enough of its kinetic energy so that it will become bound. Detailed analysis [19] confirms that $\xi_a \approx 1$ for neutral atoms, although important uncertainties remain for C^+ and other positive ions.

We compute now the rate at which the grain radius a will grow by this accretion process. If m_d denotes the mass of the grain, while the particle density, mass, and velocity of the colliding atoms are denoted by n_a, Am_H, and w_a, respectively, and if the loss processes considered below are ignored, the rate of mass accretion by the grains becomes

$$\frac{dm_d}{dt} = \pi a^2 w_a n_a A m_H \xi_a, \tag{9-28}$$

giving an increase in the grain radius at the rate

$$\frac{da}{dt}=\frac{w_a\rho_a\xi_a}{4\rho_s}=\frac{3.3\times10^4\rho_a\xi_a}{A^{1/2}\rho_s}\frac{\text{cm}}{\text{s}}, \tag{9-29}$$

where ρ_a is the interstellar mass density of atoms of type a, equal to $n_a A m_H$, and ρ_s is again the density of solid matter within the grain. For w_a the mean value for $T=80°$K has been used, computed from equation (2-19). Thus if ρ_s equals 1 g/cm^3, a typical value for dielectric grains, and if $\rho_a/A^{1/2}$ is obtained by summing values of $\rho_X/A_X^{1/2}$ for the cosmic composition shown in Table 1.1 (excluding H and the noble gases), with $n_H=20\,\text{cm}^{-3}$ [11.1a], we find

$$\frac{da}{dt}=2\times10^{-13}\xi_a\frac{\text{cm}}{\text{year}}. \tag{9-30}$$

This growth rate is sufficiently rapid to have interesting consequences. Even if ξ_a is only 0.1, the radius of a grain will grow to 2×10^{-5} cm in 10^9 years, which may be less than the mean lifetime of the grains. Thus if the condensation nuclei are themselves small spheres of refractory material, such as might be produced in stellar atmospheres, one would anticipate that mantles of icy materials, such as H_2O, NH_3, and CH_4, might form. The observational evidence discussed in the previous two chapters is probably consistent with this point of view. The weakness of the observed H_2O absorption feature at 3.1 μ [S7.2b] may perhaps be explained by chemical bonding between H_2O and other atoms, catalyzed by ultraviolet photons [16].

The mean life t_{ad} for an atom in the gas before it sticks to a grain is relatively short. For any one atom we may write

$$t_{ad}=\frac{1}{w_a n_d \sigma_d \xi_a}=\frac{1}{w_a n_H \Sigma_d \xi_a}=\frac{4\times10^7}{\xi_a}\text{ year}, \tag{9-31}$$

where equation (7-23) has again been used, and the numerical value has been obtained for $T=80°$K, $n_H=20$ cm^{-3}, and $A=16$. Thus for $\xi_a=0.1$, t_{ad} is about 4×10^8 years. This is somewhat less than the grain growth time found above; evidently for typical clouds, accretion of all the heavy atoms present in the gas would not much increase the radii of the larger grains. In a molecular cloud, where n_H may exceed 10^4 cm^{-3}, t_{ad} may be less than about 10^6 years.

b. Denudation and Disruption

If grains continued to grow by accretion until engulfed in a newly created star, there would be difficulty in understanding how any heavy atom would remain in the interstellar gas, in view of the relatively short values of t_{ad} found from equation (9-31). This difficulty is enhanced by the likelihood that most of the gas now present in molecular clouds returns again to the interstellar medium when a small fraction of this material condenses into new stars. We assume that, in fact, grains lose atoms as well as gain them, and we list below some of the main processes that may strip atoms off grains. In general, the chief four mechanisms [15] seem to be a high internal temperature, collisions with energetic atoms, absorption of ultraviolet photons, and collisions with other grains.

Molecules will tend to sublime (or evaporate) from grains at a rate given by their vapor pressure at the temperature T_s. For molecules adsorbed on the surface, this process will be significant only if T_s exceeds values somewhere between 20 and 40°K [19]. Such temperatures are anticipated [S9.1b] within H II regions if $n_H = 10$ cm^{-3}, with considerably higher temperatures at $n_H = 10^3$ cm^{-3} with r_S consequently less than a parsec. Thus mantles of icy dielectrics may tend to sublime within H II regions of moderate density; within the denser ionized clouds, such as the Orion nebula, sublimation seems rather likely. From a very small grain, atoms may sublime during the temperature rise following absorption of a single photon or surface formation of one molecule [20].

Collisions with thermal atoms and ions will knock molecules out of a grain, a process known as sputtering. The energy required to detach an H_2O molecule, for example, is only about 0.5 eV. However, the fraction η of kinetic energy transmitted from a proton to such a molecule is small. If m_a and m_m represent the masses of incident atoms and molecules, respectively, then for a single head-on collision we have

$$\eta = \frac{4m_a m_m}{(m_a + m_m)^2}, \tag{9-32}$$

which is only 0.20 for H colliding with H_2O, giving a sputtering threshold energy E_s of about 3 eV for sputtering for the most favorable impacts. For normal incidence, the accelerated molecule is moving toward the grain, and loses much of its energy inelastically before it is reflected back outward; as a result, E_s for sputtering at this angle is increased by about another order of magnitude. In addition, the overall efficiency y_s of the sputtering process, defined as the number of atoms sputtered off the surface per incident atom, is very low for atoms of incident energy E not far above E_s. If E and E_s are measured in electron volts, then very roughly $y_s = 10^{-3}$ $(E - E_s)$ [21]. As a result of this low efficiency, sputtering is usually quite unimportant in H II regions. In collisionally ionized gas, where values of 100 eV for E are typical, sputtering efficiencies are appreciable, and at a density of $10^{-3}\,\text{cm}^{-3}$ the radius of a grain will decrease by $10^{-5}\,\text{cm}$ in about 10^8 years [7].

Atoms of energies much greater than 100 eV, including cosmic rays, can knock particles out of grains in various ways [7, 15], and can completely evaporate very small grains, with a of about $10^{-7}\,\text{cm}$. However, the number of such energetic atoms present in the interstellar gas is unknown.

Absorption of ultraviolet photons, with energies between 5 and 13 eV, can lead to relatively rapid ejection of molecules from the surface of a grain. We have seen in Section 4.3 that the mean probability of an upward electronic transition in H_2 is about $5 \times 10^{-10}\,\text{s}^{-1}$, if the radiation field U_λ is assumed to have the mean interstellar values shown in Table 5.5. For other molecules the transition rates should be similar, and thus an adsorbed molecule on the surface of a dust grain will be excited in a time of about $10^9\,\text{s}$. In the excited state the molecule may be repelled rather than attracted by the grains, with ejection the immediate result [19]. If molecules can become chemically bound to the grains, they will be less subject to this effect. It is not clear to what extent photo ejection will prevent the growth of mantles in diffuse interstellar clouds, where appreciable ultraviolet radiation can penetrate.

Finally, we consider collisions between grains. If two such solid particles strike each other head-on with a velocity of a few $\text{km}\,\text{s}^{-1}$, both grains will evaporate completely [22]. The relative velocity giving sufficient energy for vaporization ranges from 3 $\text{km}\,\text{s}^{-1}$ for icy dielectrics up to 8 $\text{km}\,\text{s}^{-1}$ for Fe particles. Such impacts will result when grains pass through a shockwave such as would be produced by collisions between clouds or by an explosive stellar energy source (see Chapter 12). However, unless the grains are accelerated further within the shock, the fraction of grains evaporating in this way on passage through one shock will be small. Instead, most grains will be slowed down by collisions with the gas before they have a chance to collide with other grains.

As a measure of the fraction of grains that collide with each other, we compute the ratio of λ_{dm} to λ_{dd}, where λ_{dm} is the mean distance over which a dust grain will collide with a mass of gas equal to the grain mass, whereas λ_{dd} is the mean free path for collisions between two grains. Since H and He atoms include almost all the mass of the gas, we have in the same notation as before

$$\lambda_{dm} = \frac{4a\rho_s}{3n_H m_H(1+4n_{He}/n_H)}. \tag{9-33}$$

The quantity λ_{dm} is closely equal to the distance over which the velocity w_d of the dust grain, assumed much larger than $\langle w_H \rangle$ or $\langle w_{He} \rangle$, decreases by $1/e$.

Electrostatic forces are ignored in equation (9-33) since they are not likely to make an important contribution to the drag on fast-moving grains with a velocity w_d significantly greater than the mean proton random velocity $\langle w_p \rangle$. If $\langle w_p \rangle > w_d$, the proton drag exceeds the electron drag, given in equation (2-7), by the factor $(m_p/m_e)^{1/2}$, and is relatively large. When $w_d > \langle w_p \rangle$, however, the proton drag is reduced by an additional factor $[\langle w_p \rangle / w_d]^3$. In addition, when w_d significantly exceeds $\langle w_p \rangle$, the velocities of the protons with respect to the grains are increased above $\langle w_p \rangle$, and the fluxes of electrons and protons striking a neutral grain will become more nearly equal; if photoelectric emission is unimportant, the value of $-U$ will be less than the value obtained from equation (9-15). A reduced $|Z_d|$ reduces the electrostatic drag in proportion to Z_d^2 [S2.1]. Immediately behind a shock, when T and $\langle w_p \rangle$ are both high, electrostatic drag may be significant, but in the compressed cooler gas following an isothermal shock [S10.2a] this effect is small.

The value of λ_{dd} is set equal to $(\alpha_d n_H \Sigma_d)^{-1}$, or $1.0 \times 10^{21}/\alpha_d n_H$ from equation (7-23). Here α_d is a numerical factor that would equal 4 if all grains were the same size and grazing collisions would lead to complete vaporization. For the more realistic case in which large grains collide with a distribution of small grains, α_d is less than 4, approaching 1 as a lower limit. For $a = 3 \times 10^{-5}$ cm, $\rho_s = 1$ gm cm^{-3}, $n_{He}/n_H = 0.1$, and $\alpha_d = 2$ we obtain

$$\frac{\lambda_{dm}}{\lambda_{dd}} = 0.034. \tag{9-34}$$

The probability dP_{ev} of grain evaporation in traveling a distance dx is dx/λ_{dd}, while $d\ln w = -dx/\lambda_{dm}$. Hence if the grain velocity is initially w_0 and evaporation stops if w falls below a critical value w_c, we have [23]

$$P_{ev} = \frac{\lambda_{dm}}{\lambda_{dd}} \ln\left(\frac{w_0}{w_c}\right). \tag{9-35}$$

For shock velocities between 20 and 100 km s^{-1}, and for $w_c = 8$ km s^{-1}, equations (9-34) and (9-35) give P_{ev} in the range from 3 to 9 percent for these large grains, which are bigger than the average.

The above analysis assumes that the grains will, in fact, collide with each other as they move through the gas behind the shock. If the only force acting on the grains were the collisions discussed above, this assumption would be unrealistic, since grains of the same $a\rho_s$ would slow down at the same rate and their random velocities would be small. However, gyration around a magnetic field **B** will randomize the grain velocities perpendicular to **B**, and ensure that the grains will collide with each other at velocities up to $2w_d$. Even relatively small values of **B** and $|Z_d|$ suffice for this result. Moreover, forces on the charged grains associated with the electric field in hydromagnetic shocks will tend to accelarate the grain [24], increasing P_{ev} somewhat above the values given in equation (9-35).

Another physical process that may disrupt grains is the spin-up to high rotational speeds, discussed in Section 8.3. If the rotational speed v_r at the greatest distance from the rotational axis is 10^4 cm s^{-1}, about one-tenth the random velocity of the H atoms at 80°K, the tensile stress at the grain center will amount to about $\rho_s v_r^2$ or 100 kg/cm^2, if $\rho_s = 1$ gm cm^{-3}. At a rotational speed five times higher, the stress will be comparable with the tensile strength of some materials, and an even lower stress might fracture fragile grains.

The various processes discussed here, probably together with others including possible shattering in low-speed impacts between grains, must produce the distribution of grain radii and types present in interstellar clouds. In view of the variety of mechanisms operating, and their dependence on local conditions, it is perhaps surprising that the properties of the grains seem in many ways so nearly constant over the sky.

The reduced depletion of Mg, Si, and Fe observed in high-velocity clouds [S3.4c] is consistent with the view that atoms which are normally in grains return to the gas phase behind a high-speed shock. Of the various processes discussed here, the two which seem most likely to be associated with shocks are sputtering in the hot shocked gas and evaporative collisions between grains; both mechanisms may be responsible at least in part for this observed effect.

REFERENCES

1. J. M. Greenberg, *Astron. Astroph.*, **12**, 240, 1971.
2. C. M. Leung, *Ap. J.*, **199**, 340, 1975.
3. M. W. Werner and E. E. Salpeter, *M.N.R.A.S.*, **145**, 249, 1969.
4. L. Spitzer and J. L. Greenstein, *Ap. J.*, **114**, 407, 1951.

5. D. E. Osterbrock, *Astrophysics of Gaseous Nebulae,* 1974, W. H. Freeman (San Francisco), Section 4.3.
6. L. Spitzer, *Ap. J.,* **107**, 6, 1948.
7. J. Silk and J. R. Burke, *Ap. J.,* **190**, 11, 1974.
8. W. D. Watson, *Ap. J.,* **176**, 103, 1972.
9. M. Jura, *Ap. J.,* **204**, 12, 1976.
10. B. Feuerbacher, R. F. Willis, and B. Fitton, *Ap. J.,* **181**, 101, 1973.
11. H. C. van de Hulst, *Rech. Astron. Obs. Utrecht* **11**, Part 1, 1946; especially Fig. 2.3.
12. H. P. Gail and E. Sedlmayer, *Astron. Astroph.,* **41**, 359, 1975.
13. R. Y. Chiao and N. C. Wickramasinghe, *M.N.R.A.S.,* **159**, 361, 1972.
14. B. Wolfe, P. McR. Routly, A. S. Wightman, and L. Spitzer, *Phys. Rev.,* **79**, 1020, 1950.
15. P. Aannestad and E. M. Purcell, *Ann. Rev. Astron. Astroph.,* **11**, 309, 1973.
16. G. B. Field, *Ap. J.,* **187**, 453, 1974.
17. D. C. Morton, *Ap. J. (Lett.),* **193**, L35, 1974; **197**, 85, 1975.
18. T. P. Snow, *Ap. J. (Lett.),* **202**, L87, 1975.
19. W. D. Watson and E. E. Salpeter, *Ap. J.,* **174**, 321, 1972.
20. E. M. Purcell, *Ap. J.,* **206**, 685, 1976.
21. M. J. Barlow, *Nature, Phys. Sci.;* **232**, 152, 1971.
22. J. H. Oort and H. C. van de Hulst, *B.A.N.,* **10**, 187, 1946; No. 376.
23. M. Jura, *Ap. J.,* **206**, 691, 1976.
24. L. Spitzer, *Comments on Astroph.,* **6**, 177, 1976.

10. Dynamical Principles

Previous chapters have discussed physical processes in which individual atoms, molecules, electrons, and grains in interstellar space usually interact with each other or with photons in binary processes, with only two particles interacting at any one time. Computation of the rates of these various processes, together with use of the steady-state equations, has given some understanding of the equilibrium of the interstellar medium. In particular, this theory has made it possible to interpret a vast amount of observational data in terms of physical conditions in interstellar space.

The remainder of this book is largely devoted to dynamical problems of the interstellar gas; that is, to the acceleration and deceleration of the gas and to the possible steady equilibrium states in which the forces balance. Analysis of this topic requires knowledge of two field quantities, the macroscopic velocity $\mathbf{v}$ and the magnetic field $\mathbf{B}$. The basic research technique is the solution of the differential equations for $\mathbf{v}$ and $\mathbf{B}$ instead of the steady-state equations for population distributions, discussed in previous chapters. The present chapter is devoted to a presentation of these equations and their elementary consequences. Subsequent chapters discuss a few of the principal dynamical processes that have been studied with the use of these equations. Because of the many complications in this field, both mathematical and physical [1], the discussion is necessarily limited to a few dominant processes that can be approximated with relatively simple theoretical models.

10.1 BASIC EQUATIONS

The change of $\mathbf{v}$, the fluid velocity of the gas, is determined by the usual momentum equation, which we write in the form

$$\rho \frac{D\mathbf{v}}{Dt} = \rho\left[\frac{\partial \mathbf{v}}{\partial t} + \mathbf{v}\cdot\nabla\mathbf{v}\right] = -\nabla p - \frac{1}{8\pi}\nabla B^2 + \frac{1}{4\pi}\mathbf{B}\cdot\nabla\mathbf{B} - \rho\nabla\phi. \quad (10\text{-}1)$$

In this equation D/Dt denotes the Lagrangian time derivative, following a fluid element, whereas $\partial/\partial t$ represents the Eulerian time derivative at a fixed location. The quantities ρ and ϕ, which denote the density and the gravitational potential, respectively, are determined by the equation of continuity and Poisson's equation

$$\frac{D\rho}{Dt} = \frac{\partial \rho}{\partial t} + \mathbf{v}\cdot\nabla\rho = -\rho\nabla\cdot\mathbf{v}, \tag{10-2}$$

$$\nabla^2\phi = 4\pi G\rho. \tag{10-3}$$

The two terms involving **B**, the magnetic field, in equation (10-1) result from the ponderomotive force $\mathbf{j}\times\mathbf{B}$, when the current density **j** is replaced by $\nabla\times\mathbf{B}/4\pi$. The change of **B** with time is determined by

$$\frac{\partial \mathbf{B}}{\partial t} = \nabla\times(\mathbf{v}\times\mathbf{B}) + \frac{\eta}{4\pi}\nabla^2\mathbf{B}, \tag{10-4}$$

subject to the usual condition

$$\nabla\cdot\mathbf{B} = 0. \tag{10-5}$$

Equation (10-4), in which η represents the electrical resistivity in electromagnetic units, is only approximate [2]*, but is generally valid under most interstellar conditions. The pressure p includes both the pressure p_G of the gas and the pressure p_R of the energetic particles or cosmic rays. The gas pressure p_G is given by the usual perfect gas law,

$$p_G = \frac{\rho kT}{\mu} = \frac{\rho C^2}{\gamma}; \tag{10-6}$$

C is the sound speed in the idealized case where p is proportional to ρ^γ [see equation (10-7)]. The quantity μ in equation (10-6) is the mean mass per particle in grams, and k is the Boltzmann constant; except in shock fronts the temperature T is determined by the energy equation (6-1). Turbulence in the gas may be taken approximately into account by including a turbulent pressure in p_G. The gradient of cosmic-ray pressure perpendicular to **B** enters into equation (10-1) in the same way as ∇p_G; however, if p_R has a gradient parallel to **B**, the pressure is anisotropic, and a more complex analysis is required.

*See the additional terms in equation (2-12) and in the simplified equation (2-21) of ref. [2].

Equation (10-1) is sometimes called a "macroscopic equation," since it describes the change of the macroscopic fluid velocity **v**, defined as the average of the particle velocity **w** for all the particles within a small volume element. One can also write down the "microscopic equations," which govern the motion of an individual particle in magnetic and electric fields [2]. The most important result of the microscopic equations is the well-known helical motion of a charged particle around the magnetic lines of force, discussed in Section 9.3a [see equations (9-20) and (9-21)].

The macroscopic equations above have a number of simple consequences [2, 3]. In the absence of fluid velocities, the time required for the magnetic field to change by ohmic dissipation is of the order $4\pi L^2/\eta$, where L is the scale size of the magnetic field. Even for L as small as 1 AU this time is at least 10^9 years for a typical interstellar conductivity, and for larger scales the field is effectively constant during the age of the Universe. In the presence of material motions, equation (10-4) implies that the magnetic flux through any circuit moving with the fluid is constant, provided η is ignored. Hence the lines of magnetic force may be regarded as following the fluid in its motion and are said to be "frozen" into the fluid.

For smaller disturbances about an equilibrium condition, the equations above may be replaced by a somewhat simpler set of linear equations for the disturbances [S13.3a]. These linearized equations permit a number of well-known types of wave motions [2]. If **B** and ϕ are ignored and ρ is assumed nearly uniform, pressure disturbances propagate at the sound velocity C, given by

$$C^2 = \frac{dp}{d\rho}. \qquad (10\text{-}7)$$

For sinusoidal disturbances with a period P much less than the cooling time t_T [equation (6-2)], the compression is adiabatic, and C is given by equation (10-6), with γ equal to C_p/C_V, the ratio of specific heats at constant pressure and constant volume. For a monatomic gas, with no internal degrees of freedom, $\gamma = 5/3$. In a gas of atoms or molecules with energy levels that can be excited, C_V and C_p depend on the temperature as well as the pressure, and γ falls below $5/3$ when T is high enough to permit appreciable excitation.

If the period P of the pressure disturbances much exceeds t_T, the kinetic temperature will remain closely equal to T_E, its value in radiative equilibrium, and equation (10-7) yields

$$C^2 = \frac{kT_E}{\mu}\left(1 + \frac{\rho}{T_E}\frac{dT_E}{d\rho}\right). \qquad (10\text{-}8)$$

In H II regions T_E is independent of ρ for $n_e < 10^2\,\text{cm}^{-3}$ [S6.1b]; hence the factor in parentheses equals unity and C is the isothermal sound speed obtained from equation (10-6), with $\gamma = 1$. In H I regions also, C is often set equal to the isothermal sound speed; T_E changes with ρ in such regions but usually not very rapidly. Alternatively, the factor in parentheses may be set equal to an effective value of γ in the radiation-dominated regime, and equation (10-6) again is used with $T = T_E$.

If a uniform magnetic field is present, equations (10-1) and (10-4) may be combined to show [2] that disturbances transverse to **B** will travel along **B** at the Alfvén speed V_A, given by

$$V_A^2 = \frac{B^2}{4\pi\rho}. \tag{10-9}$$

These transverse waves traveling along the field lines are similar to a transverse wave traveling along a string under tension.

a. Virial Theorem

To derive this useful theorem, which applies to an entire system within some comoving surface S, one takes the scalar product of the position vector $\mathbf{r}$ with equation (10-1) and integrates over the volume interior to S. The enclosed mass is constant and integration by parts yields

$$\frac{1}{2}\frac{D^2 I}{Dt^2} = 2T + 3\Pi + \mathfrak{M} + W + \frac{1}{4\pi}\int_S (\mathbf{r}\cdot\mathbf{B})\mathbf{B}\cdot d\mathbf{S} - \int_S \left(p + \frac{B^2}{8\pi}\right)\mathbf{r}\cdot d\mathbf{S}, \tag{10-10}$$

where $d\mathbf{S}$ is taken to be positive in the outward direction and where the quantities I, T, Π, $\mathfrak{M}$, and W are defined by the following integrals over the volume bounded by S:

$$I = \int \rho r^2\,dV, \quad T = \tfrac{1}{2}\int \rho v^2\,dV, \quad \Pi = \int p\,dV,$$
$$\mathfrak{M} = \frac{1}{8\pi}\int B^2\,dV, \quad W = -\int \rho\mathbf{r}\cdot\nabla\phi\,dV. \tag{10-11}$$

The quantity I is a generalized moment of inertia, which in the derivation above has been expressed as the integral of r^2 over the mass element $dM = \rho\,dV$. Of the remaining integrals, T is the kinetic energy of the fluid, Π equals two-thirds of the random kinetic energy of the thermal particles in the gas plus one-third of the energy of the relativistic particles, and $\mathfrak{M}$ is the magnetic energy within the surface S.

The integral W is the total gravitational energy of the system only if any masses outside the surface S can be ignored in the computation of the potential. This result may be seen if W is written in the form

$$W = -\sum_j m_j \mathbf{r}_j \cdot \sum_k \frac{G m_k (\mathbf{r}_j - \mathbf{r}_k)}{|(\mathbf{r}_j - \mathbf{r}_k)|^3}, \tag{10-12}$$

where j is summed over all masses within the system, and k is summed over all gravitating masses. If masses outside the system can be ignored, then each interaction will be counted twice in the double sum, with $\mathbf{r}_j$ and $\mathbf{r}_k$ interchanged. As a result we have in this case

$$W = -\sum_{j<k} \frac{G m_j m_k}{|(\mathbf{r}_j - \mathbf{r}_k)|}, \tag{10-13}$$

which equals the total gravitational energy of the system.

Equation (10-10) may be applied to some components of a system, as, for example, the interstellar gas, magnetic field, and relativistic particles, but excluding the stars. In this case W is not generally equal to the gravitational energy of the components considered.

10.2 SHOCK FRONTS

When a pulse of increased pressure, with an appreciable amplitude, propagates through a gas, the front of the pulse tends to steepen because the sound velocity is higher in the compressed region; this steepening progresses until a nearly discontinuous shock front is formed. Such shock fronts generally appear whenever supersonic motions are present and must therefore be expected in the interstellar gas, where the cloud velocities are often much greater than the sound speed. We consider here the properties of such a front, idealized as a one-dimensional disturbance propagating through a homogeneous medium with a constant velocity u_1. Let the direction of motion be taken as the x axis; all quantities are then functions of x and t only.

The situation is more conveniently analyzed in a frame of reference traveling with the velocity u_1; in this frame the flow is steady, and all quantities are functions of x only. The undisturbed material enters the shock at the plane x_1, where its velocity, density, and temperature are u_1, ρ_1, and T_1, respectively. At the other side of the front, beyond the plane x_2, the corresponding quantities are u_2, ρ_2, and T_2. Between x_1 and x_2, ρ and u

are functions of x, and T may not be relevant since the gas may be far from thermal equilibrium in the shock front.

We consider first the change of physical quantities across the shock when no magnetic field is present; hydromagnetic shocks in the presence of a finite **B** are considered subsequently.

a. Perfect Gas, $\mathbf{B}=0$

The large-scale properties of a shock in a perfect gas, with $\mathbf{B}=0$, are measured by the three ratios u_2/u_1, ρ_2/ρ_1, and p_2/p_1. These ratios are determined in terms of conditions ahead of the shock (i.e., u_1, ρ_1, and p_1) by three relationships or "jump conditions," which relate physical quantities on each side of the front and result from conservation of matter, momentum, and energy. The first two of these may be obtained from equations (10-2) and (10-1), integrating these across the front to yield

$$\rho_1 u_1 = \rho_2 u_2, \tag{10-14}$$

$$p_1 + \rho_1 u_1^2 = p_2 + \rho_2 u_2^2. \tag{10-15}$$

In deriving equation (10-15), we have set $\nabla\phi$ and **B** equal to zero, and used the constancy of ρu. (Within the shock p_{xx} replaces p.)

The third jump condition expresses the conservation of energy across the front. This relation must be derived from first principles, since the energy equation (6-1), which may usually be used to determine p in equation (10-1), is not valid within a shock front, where strong irreversible processes produce an increase of entropy. We consider here the two limiting cases in which the gas does not radiate at all and in which radiation is so rapid that radiative equilibrium is established. The first case is often called an "adiabatic" shock, although in the shock front itself the gas is not adiabatic in the usual sense, since the entropy increases. If T_E, the temperature in radiative equilibrium, does not depend significantly on density, the shock in this second case may be regarded as "isothermal."

In the first case, where no radiation occurs, the energy equation states that the rate of increase of fluid energy, as a result of passage across unit area of the front per unit time, equals the corresponding rate at which work is done on the fluid by pressure forces. Hence we have

$$u_2\left(\tfrac{1}{2}\rho_2 u_2^2 + \mathcal{U}_2\right) - u_1\left(\tfrac{1}{2}\rho_1 u_1^2 + \mathcal{U}_1\right) = u_1 p_1 - u_2 p_2. \tag{10-16}$$

The quantity $\mathcal{U}$ is the internal energy of the fluid per cm^3, which must be added to $\frac{1}{2}\rho u^2$ to give the total energy content per unit volume. If the fluid behaves as a perfect gas on each side of the front, we have

$$\mathcal{U} = \frac{1}{\gamma - 1} p. \tag{10-17}$$

Use of this relationship and also of equation (10-14) in equation (10-16) gives the desired jump condition,

$$u_1^2 + \frac{2\gamma}{\gamma - 1} \frac{p_1}{\rho_1} = u_2^2 + \frac{2\gamma}{\gamma - 1} \frac{p_2}{\rho_2}. \tag{10-18}$$

From these three jump conditions, equations (10-14), (10-15), and (10-18), we obtain with some algebra [4] for a nonradiating shock,

$$\frac{p_2}{p_1} = \frac{2\gamma}{\gamma + 1} \mathfrak{M}^2 - \frac{\gamma - 1}{\gamma + 1}, \tag{10-19}$$

$$\frac{u_2}{u_1} = \frac{\rho_1}{\rho_2} = \frac{\gamma - 1}{\gamma + 1} + \frac{2}{\gamma + 1} \frac{1}{\mathfrak{M}^2}. \tag{10-20}$$

The "Mach number" $\mathfrak{M}$ of the shock in these equations is defined by

$$\mathfrak{M}^2 = \frac{\rho_1 u_1^2}{\gamma p_1} = \frac{u_1^2}{C_1^2}. \tag{10-21}$$

Evidently for large $\mathfrak{M}$, we have the asymptotic results

$$\frac{\rho_2}{\rho_1} = \frac{\gamma + 1}{\gamma - 1}, \tag{10-22}$$

$$p_2 = \frac{2\rho_1 u_1^2}{\gamma + 1}. \tag{10-23}$$

In this limiting situation $\rho_2/\rho_1 = 4$ if $\gamma = 5/3$.

We consider next the second case, in which radiative equilibrium is established. Under typical conditions prevailing in an interstellar H I cloud the cooling time given in equation (6-15) is 10^4 years for a density of some 20 H atoms/cm^3. If the shock velocity is 10 km s^{-1}, and n_H equals 20 cm^{-3} behind the shock, T will reach its equilibrium value within approximately 0.1 parsec behind the shock front. One may regard x_2 as the value of x at which this equilibrium has been reached, and include in the shock front between x_1 and x_2 not only the abrupt initial increase of temperature and

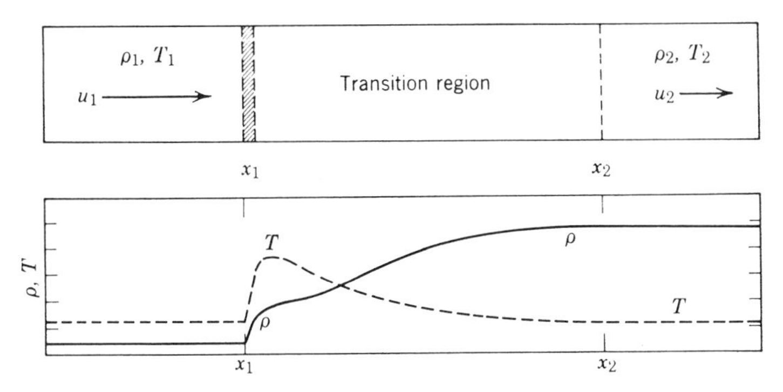

Figure 10.1 Schematic diagram of radiating shock. In the upper diagram the fluid is shown moving from left to right. At x_1 the fluid enters the shaded area, representing the nonradiative shock, followed by the transition region, where the temperature drops by radiation. The changes of density and temperature are indicated schematically in the lower figure.

density discussed above, but also the subsequent cooling and compression, as indicated in Fig. 10.1. If the hydrogen gas is neutral on both sides of the front, and the dependence of T_E on ρ is ignored [S10.1], the shock may be regarded as isothermal, and we may set $\gamma=1$. In this case p equals ρC^2, and combination of equations (10-14) and (10-15) yields

$$\frac{\rho_2}{\rho_1}=\frac{u_1^2}{C^2}=\mathfrak{M}^2, \tag{10-24}$$

where the sound speed C is given in equation (10-8), and is assumed to be the same on both sides of the shock. If u_1 is 10 km s^{-1} and C is about 1 km s^{-1}, a reasonable value for an H I cloud, ρ_2 is about 100 ρ_1. Evidently large compressions are possible in an isothermal shock.

b. Hydromagnetic Shocks

We consider next the modifications resulting when a magnetic field is present. As we shall see, the compression in an isothermal shock can be much reduced in this case. If we assume that the lines of force are straight and are everywhere parallel to the shock front (i.e., perpendicular to the x axis), the $\mathbf{B}\cdot\nabla\mathbf{B}$ term disappears from equation (10-1), and integrating this equation across the front yields, in place of equation (10-15),

$$p_1+\rho_1 u_1^2+\frac{B_1^2}{8\pi}=p_2+\rho_2 u_2^2+\frac{B_2^2}{8\pi}. \tag{10-25}$$

We have already seen that the magnetic flux remains constant through any circuit moving with a conducting fluid, and for the one-dimensional compression across a shock we have, therefore,

$$\frac{B_1}{\rho_1}=\frac{B_2}{\rho_2}. \tag{10-26}$$

In a strong nonradiating shock we have seen that if **B** is ignored, the density increases by only a factor 4, and the magnetic pressure $B^2/8\pi$ increases by only a factor 16; if **B** is less than 3×10^{-6} G in front of the shock, the magnetic pressure behind the shock is less than one-fifth the value of p_2 found from equation (10-19), if ρ_1 and u_1 have typical values for interstellar clouds. The magnetic field in this case has no great effect on the shock front, and the neglect of **B** is valid.

For an isothermal shock p equals ρC^2, with the same value of C on both sides of the front, and equations (10-25), (10-26), and (10-14) can be combined to give ρ_2/ρ_1. In the limiting case where ahead of the shock the Alfvén velocity V_{A1} much exceeds C, and ρ_2 much exceeds ρ_1, we obtain

$$\frac{\rho_2}{\rho_1}=2^{1/2}\frac{u_1}{V_{A1}}. \tag{10-27}$$

The criterion for a strong supersonic shock is now that u_1 must be large compared to V_{A1} rather than compared to C. If **B** is 3×10^{-6} G and ρ_1 in an H I cloud is 4.7×10^{-23} g cm^{-3} (see Table 11.1), V_A computed from equation (10-9) is 1.2 km s^{-1}, about equal to C; evidently, the compression given by equation (10-27) is then substantially less than is found from equation (10-24).

When energetic particles are present, gyrating around the lines of force, the pressure p_R contributed by these suprathermal particles must be included in equation (10-25). For a given u_1, the value of ρ_2/ρ_1 may be somewhat reduced by this additional effect. When **B** has a component perpendicular as well as parallel to the shock, forces appear in the shock zone parallel to the front, as a result of the $\mathbf{B}\cdot\nabla\mathbf{B}$ term in equation (10-1), and the results above are substantially altered [5].

10.3 INSTABILITIES

In many situations the equations presented in Section 10.1 possess relatively simple solutions, often with $\mathbf{v}=0$. Small disturbances of these equilibrium configurations are of interest, since under some conditions these disturbances will oscillate without steady growth, and under other conditions, the disturbance will grow, either with or without oscillations. If any disturbances are of this latter type, the equilibrium is said to be unstable, otherwise it is stable. The possible presence of instabilities for any particular type of disturbance can be examined with a relatively simple linearized theory, in which all departures from the equilibrium, or zero-order, solution are assumed infinitesimal, and all quadratic terms are ignored. If the zero-order equilibrium is independent of time, the equations for the first-order terms are all satisfied by a time variation of the type $\exp(i\omega t)$. If the first-order equations are satisfied with a real value for ω, the infinitesimal perturbations are stable, whereas if ω has a negative imaginary part, the disturbance is unstable. To examine the effects produced by such instabilities, a theory of finite amplitude disturbances is required, but this first-order analysis is very helpful in indicating when such instabilities may be important.

a. Rayleigh-Taylor Instability

One of the simplest nontrivial instabilities of astrophysical importance is that arising when a lighter fluid is pushing or accelerating a heavier one. In a uniform gravitational field this instability arises when a heavier fluid above is supported by a lighter fluid below. We give here the theory of this instability in an idealized case [6].

Let us assume that both the upper and lower fluids are incompressible, with density ρ_a above the interface and ρ_b below. The interface initially is in the plane $z=0$, but becomes deformed in the perturbation. The gravitational potential ϕ is assumed to equal gz. All quantities are independent of y. We let $\mathbf{v}_a$ and p_a denote the velocity and pressure in the fluid above the interface, with $\mathbf{v}_b$ and p_b denoting the corresponding quantities below. In this idealized situation the gravitational potential is known and equations (10-1) and (10-2) in the fluid can be solved exactly, if the quadratic term in $\mathbf{v}$ is ignored. The essential problem is then to satisfy the boundary conditions at the interface, which we do here to first order in the perturbation.

To satisfy equation (10-2), we introduce the velocity potential Ψ, defined by

$$\mathbf{v} = -\nabla\Psi. \tag{10-28}$$

Since $\partial\,(\nabla\times\mathbf{v})\partial t$ vanishes for constant ρ and negligible $\mathbf{B}\cdot\nabla\mathbf{B}$, according to equation (10-1), equation (10-28), which ensures that $\nabla\times\mathbf{v}$ remains zero, does not restrict the possible motions. Since ρ is constant above and below the interface, equation (10-2) gives

$$\nabla^2\Psi = 0. \tag{10-29}$$

A solution for this equation, which satisfies the condition that Ψ vanish for $|z|$ very large is

$$\Psi = \begin{cases} \Psi_a = K_a e^{i\omega t - \kappa z} \sin \kappa x & \text{above,} \\ \Psi_b = K_b e^{i\omega t + \kappa z} \sin \kappa x & \text{below.} \end{cases} \tag{10-30}$$

We take the real part of $\Psi/(i\omega)$ as the solution of physical interest. If we introduce equation (10-28) in equation (10-1) with $\mathbf{B}=0$ and $\mathbf{v}\cdot\nabla\mathbf{v}$ ignored, we find on integration over space

$$\rho \frac{\partial \Psi}{\partial t} = p + \rho g z, \tag{10-31}$$

which holds separately both above and below the interface. In equation (10-31) we omit a constant of integration, whose only effect is to keep p positive for the relevant range of positive gz.

The two boundary conditions to be satisfied across the interface are that both v_z and p be continuous. We may determine z_i, the value of z on the interface, by integrating v_z over time, obtaining with use of equation (10-28) and (10-30)

$$z_i = -\frac{1}{i\omega}\frac{\partial \Psi}{\partial z} = \begin{cases} \dfrac{\kappa}{i\omega}\Psi & \text{above,} \\ -\dfrac{\kappa}{i\omega}\Psi & \text{below.} \end{cases} \tag{10-32}$$

The constant of integration (which may depend on x) vanishes, since z_i and Ψ vanish together (at $t=0$ for real ω and at $t=-\infty$ for imaginary ω). Equation (10-32) is accurate only to first order in Ψ; to follow exactly the interface in its motion, the small change with time of z_i in the exponent of equation (10-30) should be considered in the integration of $\partial\Psi/\partial z$ over time, and would give terms of higher order. The condition that v_z be continuous across the interface is equivalent to the condition that z_i be the same above and below, giving

$$\kappa\Psi_a = -\kappa\Psi_b, \tag{10-33}$$

evaluated at $z=z_i$. If we again retain only terms of first order in Ψ, we may evaluate equation (10-33) at the plane $z=0$. Hence we have

$$K_a = -K_b. \tag{10-34}$$

The continuity of p across the interface may next be used to determine ω. From equation (10-31) we obtain

$$p = i\omega\rho\Psi - \rho g z. \tag{10-35}$$

We use equation (10-32) to express z again to first order in Ψ. The condition that $p_a = p_b$, evaluated again in the plane $z=0$, with use of equation (10-34), now gives

$$\omega^2 = g\kappa \frac{\rho_b - \rho_a}{\rho_b + \rho_a}. \tag{10-36}$$

If ρ_b is less than ρ_a, ω is imaginary and the instability grows. It is physically obvious that a layer of water cannot be supported in a gravitational field by an underlying layer of air, for example; the gravitational energy of the system is reduced as globules of water fall down through the air. Similarly when a cold, dense gas is accelerated by pressure from a hot, more rarefied gas, the flow is unstable; spikes and globules of the colder fluid will tend to penetrate the warmer fluid. This instability is believed to be important in the later phases of an expanding supernova shell [S12.2d].

The Rayleigh-Taylor instability is only one of the many instabilities to which a fluid is subject. A difference of tangential velocity across an interface also leads to exponentially increasing perturbations of the interface, known as the Helmholtz instability [6]. In the presence of a magnetic field, hydromagnetic instabilities may appear as the fluid seeks to minimize its magnetic energy [2]. Still other instabilities are associated with non-Maxwellian distribution functions of particle velocities. In a particular equilibrium configuration it is often difficult to know whether all relevant disturbances have been examined for possible instabilities.

REFERENCES

1. S. A. Kaplan, *Interstellar Gas Dynamics,* 2nd revised edition, F. D. Kahn, Editor, Pergamon Press (New York), 1966.
2. L. Spitzer, *Physics of Fully Ionized Gases,* 2nd revised edition, Interscience (New York), 1962.
3. L. Woltjer, *Stars and Stellar Systems,* Vol. **5**, University of Chicago Press (Chicago), 1965, p. 531.
4. L. D. Landau and E. M. Lifshitz, *Fluid Mechanics,* Pergamon Press (New York), 1976, Chapt. IX.
5. H. L. Helfer, *Ap. J.,* **117**, 177, 1953.
6. H. Lamb, *Hydrodynamics,* Dover Reprints (New York), 1945.

11. Overall Equilibrium

The basic equations of motion in Chapter 10 permit a tremendous variety of dynamical processes, many of which, in fact, occur in the interstellar gas. To make some progress in understanding the observed phenomena it is helpful to analyze individual processes in ideally simple situations, disregarding initially the other complications that are present in reality. The present chapter discusses the equilibrium of gas in interstellar space, on the assumption that all forces are in balance and the medium is motionless, with no net acceleration. In some regions, as in the exploding gases discussed in the next chapter, the assumption of pressure equilibrium is obviously incorrect, but for the general distribution of matter in the galaxy and for the interrelation between gas in the low-velocity clouds and in the intercloud regions such a model may have some relevance to reality.

11.1 PARAMETERS OF THE INTERSTELLAR GAS

In the subsequent discussion of dynamical processes, specific values for the parameters of the medium must be assumed in the calculations. This section summarizes some of the relevant information, first on the physical state of the gas and second on the sources of energy for maintaining the observed cloud motions.

a. Physical State

Table 11.1 gives values for various physical properties of the interstellar gas within about 1000 pc from the Sun. These numbers, based on the

Table 11.1. Parameters of the Interstellar Gas

Mean density ρ	3×10^{-24} gm cm^{-3}
Typical particle density	
n(H I) in diffuse clouds	20 cm^{-3}
n(H I) between clouds	0.1 cm^{-3}
n_H in molecular clouds	10^3–10^6 cm^{-3}
Typical temperature T	
Diffuse H I clouds	80°K
H I between clouds	6000°K
H II photon ionized regions	8000°K
Coronal gas between clouds	6×10^5°K
Root mean-square random cloud velocity	10 km s^{-1}
Isothermal sound speed C	
H I cloud at 80°K	0.7 km s^{-1}
H II gas at 8000°K	10 km s^{-1}
Magnetic field B	2.5×10^{-6} G
Effective thickness $2H$ of H I cloud layer	250 pc

observational data discussed in previous chapters, are mostly average values; most of the physical quantities are known to show large fluctuations. We review briefly the evidence for these various values.

The best measure of the mean ρ within the solar neighborhood appears to be the mean E_{B-V} for stars within about 1000 pc of the Sun and about 100 pc of the galactic plane [see Table 7.1 and equation (7-17)] together with the correlation between E_{B-V} and N_H determined for Lα measures [see equation (7-18)]. The mean E_{B-V} may be reduced by observational selection, since the stars behind a large cloud with $A_V=4$ mag, for example, might escape observation (see Table 7.2). On the other hand, the tendency of obscuring matter to be concentrated near early-type stars increases $\langle E_{B-V}\rangle$ above its value for a random line of sight. The mean value for n_H of 1.2 cm^{-3}, based on these arguments and given in equation (7-19), is consistent with the average values between 0.5 and 0.7 cm^{-3} found for n(H I) from 21-cm data [S3.3b; S3.4a], provided allowance is made for the increased values of n(H I) in the local Orion arm and for H atoms in molecular or ionized form. The value of ρ in Table 11.1 has been based on this mean n_H, with use of the He to H ratio in Table 1.1. The variation of ρ with z, the distance from the galactic plane, which is shown [1] in more detailed studies of $\langle E_{B-V}/L\rangle$, is ignored here.

Values of n(H I) in diffuse clouds have been obtained from the measured rotational excitation of H_2 lines [S4.3b], giving about 30 cm^{-3} as a typical value for clouds with a low value of β_0, the probability per second of H_2 excitation by photons. Clouds of higher β_0 have higher $n(\mathrm{H\ I})$, presumably as a result of dynamical effects associated with expanding gas shells near the stars. Densities determined in this way are somewhat higher than those obtained from atomic absorption lines [2], and the collisional cross section for H–H_2 collisions may be underestimated. Accordingly, we adopt the range between 10 and 30 cm^{-3} for $n(\mathrm{H\ I})$ in a diffuse cloud, with 20 cm^{-3} taken as a typical value in Table 7.2; the mean thickness of about 6 pc obtained from the mean $N(\mathrm{H\ I})$ in Table 7.2 corresponds to a spherical cloud radius of three-quarters the mean thickness or about 5 pc. (The mean thickness is $(4\pi r^3/3)/(\pi r^2)$ or $4r/3$.) Almost certainly a substantial range in $n(\mathrm{H\ I})$ must be present in individual clouds, with values ranging continuously up to those in the molecular clouds.

The value of 80°K for T in diffuse clouds is taken from equation (3-46). The distribution of T values obtained both from 21-cm absorption-emission comparisons [S3.4a] and from the $N(1)/N(0)$ rotational excitation ratio in H_2 [S3.4b] suggests that this temperature is reasonably accurate despite a large spread; most of the values lie between 40 and 120°K.

The values of T and $n(\mathrm{H\ I})$ for the gas between the clouds are also uncertain, although the agreement between several different determinations is relatively good. The value of T deduced from absorption-emission comparisons in the 21-cm line is relatively high for the most transparent clouds [S3.4a] with several determinations in the range between 4000 and 8000°K [3]. While the basis for such determinations has been criticized [4], the minimum width of 21-cm emission lines with no corresponding absorption [S3.4a] corresponds to a kinetic temperature of about 8000°K [5]. In addition, ultraviolet measures of absorption lines of the elements H, D, O, and Ar show that the velocity dispersion found from the curve of growth decreases systematically with increasing atomic weight, giving a temperature for this "warm" gas of about 6000°K [6]. Finally, direct measurements of the Doppler width of the Lα radiation scattered by H atoms streaming toward the Sun indicate a T between 5800 and 13,000°K [7].

The mean particle density $\langle n(\mathrm{H\ I})\rangle$ of this intercloud medium as determined from statistical studies of 21-cm emission [S3.4a], is 0.2 cm^{-3} [8]. Measures of 21-cm emission with a 10 arcminute beam at $b \approx 15°$ show that much of the emitting gas is distributed uniformly, giving $n(\mathrm{H\ I}) = 0.2$ cm^{-3} if the gas is assumed to extend about 300 pc along the line of sight [9]. The mean $n(\mathrm{H\ I})$ along lines of sight to unreddened stars [$E_{B-V} \leqslant 0.01$] found from Lα absorption measures is 0.1 cm^{-3}, with individual values ranging from $\leqslant 0.01$ up to 0.3 cm^{-3} [10]; allowance for possible errors in E_{B-V} might decrease this average. A mean value of about 0.1 cm^{-3} has also been obtained from interplanetary scattered Lα [11], and this value has been adopted in Table 11.1. On the other hand, the presence of H_2

absorption lines in most unreddened stars indicates that $n(\mathrm{H\ I})$ is at least 0.3 cm^{-3} in the region where these lines are being produced [12]. Clumping of the H I intercloud gas will give local densities higher than the mean values along the line of sight. Thus if this warm H I gas is assumed to occupy one-third the volume in the galactic disk, with the coronal gas filling the rest of the volume outside H I clouds and normal H II regions, the local value of $n(\mathrm{H\ I})$ is about 0.3 cm^{-3}.

Observations of O VI absorption lines [S3.4c] indicate that a hot coronal-type gas must occupy some fraction of the line of sight between clouds. The mean value of n_{H} in this collision-ionized gas, averaged along the line of sight, is uncertain, depending on T as well as on d_{O}, the depletion of oxygen, but must exceed 1.0×10^{-4} [S5.2b], the minimum value for $T = 3 \times 10^{5}$°K and no depletion ($d_{\mathrm{O}} = 1$). It seems unlikely that T is so low in this gas, and in Table 11.1 we have arbitrarily put T equal to 6×10^{5}°K, the value giving pressure equilibrium between this hot gas and the diffuse clouds if $d_{\mathrm{O}} = 1$ [S11.3], and if the coronal gas occupies most of the galactic disk. Measures of soft X-rays indicate that a spread of temperatures must undoubtedly be present. However, it is possible that the observed O VI lines may be produced primarily by hot expanding gas in the vicinity of early-type stars.

The values of n_{H} within dense clouds containing abundant molecules can, in principle, be determined from the analyses of molecular emission line formation. Present estimates range from about 10^{3} up to about 10^{6} cm^{-3} or more [S4.1d].

For H II regions the temperatures obtained from collisional excitation of metastable lines [S4.1c], from the strength of radio recombination lines [S4.2c], and from the brightness temperature observed at low radio frequencies [S3.5a] all agree in giving about 8000°K, with a range of some ± 2000°K for most measures [13]. The density of H II regions is not included in Table 11.1 since this quantity evidently extends over an enormous range. In compact H II regions n_{H} exceeds 10^{4} cm^{-3} [S3.5a]. On the other hand, in the extended H II regions which surround B stars and the nuclei of planetary nebulae, and which are believed to produce the diffuse Hα emission, rms values of 0.1 cm^{-3} for n_{H} seem plausible [S3.3a]. Within conspicuous visible H II regions, rms values of n_{H} between 10 and 10^{2} cm^{-3} are normal [S3.3a].

The rms cloud velocity is determined primarily from the dispersion of velocities among different clouds seen at 21 cm, restricting the analysis either to absorbing clouds, for which τ_0 mostly exceeds 0.2 [S3.4a], or to emitting clouds with narrow profiles (presumably again with appreciable τ_0) corresponding to an internal velocity dispersion less than 3 $\mathrm{km\,s}^{-1}$ [14]; these data agree in giving a radial-velocity dispersion (corrected for galactic rotation) of 6 $\mathrm{km\,s^{-1}}$, yielding 10 $\mathrm{km\,s^{-1}}$ for the corresponding three-dimensional dispersion on the arbitrary assumption that the velocity distribution is isotropic. The dispersion found from all the 21-cm emission data is higher, presumably in part because of high random thermal velocities in the intercloud medium. For the Ca II lines the presence of high-speed clouds of relatively low mass increases the velocity dispersion somewhat if all clouds are given the same weight, but for the Na I D lines the dispersion of radial velocities is again about 6 $\mathrm{km\ s^{-1}}$ [15].

The value of the magnetic field B in Table 11.1 is obtained entirely from pulsar data [S3.6c], and represents a rough average between the mean $\mathbf{B}$, given in equation (3-73), and the rms field, which may be more nearly 3×10^{-6} G. The appreciably greater values of B in some dense clouds obtained from 21-cm Zeeman effect measures [S3.4a] presumably represent field compression in a contracting cloud [S11.3b].

The effective thickness $2H$ has been set equal to the value obtained from the 21-cm emission line [S3.3b]. The closely equal value of 240 pc has been obtained [16] for the Ca II layer, dividing $2N$(Ca II) for high latitude stars by a mean n(Ca II) in the galactic plane. A similar value of 200 pc has been obtained for dust [S7.2a]. For other components of the gas the effective thickness differs markedly from these values. The effective thickness of the electron-ion gas is much greater, about 700 pc [S3.6c], whereas for the CO molecules, produced in the denser, nearly opaque clouds, $2H\approx120$ pc [S4.1d]. For the warm H I gas between the H I clouds a value of about 400 pc for $2H$ seems indicated [15].

b. Energy Source for Cloud Motions

The large velocities of interstellar clouds form a conspicuous feature of the interstellar gas. Explosions of H II regions and of supernovae, the two most powerful mechanisms for accelerating gas, are analyzed in the next chapter. Here we discuss the power available from these two sources for maintaining the kinetic energy of the gas and compare this power with that dissipated by the cloud motions.

We define P_u as the total average power available per $\mathrm{cm^3}$ in the form of kinetic energy imparted to photoelectrons in H II regions. Since the rate of photoemission from H atoms must closely balance the rate of radiative capture by protons, we have for P_u

$$P_u=\alpha^{(2)}\langle n_e^2\rangle kT_c\langle\psi\rangle, \tag{11-1}$$

where we have used equation (6-7) to evaluate $\bar{E}_2$, the mean photoelectron energy, and where the presence of helium has been ignored. To evaluate this expression we make use of the observed value of about 0.1 cm^{-6} for $\langle n_e^2 \rangle$ [S3.5a], let $\alpha^{(2)} = 3.1 \times 10^{-13}\,\mathrm{cm}^3\,\mathrm{s}^{-1}$ from equation (5-14) and Table 5.2, for $T = 8000°\mathrm{K}$, and set $T_c = 32{,}000°\mathrm{K}$, a representative value for the color temperature of the radiation shortward of the Lyman limit in O stars. With $\langle \psi \rangle = 1.38$ from Table 6.1, we obtain

$$P_u = 1.9 \times 10^{-25}\,\mathrm{ergs\,cm^{-3}\,s^{-1}}. \tag{11-2}$$

The kinetic energy of mass ejected from these early-type stars is uncertain, but may be comparable with P_u [17], increasing somewhat the power going into exploding H II regions.

For supernova explosions we define P_s as the average power available initially in the form of kinetic energy, again per cm^3. For a supernova of type I, with an initial expansion velocity of 20,000 $\mathrm{km\,s^{-1}}$ [18], an ejected mass of 0.1 $M_\odot$ gives an initial kinetic energy of 4×10^{50} ergs, the value of E_0 obtained from the X-rays emitted by old supernova remnants (see Table 1.2). For supernovae of type II, associated with young stars and spiral arms, the expansion velocity is somewhat less [18], but the mass is believed to be greater, and we assume the same value for E_0. The supernova frequency in our Galaxy is about one per 25 years [19]. These numbers are all uncertain by a factor 2, but if the supernovae are assumed to occur within a galactic disk 10,000 pc in radius and 250 pc in thickness, we obtain

$$P_s = 2.2 \times 10^{-25}\,\mathrm{ergs\,cm^{-3}\,s^{-1}}. \tag{11-3}$$

According to the theoretical calculations, the efficiencies with which these power inputs are converted into observed cloud motions are about 1 and 3 percent, respectively [S12.1c and 12.2c]. Hence the total power input available for maintaining cloud motions is about $8 \times 10^{-27}\,\mathrm{ergs\,cm^{-3}\,s^{-1}}$, with supernovae contributing about four times as much as ultraviolet photons from early-type stars. These numerical results are uncertain by at least factors of 2.

For comparison we now consider the power P_c dissipated per cm^3 in cloud collisions. According to Table 7.1, the material intersected by a line of sight in the galactic plane per kpc can be idealized as 6.2 standard clouds and 0.8 large clouds; the mass of this material is 62 percent in standard clouds and 38 percent in large clouds, giving densities averaged over the galactic disk of 1.9×10^{-24} and $1.1 \times 10^{-24}\,\mathrm{gm\,cm^{-3}}$, respectively, in these two types of clouds. The rms velocity of the standard clouds is 10 $\mathrm{km\,s^{-1}}$, according to Table 11.1, giving a relative velocity of 14 $\mathrm{km\,s^{-1}}$, and collisions of one standard cloud with another every 1.1×10^7 years. On the average, the kinetic energy of relative motion, which we assume is entirely lost in each collision, is half the total kinetic energy of the two identical colliding clouds. If the velocity dispersion of the large clouds is assumed to be the same as for the standard clouds, collisions of these large clouds with each other and with the smaller clouds will increase the total dissipation rate by about 60 percent, and we obtain

$$P_c = 2.1 \times 10^{-27}\,\mathrm{ergs\,cm^{-3}\,s^{-1}}. \tag{11-4}$$

This computation is evidently highly idealized, and neglects, for example, the energy dissipated by high-velocity clouds, which are ignored in the consideration of rms velocities. Also the distribution of clouds is known not to be entirely random. Subject to these uncertainties the available energy input, which is four times the energy dissipation rate P_c, seems clearly sufficient to maintain the velocities of the observed clouds.

11.2 GALACTIC EQUILIBRIUM

We now consider model systems in which the interstellar gas is in equilibrium in the gravitational potential field ϕ produced by the stars. Two simplified situations will be treated; first, a spherically symmetrical potential, as in a spherical galaxy or globular cluster, and second, a potential which is a function only of z, the distance from the galactic plane. In the first situation the magnetic field will be ignored, whereas in the second the magnetic field will be taken into account, both in the case where **B** is assumed parallel to the plane and where large distortions of **B** perpendicular to the plane are allowed.

a. Spherically Symmetric System

The virial theorem may be used to find $\langle r_G^2 \rangle$, the mean square distance of the gas from the center of the system. With $\mathbf{v}_G$ and **B** both set equal to zero, we see from equation (10-10) that $3\Pi_G = -W_G$. We assume that the gravitational potential is produced by stars, whose density ρ_S, per unit volume of the stellar system, is taken to be constant in the region occupied by the gas. Hence in equation (10-3) we can set ρ equal to a constant ρ_S, and in equation (10-11) for W_G, the integral of $\rho_G r^2 dV$ becomes $M_G \langle r_G^2 \rangle$; equation (10-10) now gives

$$\langle r_G^2 \rangle = \frac{3\langle w_G^2 \rangle}{4\pi G \rho_S}, \tag{11-5}$$

since in equation (10-11) for Π, $p=\rho_G\langle w_G^2\rangle/3$, where $\langle w_G^2\rangle$ equals the mean square value of the random gas velocities. A similar relation holds also for the stars, with r_G and w_G replaced by r_S and w_S, with r_S interpreted as the radius of the inner core of the system, and with a numerical constant of order unity appearing. Dividing these results, we have the approximate relation

$$\frac{\langle r_G^2\rangle}{r_S^2}\approx\frac{\langle w_G^2\rangle}{\langle w_S^2\rangle}. \tag{11-6}$$

For a gas at densities of a few atoms or more per cm^3, $\langle w_G^2\rangle^{1/2}$ is not likely to much exceed 20 km s^{-1}, its value for an ionized gas at 10^4°K. Stellar velocities in galaxies are normally one order of magnitude greater. It follows that a system of gas in hydrostatic equilibrium within a spherical galaxy must be strongly concentrated toward the center, unless its temperature is somehow much higher than normal, or unless the energies of cosmic rays and magnetic fields are comparable with $-W_G$.

If ρ_G is comparable with ρ_S, the gravitational potential is increased, and the self-gravitational energy of gas increases $|W_G|$. In this case it is readily shown [20] that if the ratio $\langle w_G^2\rangle/\langle w_S^2\rangle$ is small and constant, the maximum fraction of the mass of the system which can be in equilibrium at the center is about equal to the 3/2 power of this ratio, and in typical spherical galaxies is about 10^{-3}. In globular clusters the ratio of mean square velocities can be unity or more, in which case the gas can escape by evaporation and outward flow. The general conclusion is that in a spherical system, where neither stars nor gas have appreciable angular momentum, the fraction of gas present in pressure equilibrium must be small.

This conclusion is completely altered, of course, if the gas has appreciable angular momentum. In this case the large velocities required for gas to be in equilibrium at moderate r_G can be mostly systematic velocities of rotation, which are not subject to dissipation by inelastic collisions as are random velocities.

b. Plane One-dimensional System

We consider now the equilibrium of the interstellar gas in the galactic disk, with all quantities assumed to be functions only of z, the distance from the galactic plane. As shown in Table 11.1, the effective thickness of this layer is taken to be 250 pc. We investigate the density distribution in this layer, using the equation of hydrostatic equilibrium obtained from equation (10-1) by setting the time derivatives equal to zero.

The pressure p in this equation may be expressed as the sum of the gas pressure $p_G(z)$ and the cosmic ray pressure $p_R(z)$. Since the mean square random cloud velocities $\langle v_c^2 \rangle$ much exceed the random thermal velocities, the thermal pressure of the gas may be neglected compared to the macroscopic turbulent pressure associated with these cloud motions, and for $z=0$, p_G becomes

$$p_G(0) = \tfrac{1}{3}\rho\langle v_c^2 \rangle = 1.0\times 10^{-12}\,\text{dynes cm}^{-2}, \tag{11-7}$$

where $\langle v_c^2 \rangle$ and the mean density ρ have been evaluated from Table 11.1. The cosmic ray pressure is equal to one-third the energy density U_R for these relativistic particles, and from the values of U_R given in Section 1.4, we obtain

$$p_R(0) = 0.4\times 10^{-12}\,\text{dynes cm}^{-2}. \tag{11-8}$$

We assume here that the magnetic lines of force are straight and parallel. The $\mathbf{B}\cdot\nabla\mathbf{B}$ term, which represents the net force produced by the tension along $\mathbf{B}$, then vanishes, and the remaining term contributes effectively a magnetic pressure p_B, which at $z=0$ becomes

$$p_B(0) = \frac{B^2}{8\pi} = 0.25\times 10^{-12}\,\text{dynes cm}^{-2}, \tag{11-9}$$

where B in the galactic plane has been evaluated from Table 11.1.

To obtain a simple solution we assume arbitrarily that all three pressures vary in the same way with z. Thus $\alpha \equiv p_B/p_G$ and $\beta \equiv p_R/p_G$ are both constant. We assume also that $d\phi/dz = -z(dg_z/dz)$, where the slope of g_z is taken to be constant, a valid assumption for the stellar gravitational field at $z \leqslant 250$ pc (see Fig. 1.7). Equation (10-1) now becomes

$$(1+\alpha+\beta)\frac{1}{3\rho}\frac{d}{dz}\left(\rho\langle v_c^2 \rangle\right) = z\frac{dg_z}{dz}. \tag{11-10}$$

If $\langle v_c^2 \rangle$ is assumed constant, this equation has the solution

$$\rho(z) = \rho(0)e^{-(z/h)^2}, \tag{11-11}$$

where

$$h = \left[\frac{2(1+\alpha+\beta)\langle v_c^2 \rangle}{-3\,dg_z/dz}\right]^{1/2} = 100(1+\alpha+\beta)^{1/2}\,\text{pc}. \tag{11-12}$$

The numerical value in equation (11-12) has been determined with $dg_z/dz = -2.2 \times 10^{-11}\,\mathrm{cm\,s^{-2}\,pc^{-1}}$. The effective thickness $2H$ of the gaseous layer is defined as the total mass per cm^2 divided by $\rho(0)$, giving

$$2H = \pi^{1/2}h = 180(1+\alpha+\beta)^{1/2}\,\mathrm{pc}. \tag{11-13}$$

To apply this result we insert the values of 0.25 and 0.4 for α and β obtained from equation (11-7), (11-8), and (11-9) and find a theoretical value of 230 pc for $2H$; the agreement with the value in Table 11.1 is closer than one would expect from the uncertainty in the observational quantities used. Since the agreement is only slightly worse if α and β are zero, this discussion provides a useful check on the self-consistency of $2H$, $\langle v_c^2\rangle$, and dg_z/dz, but gives no conclusive information on whether or not cosmic rays and the magnetic field are largely confined to the Galaxy. The chief argument for such confinement is the problem of an energy source for cosmic rays and magnetic fields if these have throughout the Universe about the same values observed near the Sun.

This discussion indicates that mean magnetic fields as great as 10^{-5} G, required for explaining the observed synchrotron radiation with the observed number of relativistic electrons reaching the Earth, would give $\alpha = 4$ and could not easily be confined to the galactic plane. As suggested in Section 1.4, regions in which both B and the particle density of relativistic electrons are greater than average probably emit the observed radiation.

c. Equilibrium in a Plane Gravitational Potential

If **B** is assumed parallel to the galactic plane, the equilibrium found above is unstable under some conditions [21]. In equilibrium the magnetic field is pushing the matter away from the galactic plane, while the gravitational field is pulling it back. If a line of force is assumed curved, or bowed, with some sections further from the galactic plane than others, the clouds are free to slide down the lines of force, leaving the outer parts of these lines relatively unencumbered with matter, and therefore free to move outward because of the magnetic pressure. Calculations show that if $\langle v_c^2\rangle$ is independent of density ($\gamma = 1$), the instability appears for wavelengths greater than some critical value λ_c, where λ_c is approximately $4\pi h$, if α and β are comparable and small with respect to unity [21]; h is again the value of z at which $\rho(z)$ falls to $\rho(0)/e$. Thus λ_c near the Sun is in the neighborhood of 1600 pc. This instability develops slowly, and the increase of the star formation rate with density may increase $\langle v_c^2\rangle$ with ρ, giving an effective γ greater than unity, tending to stabilize the equilibrium [22]. However, the possibility remains that the plane equilibrium described above may in fact be unstable.

Analysis of this more general situation, in which the lines of force are curved in vertical planes, shows that such instabilities can lead to a new equilibrium, in which the tension along the curved lines of force balances the magnetic pressure outward [23]. We derive here the relations which lead to this result, with the assumption that the gas is isothermal ($\gamma=1$) and that the cosmic ray pressure p_R vanishes ($\beta=0$). To begin, we transform equation (10-1) in two ways. First, we replace the two magnetic terms by $\mathbf{j}\times\mathbf{B}$, where the current density $\mathbf{j}$ (in emu) is given by

$$4\pi\mathbf{j}=\nabla\times\mathbf{B}. \tag{11-14}$$

Second, we eliminate the term in $\nabla\phi$ with the substitution

$$q\equiv pe^{\phi/C^2}, \tag{11-15}$$

where C is the isothermal sound speed in the equation of state (10-6), which has been used to eliminate ρ. For an equilibrium state in which all quantities are independent of time and $\mathbf{v}=0$, equation (10-1) becomes

$$\nabla q=e^{\phi/C^2}\mathbf{j}\times\mathbf{B}. \tag{11-16}$$

Since $\mathbf{B}\cdot\nabla q$ vanishes as a result of equation (11-16), it follows that q must be constant along a line of force. It is evident physically that in equilibrium the usual equation of hydrostatic equilibrium, $\nabla p=-\rho\nabla\phi$, must hold along each line of force, since the magnetic forces are always perpendicular to $\mathbf{B}$. Thus along each line of force p follows the usual exponential law for an isothermal gas and varies as $\exp(-\phi/C^2)$.

We wish to solve equation (11-16) in the two-dimensional situation where all physical quantities are independent of y, and $\mathbf{B}$ is in the xz plane, where z is the distance from the galactic plane. Since B_y, $\partial B_x/\partial y$, and $\partial B_z/\partial y$ are all zero, it follows from equation (11-14) that $\mathbf{j}$ is parallel to the y axis. Let us now introduce the vector potential A, such that

$$\mathbf{B}=\nabla\times\mathbf{A}, \tag{11-17}$$

which guarantees that equation (10-5) is satisfied. Since $B_y=0$ and $\mathbf{A}$ is not a function of y, it is readily seen that $\mathbf{B}$ depends only on A_y, and we take $\mathbf{A}$, like $\mathbf{j}$, to be parallel to the y axis; that is, we set $A_x=A_z=0$.

In this two-dimensional problem A_y is constant along each line of force, since from equation (11-17),

$$\mathbf{B}\cdot\nabla A_y=-\frac{\partial A_y}{\partial z}\frac{\partial A_y}{\partial x}+\frac{\partial A_y}{\partial x}\frac{\partial A_y}{\partial z}=0. \tag{11-18}$$

Moreover, the difference of A_y between two lines of force in a particular xz plane is proportional to the magnetic flux between them, as may be seen by applying Stokes' theorem to a rectangular area in the zy plane with the two vertical sides separated by unit distance in the y direction. If we stipulate that in the galactic midplane, where $z=0$, **B** is a straight line in the x direction, and take $A_y=0$ on this line, it follows that $A_y(x,z)$ equals the total magnetic flux between the midplane and z, per unit distance along the y axis. In the subsequent discussion we omit the subscripts y from both j_y and A_y.

If now we substitute $\nabla\times\mathbf{A}$ for **B** in equation (11-16), we find

$$\nabla q = e^{\phi/C^2} j \nabla A. \tag{11-19}$$

The current density j can be expressed in terms of $\nabla^2 A$ by substituting equation (11-17) into equation (11-14), yielding the final equation

$$\nabla^2 A = -4\pi \frac{dq}{dA} e^{-\phi/C^2}, \tag{11-20}$$

where we have let $\nabla q/\nabla A$ equal dq/dA; since q and A are both constant along each line of force, q is a function of A only. We assume that the gravitational potential ϕ is a known function of z only.

Equation (11-20) may be solved numerically, giving $A(x,z)$ and the topography of the magnetic field, if dq/dA is known as a function of A. To determine this function, we use the fact that the lines of force are frozen in the fluid, and therefore the total mass between two lines of force is constant during all motions, although the distribution along the lines of force may change. The value of A on a particular line of force is also constant during any motion, since A equals the flux through a strip of unit width between $z=0$ and the particular line of force being considered. Hence the total mass between two lines of force will be an invariant function of A during the motion. To express these ideas quantitatively, we let $m(A)dA$ be the total mass in a volume of unit width in the y direction, bounded in z by force lines with vector potentials A and $A+dA$, and of length X in the x direction; we shall assume that the pattern of **B** is periodic and shall set X equal to the wavelength of this pattern. The total mass may be obtained by integrating ρdV over this volume. If we let $z_A(x)$ denote the z value, for each x, of the line of force with a vector potential A, then $(\partial z_A/\partial A)dA$ is the thickness of the volume in question. If we let $\rho = p/C^2$, and eliminate p by equation (11-15), we obtain

$$m(A) = \frac{1}{C^2}\int_0^X dx\left(\frac{\partial z_A(x)}{\partial A}\right) q e^{-\phi/C^2}. \tag{11-21}$$

The quantity q is a function of A only, and may be taken out of the integral, giving a direct relationship between $m(A)$ and $q(A)$.

Equation (11-21) gives $q(A)$ if $m(A)$ and the magnetic field topography [i.e., $z_A(x)$] are known, while equation (11-20) gives $A(x,z)$ and $z_A(x)$ if $q(A)$ is known. A solution may be sought by iteration, assuming a particular $z_A(x)$, computing $q(A)$ and thus giving the right-hand side of equation (11-20) as a function of x and z; integration of this equation then yields a revised z_A. If the process converges, it will give an equilibrium solution.

Such solutions have been obtained [23] with $m(A)$ computed for an initially one-dimensional equilibrium configuration [S11.2b], except that ϕ was assumed proportional to $|z|$ instead of z^2. The initial magnetic field was then given a small sinusoidal perturbation with wavelength λ in the x direction. Empirically it was found that the character of the solution obtained in subsequent iterations depended on whether λ was less or greater than λ_c, where the critical wavelength λ_c obtained from the linearized instability theory [21] is equal to

$$\lambda_c = \frac{4\pi h}{(1+2\alpha)^{1/2}}. \tag{11-22}$$

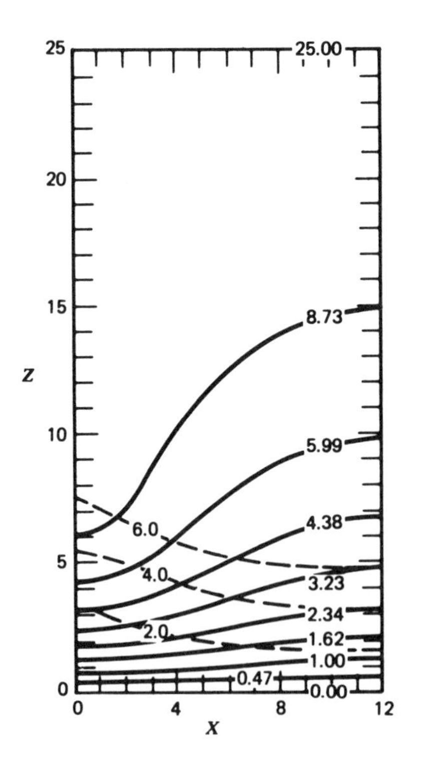

Figure 11.1 Galactic equilibrium of magnetized interstellar gas [23]. The solid lines show magnetic lines of force in equilibrium, with magnetic stresses and pressure gradients balanced by a uniform gravitational field g toward the galactic midplane, $z=0$. The dashed lines represent isodensity contours on which $\rho(z)/\rho(0)=1/e$, $1/e^2$, and $1/e^3$. The number shown for each curve represents the height above the galactic plane for the corresponding line in the initial one-dimensional but unstable equilibrium.

For $\lambda < \lambda_c$, the computations converged back to the plane equilibrium, with straight lines of force. For longer wavelengths, the computations converged to a new equilibrium. A plot of z_A along several lines of force for a typical such solution, with $\alpha = 1$, is shown in Fig. 11.1. Distances are measured in units of $C^2/g = h/2$, with $\lambda_c/2 = 7.26$ in these units. The curving lines of force are suggestively similar to the curving filaments observed in a survey of 21-cm emission [24]. If cosmic-ray pressure had been taken into account, the solutions would be quantitatively different, but probably not qualitatively [25].

The theoretical model is certainly very idealized in that irregularities in the magnetic field produced by cloud motions are ignored. Such magnetic irregularities are inevitable since the kinetic energy of the H I clouds exceeds the magnetic energy density, and are in any case observed [S3.6c, S8.2c]. It is not clear what detailed effects such large transient irregularities might have on the solution. Nevertheless, the existence of an equilibrium solution in this idealized situation is of interest. The large mass concentration in the magnetic "valleys" and the curving loops of **B** on the magnetic "summits" may help to explain some observational features of spiral arms and star formation.

11.3 EQUILIBRIUM OF CLOUDS

It is clear from the basic equations that interstellar motions tend toward pressure equilibrium. Any pressure gradients too great to be balanced by gravitational forces will produce motions which expand the high pressure regions and contract those at low pressure. The magnitude of the pressure differences present, on the average, will depend, of course, on the rate at which such pressure differences are produced by various energy sources, including newly formed hot stars and supernovae. Since pressure differences in a region tend to equalize in the time taken for sound to travel across the region (see the values of C in Table 11.1), one might expect that diffuse H I clouds, with radii of roughly 5 pc, might be expected to come to pressure equilibrium with their surroundings in about 7×10^6 years. For comparison, the mean time until any one cloud collides with another cloud is about 1.2×10^7 years, if we take six clouds per kpc and a cloud velocity of 10 km s^{-1}. Thus one might expect these clouds to approach pressure equilibrium but to reach it only partially.

The observational evidence seems generally consistent with this theoretical expectation. A value of 0.3 cm^{-3} seems plausible for the warm H I component of the intercloud medium, with $T = 6000°$K, if this gas occupies one-third of the volume in the galactic disk [S11.1a]. The diffuse H I clouds with a temperature of about 80°K would then be in pressure equilibrium with their surroundings if n_H were about 20 cm^{-3}, in agreement with the value adopted in Table 11.1. The equilibrium pressure, including one He atom for every 10 H atoms, is then

$$p = 2.6 \times 10^{-13} \text{dynes cm}^{-2}, \tag{11-23}$$

corresponding to $n_H T = 1700$°K cm^{-3}. Some diffuse clouds may well be at lower pressure, with others at a much higher pressure, especially those compressed by passing shocks produced by supernovae or expanding H II regions. Nevertheless, the possibility of this agreement is suggestive, and appears generally consistent with a two-phase model for heating the interstellar gas [S6.2b].

In extended ionized regions the pressure may be not greatly different from that in equation (11-23), although the uncertainties are large. In an H II gas at 8000°K the corresponding value of n_H consistent with equation (11-23), taking into account the presence of 2.1 free particles per H nucleus, instead of 1.1 in neutral H I regions, is 0.11 cm^{-3}. The rms value of n_e in the extended H II regions around B stars and planetary nuclei is about 0.1 cm^{-3}, and a local density of 0.3 cm^{-3} in 10 percent of the volume would be consistent both with this observed rms value and with the value of 0.03 cm^{-3} for $\langle n_e \rangle$ obtained from pulsar dispersion measures [S3.6a]. These numbers suggest a pressure in these H II regions about 3 times that given in equation (11-23), consistent with some dynamical expansion. In the conspicuous H II regions around O associations the excess pressure is usually greater by several orders of magnitude.

The minimum pressure within the coronal gas is about 0.1×10^{-13} dynes cm^{-2} if the temperature is as low as 3×10^5°K [S5.2b]. If this gas is assumed to occupy two thirds of the galactic disc, for example, and its temperature is taken to be 6×10^5°K, with no depletion of O, its pressure is 2.7×10^{-13} dynes cm^{-2}, rising to 21×10^{-13} dynes cm^{-2} if $T = 10^6$°K. Evidently a pressure similar to that in equation (11-23) is entirely possible, although higher pressures would seem required if the observed collision-ionized gas were mostly concentrated within the central cores of H II regions, heated by stellar winds.

While the observations are evidently not precise, they clearly indicate that the pressures in three main components of the interstellar gas (the diffuse H I clouds with low β_0, the warm H I intercloud gas, and the coronal gas, if this fills much of the galactic disk) are of roughly the same order of magnitude, despite temperature and density differences amounting to four orders of magnitude. Thus it seems likely that pressure equilibrium has some relevance to the observed distribution of interstellar matter. In the rest of this section we shall consider an idealized cloud which is assumed to be in equilibrium with the surrounding medium, taken to be much less dense and at a uniform pressure p_0. We investigate primarily the overall relations between p_0 and the cloud mass M and radius R in equilibrium, as well as the range of conditions for which hydrostatic equilibrium is possible. Again the magnetic field will be ignored initially, but considered in the later discussion.

a. Spherical Cloud, $\mathbf{B}=0$

We consider an isothermal sphere, in which $\mathbf{v}=\mathbf{B}=0$. The virial theorem gives one overall condition that must be satisfied in equilibrium, when $d^2I/dt^2=0$. Since the mass outside the sphere may be ignored, equation (10-13) is applicable and W is the total gravitational energy. If we approximate W by the usual formula for a sphere of uniform density, equation (10-10) yields

$$4\pi R^3 p_0 = \frac{3MkT}{\mu} - \frac{3GM^2}{5R}, \tag{11-24}$$

where μ is the mean mass per particle.

When M, T, and μ are all fixed, this equation yields a family of solutions in which R is a function of p_0. When R is sufficiently large the gravitational term is negligible, and the external and internal pressures are in balance; the product of p_0 and the volume V is then nearly constant. As p_0 increases, R decreases until finally the gravitational term becomes large, decreasing R somewhat further. Finally the gravitational term becomes so great that further decreases in R would require a decrease in p_0; that is, no solutions are possible with p_0 increased still further. It follows directly from equation (11-24) that the upper limit on p_0, which we designate as p_m, is given by

$$p_m = 3.15\left(\frac{kT}{\mu}\right)^4 \frac{1}{G^3M^2}. \tag{11-25}$$

Solutions for which p_0 decreases with decreasing R are unstable; for a fixed p_0 a decrease in R will decrease the internal pressure, resulting in motions which decrease R further. If p_0 exceeds p_m no equilibrium is possible, and the sphere will collapse.

When a sphere of constant M, T, and μ is forced to contract with increasing p_0, density gradients appear, and the ratio of the central density $\rho(0)$ to the mean density $\bar{\rho}$ increases. As a result, the gravitational term is greater than assumed in equation (11-24) and p_m is decreased. A detailed solution [26] for the structure of an isothermal sphere surrounded by a medium at constant pressure but negligible density yields a numerical constant 1.40 instead of 3.15 in equation (11-25); when $p=p_m$ the value of $\rho(0)/\bar{\rho}$ equals 5.8.

b. Magnetized Cloud

The effect of a magnetic field in equation (10-10) may be computed approximately. We assume that the field within the cloud is uniform, with a strength B, and that outside the cloud the mean square field strength for each value of r equals $B^2(R/r)^6$. If the closed surface S in equation (10-10) is taken to be far outside the cloud, the magnetic surface terms are negligible, and the fields inside and outside the cloud contribute equally to $\mathfrak{M}$. Calculations based on more realistic assumptions, with continuous field derivatives, show [27] that the total $\mathfrak{M}$ is not greatly altered. The sum of the other terms is the same as when the surface S coincides with the cloud surface, since the increase of Π, including the integral over the low-density surrounding medium, just offsets the increase of the surface integral involving p_0. In place of equation (11-24) we find

$$4\pi R^3 p_0 = \frac{3MkT}{\mu} - \frac{1}{R}\left[\frac{3}{5}GM^2 - \frac{1}{3}R^4B^2\right]. \qquad (11\text{-}26)$$

If the field is assumed to be frozen in the material, the total flux Φ through the cloud, equal to $\pi R^2 B$, must remain constant as the cloud contracts; consequently the ratio of gravitational-to-magnetic energies remains constant as the cloud is compressed. If magnetic forces do not prevent initial contraction, they will not prevent a spherical collapse at any stage. This same conclusion also follows from comparing the magnetic force perpendicular to $\mathbf{B}$ and the gravitational force; if expressed per cm^3, both forces vary as $1/R^5$.

In equation (11-26) the gravitational term exceeds the magnetic one if M exceeds M_c, where

$$M_c = \frac{c_1}{\pi}\left(\frac{5}{9G}\right)^{1/2}\Phi; \qquad (11\text{-}27)$$

Φ is again the magnetic flux through the cloud and c_1 is a correction factor inserted so that equation (11-27) may be applied to clouds which are initially spherical, with radius R_0, and with uniform ρ_0 and $\mathbf{B}_0$, but become centrally condensed and flattened parallel to $\mathbf{B}$ as they contract (see below). To compute M_c as a function of ρ_0 and B_0 we must allow for the dependence of Φ on M_c, with $R_0^2=(3M_c/4\pi\rho_0)^{2/3}$; equation (11-27) gives

$$M_c=\frac{c_1^3 5^{3/2}}{48\pi^2}\frac{B_0^3}{G^{3/2}\rho_0^2}. \tag{11-28}$$

Gravitational collapse can occur only if $M>M_c$ and also if p_0, the external pressure, exceeds p_m, where

$$p_m=3.15c_2\frac{(kT/\mu)^4}{G^3M^2\left\{1-(M_c/M)^{2/3}\right\}^3}, \tag{11-29}$$

with insertion of another correction factor c_2. For the radius R_m of this critical state, on the verge of collapse, we obtain

$$\frac{GM^2}{R_m^4}=\frac{25p_m}{1-(M_c/M)^{2/3}}. \tag{11-30}$$

The value of the constant factor in equation (11-30) has been taken from the exact solution for no magnetic field.

Equation (11-26) neglects the tendency of the magnetic field to flatten a cloud in a direction parallel to **B**. If the equatorial radius remains unchanged, this flattening will increase the absolute value of the gravitational energy somewhat above the value assumed in equations (11-24) and (11-26), whereas the magnetic energy will be decreased; as a result M_c is decreased below the value computed for spherical contraction [27].

This effect of nonspherical contraction and also that of nonuniform distributions of ρ and **B** have been explored with exact computations of equilibrium configurations for gaseous isothermal clouds, subject to an external gas pressure p_0 [28]. The clouds are assumed to be initially spherical, with a uniform magnetic field B_0 extending both through the cloud and the intercloud medium. These systems are initially far from equilibrium, hence in the computations they contract and deform, with the magnetic lines of force frozen in the gas. Equilibrium states are found with methods similar to those used above for equilibrium in a plane potential field [S11.2c]. In this case, the magnetic configuration has an axis of symmetry; **j** and **A** are now vectors in the direction of θ, the azimuthal angle about the symmetry axis.

The computational results indicate that because of the flattening parallel to **B**, the magnetic field increases less rapidly than $\rho^{2/3}$, with the observed $d\ln B/d\ln\rho$ between $\frac{1}{2}$ and $\frac{1}{3}$ for the models investigated. It is clear physically that for $M/M_c \ll 1$, the magnetic forces are so much stronger than the gravitational ones that a cloud will scarcely contract at all transverse to **B**, but will flatten parallel to **B** as p_0/p_m increases or as the gravitational self-attraction increases. For $M/M_c \gg 1$, however, a cloud can contract isotropically with little effect produced by B; in this limit B will vary as $\rho^{2/3}$. Thus if for the observed clouds n(H I) and M are assumed proportional, the variation of B with n(H I) found from Zeeman effect observations [S3.4a] can be understood theoretically.

The exact computations also make it possible to compute the correction factors c_1 and c_2, in equations (11-27) and (11-29), to take account of the difference between the idealized uniform spherical clouds considered in the virial theorem and more realistic clouds, centrally condensed and moderately flattened along **B**. The following values are found [29] for models on the verge of collapse:

$$c_1 = 0.53, \qquad c_2 = 0.60. \tag{11-31}$$

With this value of c_1 inserted in equation (11-28), we obtain

$$M_c = \frac{0.0035}{G^{3/2}} \frac{B_0^3}{\rho_0^2} = 1.9 \times 10^4 \frac{\{B_0(\mu G)\}^3}{n_{H0}^2} M_\odot, \tag{11-32}$$

where the numerical value has been computed with B_0 in μG and n_{H0} equal to $\rho_0/1.4m_H$. Thus for n_{H0} and B_0 equal to 20 cm^{-3} and 2.5 μG, typical values for a diffuse cloud (see Table 11.1), M_c equals 740 $M_\odot$, increasing to 2×10^5 $M_\odot$ if n_{H0} is decreased to 1.2 cm^{-3}, representative of the interstellar medium as a whole.

REFERENCES

1. T. Neckel, *Zeits. f. Astroph.*, **63**, 221, 1966.
2. J. F. Drake and S. R. Pottasch, *Astron. Astroph.*, **54**, 425, 1977.
3. R. D. Davies and E. R. Cummings, *M.N.R.A.S.*, **170**, 95, 1975.
4. E. W. Greisen, *Ap. J.*, **184**, 379, 1973.
5. G. B. Field, *Molecules in the Galactic Environment*, M. G. Gordon and L. E. Snyder, Editors, Wiley (New York), 1973, p. 21.
6. D. G. York, in preparation.
7. J. L. Bertaux, J. E. Blamont, N. Tabarié, W. G. Kurt, M. C. Bourgin, A. S. Smirnov, and N. N. Dementeva, *Astron. Astroph.*, **46**, 19, 1976.
8. P. L. Baker and W. B. Burton, *Ap. J.*, **198**, 281, 1975.
9. C. Heiles, *Ap. J. Supp.*, **15**, 97, 1967; No. 136.
10. R. C. Bohlin, *Ap. J.*, **200**, 402, 1975.

11. L. Spitzer and E. B. Jenkins, *Ann. Rev. Astron. Astroph.,* **13**, 133, 1975.
12. D. G. York, *Ap. J.,* **204**, 750, 1976.
13. D. E. Osterbrock, *Astrophysics of Gaseous Nebulae,* W. H. Freeman (San Francisco), 1974, Chapt. 5.
14. K. Takakubo, *B.A.N.,* **19**, 125, 1967.
15. E. Falgarone and J. Lequeux, *Astron. Astroph.,* **25**, 253, 1973.
16. P. J. Van Rhijn, *Publ. Kapteyn Astron. Lab.*, No. 50, Groningen, 1946.
17. J. Castor, R. McCray, and R. Weaver, *Ap. J. (Lett.),* **200**, L107, 1975.
18. L. Searle, *Supernovae and Supernova Remnants,* C. B. Cosmovici, Editor, Reidel Publ. (Dordrecht), 1974, p. 125.
19. G. A. Tammann, *Supernovae and Supernova Remnants,* C. B. Cosmovici, Editor, Reidel Publ. (Dordrecht), 1974, p. 155.
20. L. Spitzer, *Ap. J.,* **95**, 329, 1942.
21. E. N. Parker, *Ap. J.,* **145**, 811, 1966.
22. E. G. Zweibel and R. M. Kulsrud, *Ap. J.,* **201**, 63, 1975.
23. T. Ch. Mouschovias, *Ap. J.,* **192**, 37, 1974.
24. C. Heiles and E. B. Jenkins, *Astron. Astroph.,* **46**, 333, 1976.
25. T. Ch. Mouschovias, *Astron. Astroph.,* **40**, 191, 1975.
26. L. Spitzer, in *Stars and Stellar Systems,* Vol. 7, University of Chicago Press (Chicago), 1968, p. 1.
27. P. Strittmatter, *M.N.R.A.S.,* **132**, 359, 1966.
28. T. Ch. Mouschovias, *Ap. J.,* **206**, 753, 1976; **207**, 141, 1976.
29. T. Ch. Mouschovias and L. Spitzer, *Ap. J.,* **210**, 326, 1976.

12. Explosive Motions

As stars are born, develop, and die they can produce large transient pressure increases in the surrounding interstellar gas, giving rise to a rapid outward expansion or explosion of this medium. These explosions may be driven by ultraviolet photons which ionize and heat the gas in H II regions, for example, or by high-velocity expulsion of mass from a star, as from a supernova or bright early-type star. The energies available in photon-driven expansions of H II regions and in energetic supernova shells have been discussed above [S11.1b]. Here we present a very simplified theory of such explosions, neglecting such complications as initial density inhomogeneities and magnetic fields. Expansion of the interstellar medium caused by mass loss from early-type stars involves the processes present in both the other types of explosions and is not considered here. Successive sections below discuss the expansion of photon-driven H II regions, the expansion of supernova shells, and the dynamics of individual H I clouds engulfed by an ionization or shock front.

12.1 H II REGIONS

The development of an H II region is conditioned by the transitory nature of the energy source. During the lifetime of an O star, which is less than 10^7 years, interstellar gas moving at 10 km/s will travel less than 100 pc, which is comparable with the diameter of the larger H II regions. Thus before an H II region has expanded very far, its central energy source will be extinguished.

A new O star is presumably born within clouds of relatively dense cold gas. The appearance of a source of ultraviolet photons will have two effects. First, the gas surrounding the new star will become ionized. Since the mean free path of an ultraviolet photon is very short in neutral hydrogen, the photons, after traveling freely through the ionized gas immediately surrounding the star, will be absorbed in a relatively thin surrounding shell of neutral hydrogen, producing new ionization. Thus the ionized and neutral gases are separated by an ionization front, which moves rapidly outward as more and more atoms become ionized by the stream of photons. Second, the heated gas will be at a much higher pressure than the surrounding cool gas, and will tend to expand. Since the expansion velocity is likely to exceed the sound velocity in the surrounding H I region, a shock front may be expected to form, moving out through the neutral gas. The dynamical analysis of H II regions must consider the interactions between the ionization front and the shock front, together with the equations of motion for the gas behind the two fronts.

The relevant equations for an ionization front, analogous to equations (10-14) through (10-18) for a shock front, are derived below. These relations are then used in a discussion of the dynamical evolution of an H II region.

a. Ionization Fronts

As in the discussion of shock fronts, we consider a one-dimensional front propagating through a homogeneous medium with constant velocity u_1, and we adopt a frame of reference moving at this same velocity. Figure 10.1 is again applicable, except that now between x_1 and x_2 there is not only a change in velocity and density but also a change of ionization state and kinetic temperature produced by absorption of ultraviolet photons. The thickness of the ionization front must somewhat exceed the mean free photon path, $1/s_{f\nu}n(\mathrm{H\ I})$ [equation (5-6)]; if the change in density of neutral H within the front is ignored, this pathlength at $\nu=\nu_1$ equals $0.04/n(\mathrm{H\ I})$ pc. The distance traveled by the front during the time t_T required to establish thermal equilibrium in the H II region behind the front may be computed from equation (6-10), and determines the thickness of the transition region; this distance equals about $0.2/n_p$ pc if the front velocity is 10 km s^{-1} and if n_p is taken to be constant.

Equations (10-14) and (10-15) are valid as before, with the one difference that the mass flow through the front is no longer arbitrary. Instead, the number of H atoms flowing through the front per unit area per second must equal J, the corresponding number of ionizing photons reaching the front. Hence equation (10-14) becomes

$$\rho_1 u_1 = \rho_2 u_2 = \mu_i J, \tag{12-1}$$

where μ_i is defined as the mean mass of the gas per newly created positive ion; if the abundance of helium relative to hydrogen is again taken to be 0.10 by particle number (see Table 1.1), μ_i equals 1.40 m_H if the helium remains neutral, and 1.27 m_H if the helium becomes singly ionized in the front.

In an isothermal shock front the sound speeds C_1 and C_2 ahead of and behind the front are equal. In an ionization front C_2 and C_1 are markedly different, each value being determined by the equilibrium temperature and mean molecular weight in H II and H I regions (see Table 11.1). The general solution of equations (12-1) and (10-15) together with equation (10-6) (with $\gamma = 1$) is

$$\frac{\rho_2}{\rho_1} = \frac{(C_1^2 + u_1^2) \pm \left\{(C_1^2 + u_1^2)^2 - 4u_1^2C_2^2\right\}^{1/2}}{2C_2^2}. \tag{12-2}$$

The condition that equation (12-2) give a real value places a restriction on u_1. Either u_1 must exceed an upper critical value u_R, given by

$$u_R = C_2 + (C_2^2 - C_1^2)^{1/2} \approx 2C_2, \tag{12-3}$$

or u_1 must be less than a lower critical value u_D, given by

$$u_D = C_2 - (C_2^2 - C_1^2)^{1/2} \approx \frac{C_1^2}{2C_2}. \tag{12-4}$$

The approximate equality in equations (12-3) and (12-4) holds because C_2 exceeds C_1 by about one order of magnitude in an interstellar ionization front.

The nature of the ionization front depends on the value of u_1, which is determined by equation (12-1). If u_1 exceeds u_R, the solution is called "R type"; R refers to a rarefied gas, since for ρ_1 sufficiently low, u_1 will always exceed u_R. In this case ρ_2 exceeds ρ_1, and u_1 exceeds $2C_2$. If we take the negative sign in equation (12-2), the motion of the front is supersonic relative to the gas behind as well as ahead of the front. This type of front is referred to as "weak R," since the relative change of density is small. For an expansion driven by stellar photons, the "strong R" fronts, corresponding to the positive sign in equation (12-2) and to a larger density increase across the front, are not usually relevant, since these solutions require some mechanism to maintain a large outward velocity of the compressed hot gas behind the front. For the extreme case of a weak R front in which u_1 much exceeds $2C_2$, the changes in density and relative velocity across the front are given by

$$\frac{\rho_2}{\rho_1} = 1 + \frac{C_2^2}{u_1^2}, \tag{12-5}$$

$$u_1 - u_2 = \frac{C_2^2}{u_1}, \tag{12-6}$$

to first order in C_2^2/u_1^2.

If u_1 is between u_R and u_D, an ionization front of this simple type is not possible. In this case a shock front must precede the ionization front, increasing ρ_1 and thus reducing u_1 to u_D or below. If u_1 is exactly equal to u_D, the front is called "D critical," u_2 is exactly equal to C_2 and the density change is given by

$$\frac{\rho_2}{\rho_1} = \frac{C_1^2}{2C_2^2}, \tag{12-7}$$

giving p_2 equal to $p_1/2$. More generally, under interstellar conditions the shock ahead of the ionization front will compress the gas sufficiently so that u_1 found from equation (12-1) is less than u_D; this solution is called "D type," with D referring to a dense gas. In a D-type front ρ_2 is always less than ρ_1.

The "weak D" front, with the smaller relative decrease in density, now corresponds to the positive sign in equation (12-2). Strong D fronts, in which u_2 exceeds C_2, may also occur. These fronts may be present under some interstellar conditions [1], especially for brief periods in the development of H II regions.

b. Initial Ionization of the Gas

When an O star is first formed, the ultraviolet luminosity may be expected to rise rather abruptly. For example, calculations for a model star of 30 $M_{\odot}$, as it approaches the main sequence, show that during its final increase $S_u(0)$, the emission rate of photons beyond the Lyman limit, doubles in 4000 years [1]. Thus the initial growth of an H II region may be computed on the assumption that the central star abruptly starts to shine at its full luminosity.

Close to the star the ionization front will move away at a very rapid speed. Consider, for example, an ionization front at a distance of 5 pc from an O7 star, with a hydrogen density of 10 atoms/cm^3. From equation

(12−1), with J evaluated from $N_u = S_u(0)$ for an O7 star in Table 5.3, we find u_1 equal to 2400 km s^{-1}. Evidently u_1 exceeds u_R, and the front is of weak R type. If r_i denotes the radius of the ionization front we may write

$$\frac{dr_i}{dt} = \frac{S_u(r_i)}{4\pi n_H r_i^2} = \frac{1}{4\pi n_H r_i^2}\left\{ S_u(0) - \frac{4}{3}\pi r_i^3 n_H^2 \alpha^{(2)} \right\}, \qquad (12\text{-}8)$$

where the effect of absorption on S_u has been evaluated by means of equation (5-20), and where the helium atoms have been assumed neutral. An integration yields

$$r_i^3 = r_S^3 \left\{ 1 - \exp\left(- n_H \alpha^{(2)} t \right) \right\}, \qquad (12\text{-}9)$$

where equation (5-21) has been used for r_S, the equilibrium radius of the H II region at the density n_H, which we have assumed uniform throughout the region.

Equations (12-8) and (12-9) are valid only as long as dr_i/dt exceeds u_R or about $2C_2$ [equation (12-3)]. When the velocity of the ionization front falls below this critical value, a shock front appears and the ionization front is no longer R type. From equations (12-8) and (5-21) it follows directly [2] that dr_i/dt equals $2C_2$ when

$$\frac{r_S^3}{r_i^3} = 1 + \frac{6C_2}{n_H \alpha^{(2)} r_i}. \qquad (12\text{-}10)$$

If we replace r_i on the right-hand side of equation (12-10) by r_S, equal to 12 pc for an O7 star with n_H equal to 10/cm^3, and substitute from equation (5-14) for $\alpha^{(2)}$, with T equal to 8000°K, we find that the right-hand side of equation (12-10) equals 1.05. Hence the weak R-type front will persist until r_i is within about 2 percent of r_S.

Until r_i is nearly equal to r_S and S_u is substantially reduced by absorption, u_1 will much exceed C_2. Hence the velocity change Δu on passing through the ionization front will generally be much less than C_2, in accordance with equation (12-6). Throughout the ionized gas, velocities and corresponding perturbations will be present, but these are correspondingly small until r_i approaches r_S [2].

The detailed structure of the moving ionization front has been analyzed [3] numerically, including all the chief physical effects; the diffuse radiation field was handled with the "on-the-spot" approximation [S5.1c]. For a star of effective temperature 50,000°K the results show a rapid temperature rise to about 17,000°K in the ionization front, followed by a more gradual decline to the equilibrium temperature of about 8000°K.

c. Expansion of the Ionized Gas

The first phase in the development of an H II region, which has been discussed in the preceding paragraphs, is essentially the ionization of a large sphere of gas by the O star; motions are of small importance throughout most of this phase. In the second phase the heated sphere of gas tends to expand into the surrounding neutral region, where the density is presumably about the same as the initial H II density, but the pressure is less by some two orders of magnitude. Since the equilibrium radius r_S of the ionized region varies as $n_H^{-2/3}$, in accordance with equation (5-21), $n_H r_S^3$ varies as $1/n_H$; hence the amount of material ionized increases as the gas expands. This is because with the same number of photons more hydrogen atoms can be kept ionized if n_H is reduced, since the rate of recombination per cm^3 varies as n_H^2. As a result, when the H II gas expands, the ionization front will move outward through the gas. If the brightness of the O star were constant and the cloud of gas were sufficiently large, this second phase would terminate when pressure equilibrium was established. In fact, the ultraviolet luminosity of the brighter stars will drop after a few million years.

The transition between these two phases is difficult to follow analytically. During this period the changes in velocity through the ionization front are comparable with the sound velocity in the ionized region, which we denote here by C_{II}. A shock forms at the ionization front and moves away, and the ionization front changes from weak R type at the beginning to weak D type at the end. This complex dynamical process can best be followed by direct numerical integration of the dynamical equations [1]. While these computations do not include cooling of the neutral gas behind the shock front, they take most other features of the problem into account and show clearly the formation of the shock when the radius r_i of the ionization front approaches r_S, and dr_i/dt falls below $2C_2$.

After the shock has formed and has moved away from the ionization front, both fronts slow down somewhat and the ionization front becomes of weak D type and completely subsonic. The rate of expansion of the H II region during the ensuing second phase will now be deduced approximately from the known properties of the shock and ionization fronts.

We make the simplifying assumptions that the H II region inside the ionization front is in a state of uniform expansion with a density ρ_{II} which is independent of radius and that the radius r_i of the ionized zone equals its static value r_S [S5.1b]. We let v_i be the velocity, relative to the star, of the ionized gas immediately behind the ionization front. If we follow a particular shell of ionized gas, of radius r, then $r^3\rho_{II}$ will be constant during the expansion, giving

$$\frac{1}{r}\frac{dr}{dt}=\frac{v_i}{r_i}=-\frac{1}{3\rho_{\text{II}}}\frac{d\rho_{\text{II}}}{dt}. \tag{12-11}$$

The velocity V_i of the ionization front relative to the star may be expressed in terms of $d\rho_{\text{II}}/dt$ since $r_i^3\rho_{\text{II}}^2$ is constant [see equation (5-21)], provided that the ultraviolet luminosity of the central star is constant and that extinction by grains may be ignored. Hence we have

$$\frac{1}{r_i}\frac{dr_i}{dt}=\frac{V_i}{r_i}=-\frac{2}{3\rho_{\text{II}}}\frac{d\rho_{\text{II}}}{dt}. \tag{12-12}$$

Combining these results we obtain

$$v_i=\tfrac{1}{2}V_i. \tag{12-13}$$

Thus the velocity u_{i2} of the ionized gas relative to the front is also equal to $\frac{1}{2}V_i$.

To determine the shock velocity V_s, relative to the undisturbed H I gas ahead, we neglect the small difference between V_s and V_i, [see equation (12-18) below] and apply the pressure jump condition across the two fronts, from in front of the shock to behind the ionization front. With use of equation (12-13) we obtain in place of equation (10-15)

$$\rho_{\text{I}}V_s^2=p_{\text{II}}+\rho_{\text{II}}\left(\tfrac{1}{2}V_s\right)^2, \tag{12-14}$$

where ρ_{I} denotes ρ in the H I region ahead of the shock, whereas p_{II} and ρ_{II} denote the uniform pressure and density within the H II region; we have ignored p_{I}, the pressure in the H I region. Use of the isothermal equation of state (10-6) within the H II region gives

$$V_s^2=C_{\text{II}}^2\frac{\rho_{\text{II}}/\rho_{\text{I}}}{1-\rho_{\text{II}}/4\rho_{\text{I}}}. \tag{12-15}$$

To obtain the difference between V_s and V_i, and the amount of gas accumulating in the front, we must analyze in more detail the velocity changes across the front. As before, we denote by subscripts s and i quantities referring to the shock and ionization fronts, respectively, with subscripts 1 and 2 denoting quantities in front of and behind each of these two fronts. With v denoting an outward velocity with respect to the star, while u_s and u_i denote inward velocities with respect to the shock and ionization fronts, we have

$$-u_{s1}=0-V_s, \quad -u_{s2}=v_{s2}-V_s,$$

$$-u_{i1}=v_{i1}-V_i, \quad -u_{i2}=\tfrac{1}{2}V_i-V_i, \tag{12-16}$$

where we have used equation (12-13). We now assume that the gas density between the shock and ionization fronts is constant, which has the consequence that $v_{s2}=v_{i1}$. (The spherical divergence between these two fronts is ignored.) The ratio of u_{s2} to u_{i1} may now be found from the density jump condition [equation (10-14)] applied across each of the two fronts, giving

$$\frac{u_{s2}}{u_{i1}}=\frac{2\rho_{\mathrm{I}}}{\rho_{\mathrm{II}}}\frac{V_s}{V_i}. \tag{12-17}$$

Since ρ_{II} is generally less than ρ_{I}, in the ideal case of expansion in an initially uniform medium, and since V_s/V_i is nearly unity, u_{s2} is more than twice u_{i1}, and it follows that the rate at which mass leaves the compressed shell between the two fronts is less than half the rate at which it enters. Evidently much of the H I gas in front of the shock accumulates in this compressed shell.

Finally we compute V_s-V_i, which from equations (12-16) for u_{i1} and u_{s2} equals $u_{s2}-u_{i1}$, since again we set $v_{s2}=v_{i1}$. If we express u_{i1} in terms of u_{s2} with use of equation (12-17), and let $u_{s2}=\rho_{\mathrm{I}}V_s/\rho_{s2}$, with ρ_{s2} evaluated from equation (10-24), we find

$$\frac{V_s-V_i}{V_s}=\frac{C_{\mathrm{I}}^2}{V_s^2}\left(1-\frac{\rho_{\mathrm{II}}}{2\rho_{\mathrm{I}}}\frac{V_i}{V_s}\right). \tag{12-18}$$

Since $V_s^2/C_{\mathrm{I}}^2 \approx p_{\mathrm{II}}/p_{\mathrm{I}}$ according to equations (12-15) and (10-6), V_i/V_s is nearly unity as long as p_{II} much exceeds p_{I}.

Equation (12-15) may be integrated directly if $\rho_{\mathrm{II}}/4\rho_{\mathrm{I}}$ in the denominator is ignored; this result becomes exact as $\rho_{\mathrm{II}}/\rho_{\mathrm{I}}$ becomes small. Since ρ_{II} varies as $r_i^{-3/2}$, for constant stellar luminosity, we define r_{i0} by the relationship

$$\frac{\rho_{\mathrm{II}}}{\rho_{\mathrm{I}}}=\left(\frac{r_{i0}}{r_i}\right)^{3/2} \tag{12-19}$$

The radius r_{i0} is nearly equal to the initial value of r_S, the radius of the ionization front in radiative equilibrium. Since the difference between V_s and V_i is small until pressure equilibrium is approached, we may set V_s

equal to dr_i/dt, and obtain from equation (12-15)

$$\frac{r_i}{r_{i0}} = \left(1 + \frac{7}{4}\frac{C_{\rm II}t}{r_{i0}}\right)^{4/7}. \qquad (12\text{-}20)$$

The validity of the assumptions leading to equations (12-15) and (12-20) may be verified by examination of the numerical results obtained for old H II regions [1, 4]. In the numerical computations the full dynamical equations were integrated for times up to 2×10^6 years after the birth of an O7 star in a uniform H I region. The basic assumption of uniform expansion of the ionized gas, leading to equation (12-13), appears to be correct on the average, since the mean value of $2u_{i2}/V_i$ after the shock separates from the ionization front equals 1.00 in the computer results, with an average deviation of ±0.05. For an initial H density $n_{\rm I}$ equal to 6.4 cm^{-3}, the value of $d\log r_i/d\log t$ in the computations falls to 0.55 for t equal to 2×10^6 years, at which time $\rho_{\rm II}/\rho_{\rm I}$ has decreased to about 0.10; the theoretical asymptotic value of this slope should be 0.57, according to equation (12-20). Since the cooling of neutral hydrogen behind the shock front was not included in the numerical computations, the shell between the two fronts was found to be relatively thick, and equation (12-18) is not applicable. However, the computer results show clearly that equations (12-15) and (12-20) are reasonably accurate for the idealized problem under consideration.

The expansion of actual H II regions is likely to differ substantially from this simple model. Density irregularities in the H I cloud can alter the appearance of the ionization front. Even if the surrounding H I cloud were entirely uniform, the thermal instability of an isothermal shock, discussed at the end of the next section, would create irregularities, although the ionization front itself is apparently stable [1]. A magnetic field can destroy the spherical symmetry and reduce the amount of compression in the shock front [see equation (10-27)]. The lines of magnetic force will tend to remain straight when density irregularities develop, perhaps giving these a filamentary structure parallel to **B**.

These modifications are not likely to affect very greatly the efficiency ε_u, with which an H II region converts the available radiant energy into kinetic energy; this efficiency is small in any case. We define ε_u in terms of the total kinetic energy of the photoelectrons produced. (If the total energy radiated by the star were taken as a reference, ε_u would be smaller by about an order of magnitude.) The inefficiency of an H II region as a mechanical heat engine is due to the rapid rate of radiation, especially by impurity atoms. As a result, only a small fraction of the photoelectron kinetic energy is available for offsetting the heat loss resulting from adiabatic expansion, and ε_u is about 1 percent for a typical H II region around an O star [4].

12.2 SUPERNOVA SHELLS

Of the vast energy released in a supernova explosion, some will be radiated relatively early, while some will be converted into kinetic energy of the expanding gas, which will sweep up, accelerate, compress, and heat the surrounding interstellar material. Some gas is likely to escape from the Galaxy entirely, especially if the supernova is far from the galactic plane. The gas that is slowed down within the Galaxy will ultimately radiate most of its energy in some region of the electromagnetic spectrum. In this section we attempt to follow theoretically the successive stages that may be expected. The discussion is necessarily confined to an idealized situation, in which the interstellar gas is homogeneous. Any numerical values computed will refer to a supernova with a total kinetic energy E of 4×10^{50} ergs for the ejected envelope [S11.1b].

a. Initial Expansion of Supernova Material

When a supernova ejects a large mass of gas, the density near the star is much greater than the interstellar density, and the mass expands essentially into a vacuum. Analogy with terrestrial explosions suggests that a shock will begin to form when the material has swept over a distance about equal to the mean free path. If this were the correct criterion for material ejected from a supernova, a shock would never form, since the computed range is comparable with the dimensions of the Galaxy. Protons moving through neutral hydrogen atoms at a velocity of 2×10^9 cm/s, corresponding to an energy of 2 MeV, have a range of roughly 500 pc if the mean $n(\mathrm{H\ I})$ is 1.2 cm^{-3}; the mean travel time is about 10^{12} s. Evidently a conventional shock cannot be formed around a type I supernova.

However, this simple picture must be modified by several effects. According to equations (9-20) and (9-21) the charged particles will gyrate around the magnetic field with a radius of gyration of only 10^{11} cm, for protons with $w_\perp$ equal to 20,000 km/s in a field of 3×10^{-6} G. While the magnetic field has much too small an energy density to hinder the outward expansion, it provides a massless barrier between the supernova gas and the ionized interstellar material; a hydromagnetic shock will form, and move outward. Moreover, as the gas expands, the magnetic field carried along by the gas may weaken, in which case the kinetic energy of gyration around the lines of force will decrease adiabatically. Thus the mean energy of the particles initially in the envelope may decrease as the shock sweeps up more and more material.

The motion of fluid parallel to **B** is unhindered by the magnetic field. Thus one might expect that one-third of the kinetic energy would be unaffected by phenomena in the neighborhood of the supernova, and would appear in the form of energetic particles several hundred parsecs away. However, there are a number of additional mechanisms by which particles can lose energy even if their initial velocity is parallel to **B**. Penetration of the positive ions through the surrounding interstellar gas excites Alfvén waves [5], scattering the relativistic particles and keeping them within a few parsecs of the supernova shell.

It seems likely that much of the initial energy of the supernova shell is communicated to the interstellar medium through a hydromagnetic shock, moving across the interstellar magnetic field. To follow the evolution of this energy we shall adopt the approximation of a spherical shock and shall ignore the effect of the magnetic field on the jump conditions across the shock; despite the obvious physical shortcomings of this picture, the results which it yields may reproduce in a general way the physical effects associated with the actual situation. The remainder of the section is devoted to an analysis of this idealized spherical expansion of a supernova envelope.

Three different stages in the expansion may be distinguished. In the first stage the interstellar material has little effect because of its low density, and the velocity of expansion of the supernova envelope will remain nearly constant with time. This phase terminates when the mass of gas swept up by the outward moving shock is about equal to the initial mass M_s expelled from the supernova; that is, when

$$\frac{4\pi r_s^3 \rho_1}{3} = M_s, \tag{12-21}$$

where ρ_1 is the density of the gas in front of the shock, and r_s is the radius of the shock front. For M_s equal to 0.25 $M_\odot$ and ρ_1 equal to 2×10^{-24} g/cm^3, equation (12-21) will be satisfied when r_s equals 1.3 pc, which will occur about 60 years after the initial explosion.

During the second phase of expansion the mass behind the shock is determined primarily by the amount of interstellar gas swept up, but the energy of this gas will be constant; radiation during this period is unimportant. This phase is discussed immediately below. In the third phase, which is treated subsequently, radiation becomes dominant and the shock becomes effectively isothermal.

b. Intermediate Nonradiative Expansion

When the interstellar mass swept up by the outgoing shock exceeds the initial mass, the shock velocity will begin to drop. The temperature of the gas is so high at this stage that the radiation rate will be small and will be neglected here to give a first approximation. More specifically, the electrons will be heated initially to roughly $10^{7\circ}$ K on passage through a shock at 20,000 km s^{-1}, and interaction with the positive ions will raise the electron temperature even higher. Thus we may assume that the total energy of the gas within the shock front is constant, and equal to the initial energy E. This second stage is often referred to as the adiabatic or Sedov [6] phase.

While a detailed determination [6] of the motions in this phase is somewhat involved, an equation for the outward velocity of the shock can be computed readily from simple physical principles. Let a fraction K_1 of the total energy E be in heat energy, with the balance in kinetic energy, and let p_2, the pressure immediately behind the shock front, equal K_2 times the mean pressure of the heated gas within the spherical volume enclosed by the shock. If the material is assumed to be a perfect gas, with γ equal 5/3, then the mean pressure is two-thirds the mean density of thermal energy, giving

$$p_2 = K_2 \times \frac{2}{3} \times \frac{3K_1 E}{4\pi r_s^3} = \frac{KE}{2\pi r_s^3} \tag{12-22}$$

where K equals $K_1 K_2$, and r_s is again the radius of the shock. Since the Mach number $\mathfrak{M}$ is large, p_2 may be obtained from equation (10-23), giving immediately for the shock velocity,

$$u_1^2 = V_s^2 = \frac{2KE}{3\pi\rho_1 r_s^3}. \tag{12-23}$$

According to the detailed dynamical theory, a similarity solution exists for which the structure of the expanding shell is constant with time. From the exact computations for this solution, in the case where γ equals 5/3, the numerical values of K_1 and K_2 obtained are 0.72 and 2.13, respectively [7]; hence K equals 1.53.

The temperature immediately behind the shock front may be obtained from p_2/ρ_2, using the usual formula for a perfect gas. In accordance with equation (10-20), ρ_2 equals $4\rho_1$, the usual result for a strong shock with γ equal to 5/3, and with use again of equations (10-19) and (10-21) we obtain for T_2

$$T_2 = \frac{3\,\mu}{16k} V_s^2 = \frac{0.061\,\mu}{k} \frac{E}{\rho_1 r_s^3}, \tag{12-24}$$

where K has been replaced by its value 1.53, and where μ is the mean mass per particle, equal to 0.61 m_{H} for the helium-hydrogen ratio of 0.10 given in Table 1.1, if full equipartition of thermal energy is assumed between electrons and ions; such equipartition is certainly not valid at the outset, but should become so as the shock gradually slows down.

Equation (12-23) may be integrated at once, since V_s equals dr_s/dt and we have

$$r_s = \left(\frac{2.02E}{\rho_1}\right)^{1/5} t^{2/5} = \frac{0.26\left[t\text{ (years)}\right]^{2/5}}{n_{\mathrm{H}}^{1/5}} \text{ parsecs,} \tag{12-25}$$

where the numerical value has been computed for E equal to 4×10^{50} ergs, and ρ_1/n_{H} has been computed with $n_{\mathrm{He}}/n_{\mathrm{H}}$ again equal to 0.10. If we substitute this result in equation (12-24), we obtain

$$T_2 = \frac{3\,\mu}{100k}\left(\frac{2.02E}{\rho_1}\right)^{2/5} t^{-6/5} = \frac{1.5\times 10^{11}}{\left[t\text{(years)}\right]^{6/5} n_{\mathrm{H}}^{2/5}}\ {}^{\circ}\mathrm{K}. \tag{12-26}$$

At any time the temperature behind the shock front increases with decreasing r, since at earlier time, when the inner gas passed through the shock, T_2 was greater. According to the exact computations [7], T/T_2 varies very nearly as $(r/r_s)^{-4.3}$. The density ρ/ρ_2 decreases even more steeply than T_2/T as r decreases, since p/p_2 has a sharp drop right behind the shock, as shown in Table 12.1, computed for $\gamma=5/3$. As a result, half the mass of gas inside the expanding shock is in the outer 6.1 percent of the radius, and three-fourths in the outer 12.6 percent. The X-rays from supernova remnants originate mostly in this relatively thin layer; the analysis of the observations [S3.5a] is based on the theoretical results given above.

We have already seen that this similarity solution is valid only for t greater than some hundred years, when the swept-up mass much exceeds the mass of the original supernova envelope [see equation (12-21)]. In addition, these results are valid only as long as radiative cooling of the gas remains unimportant. When T falls below $10^{6}\,{}^{\circ}$K, corresponding to mean electron energies of about 100 eV, the abundant ions C, N, and O will begin to acquire bound electrons, and excitation of these will radiate

Table 12.1. Pressure Variation behind Adiabatic Supernova Shock Front

r/r_s	1.00	0.99	0.98	0.97	0.96	0.95	0.94
p/p_2	1.000	0.926	0.860	0.802	0.751	0.706	0.666
r/r_s	0.93	0.92	0.90	0.88	0.86	0.84	0.82
p/p_2	0.630	0.598	0.545	0.502	0.467	0.439	0.415
r/r_s	0.80	0.75	0.70	0.60	0.50	0.40	0
p/p_2	0.396	0.362	0.340	0.318	0.310	0.306	0.306

energy rapidly [8]. At these temperatures electrons and ions will be close to equipartition, since the equipartition time $t_s/2$ found from equation (2-7) becomes relatively short—less than $500/n_e$ years for protons. We may assume that radiation cooling becomes dominant for T about equal to 10^{6}°K, corresponding to t about 2×10^4 years, and to a radius of about 14 pc, again computed for n_H equal to unity. For other values of n_H [see Table 1.2], t and r_s both vary as $n_H^{-1/3}$.

c. Late Isothermal Expansion

When radiative cooling becomes important, the temperature of the gas will fall to a relatively low value. The outward progress of the shock is now maintained not primarily by the stored thermal energy, which is being radiated, but by the momentum of the outward moving gases. The shock itself may be regarded as isothermal. According to equation (10-24), the compression across the shock will be very large, and from equation (10-14) we see that the velocity u_2 of the gas away from the front is correspondingly very low. Material that has crossed the front will stay close to the front for a long time, forming a thin compressed shell of outward moving gas. The distance from the shock to this compressed shell will be proportional to the cooling time for gas that has been heated in the initial nonradiating shock (see Fig. 10.1).

The velocity of the shell during this phase may be computed from the condition that the outward momentum remains constant. This simple picture is sometimes referred to as the "snowplow model," since the interstellar gas is swept up by the expanding shock in somewhat the same manner as snow accumulates on the front of a snowplow.

If M_t and v_t are the mass and velocity of the shell when this model first becomes applicable, when the shell is in transition from the nonradiative to the radiative model, we have

$$Mv = M_t v_t. \qquad (12\text{-}27)$$

By the time equation (12-27) is applicable, the mass in the shell will be largely interstellar material swept up in an earlier phase. Hence M varies as r_s^3, where r_s is the shell radius, and an integration yields

$$t=\frac{\pi\rho_1}{3M_t v_t} r_s^4=\frac{M r_s}{4M_t v_t}. \qquad (12\text{-}28)$$

The retardation of the shell is now more rapid than before, with dr_s/dt varying as $1/r_s^3$, as compared with the $1/r_s^{3/2}$ variation given in equation (12-23).

The identified supernova remnants are generally smaller than the radii at which the snowplow model becomes applicable. However, if the energy required for maintaining interstellar motions comes from expanding supernova shells, this model may be used for computing ε_s, the efficiency with which supernova energy is converted into cloud motions. If a supernova shell finally produces a total mass M_c moving at the velocity v_c, then ε_s is given by

$$\varepsilon_s=\frac{M_c v_c^2}{2E}=\frac{M_c v_c^2}{M_t v_t^2}\times\frac{M_t v_t^2}{2E}, \qquad (12\text{-}29)$$

where M_t and v_t are again the mass and velocity at which the shock becomes isothermal instead of nonradiating. We have already seen that behind the nonradiating shock the kinetic energy is about one-fourth of the total energy. If we use equation (12-27) we obtain

$$\varepsilon_s=\frac{v_c}{4v_t}. \qquad (12\text{-}30)$$

To maximize ε_s we take a minimum value for v_t, and assume that the transition between the two models occurs when equation (12-24) predicts a temperature of 10^5°K. Then v_t equals 85 km s^{-1}, and if we set v_c equal to the rms cloud velocity of 10 km s^{-1} in Table 11.1, then

$$\varepsilon_s=0.029. \qquad (12\text{-}31)$$

The snowplow model is highly approximate, especially in its neglect of the heated gas far inside the expanding shell. As pointed out above, T/T_2 is least and ρ/ρ_2 greatest for the gas which has cooled by radiation a short distance behind the shock. The inner regions, which are too hot for appreciable radiative cooling, stay hot for a long time, their temperature dropping adiabatically as the entire heated sphere continues to expand.

The heated inner gas is at relatively high pressure and can continue to push on the outer cold shell, increasing somewhat its outward expansion.

d. Numerical Solutions

Detailed integrations [7, 9] of the basic equations have given information both on the details of the late phases in supernova shell expansion and on the rate of radiation throughout. The cold, neutral shell forms as expected, but its outward momentum increases somewhat with time. As a result, the shock radius r_s is proportional not to $t^{0.25}$, as predicted by equation (12-28), but rather to $t^{0.31}$, as compared with $t^{0.4}$ found from equation (12-25) for the nonradiative case. The transition between the adiabatic and the isothermal phases is accompanied by a number of rather complex dynamical phenomena which propagate inward and outward. Similar effects appear also at the transition between the first and second phases of the explosion [9], when the swept-up mass first approaches the mass of material ejected from the exploding star.

The efficiency with which the initial total energy E is converted into kinetic energy at a velocity of about 10 km s^{-1} depends on the relative importance of radiation and adiabatic expansion, which may be measured by the ratio of the expansion time r_s/V_s to the radiative cooling time t_T, proportional to $1/\rho$ [S6.2a]. If V_s and the resultant temperature behind the shock are fixed, the density ratio ρ/ρ_1 across the shock is also fixed; if r_s is eliminated with the use of equation (12-23), $r_s/V_s t_T$ is then proportional to $(n_{\text{H}}^2 E)^{1/3}$, where n_{H} is again evaluated in front of the shock. More generally, it may be shown from the equations of motion that solutions with constant $n_{\text{H}}^2 E$ may be derived from each other to a high approximation by simple scaling laws. With $E = 3 \times 10^{50}$ ergs and $n_{\text{H}} = 1$ cm^{-3}, the numerical computations show [7] that kinetic energy is 3 percent of E when $V_s = 10$ km s^{-1}. With the same E, decreasing n_{H} to 10^{-2} cm^{-3} more than doubles the kinetic energy, with a corresponding reduction if n_{H} is increased. These values agree about as well as might be expected with the crude estimate in equations (12-30) and (12-31).

Most of the harder X-rays, in the kilovolt range, are emitted in the adiabatic stage, before the cold shell forms. In the radiative stage the hot gases just behind the shock front and between the cold shell and the very hot interior both radiate copiously in the far ultraviolet, at energies between 13.6 and 50 eV [7, 9]. In this late phase, most of the initial E is radiated, with much of it appearing in visible wavelengths and, as a result of grain absorption, in the infrared as well.

The assumption of spherical symmetry in all these solutions is certainly not valid in detail. Even if the surrounding medium were uniform, the spherical flow is unstable after the cold shell develops. When the heated interior gas exerts pressure on the cold shell, increasing its outward momentum, a low-density fluid is accelerating a heavier one, and Rayleigh-Taylor instabilities must result [S10.3]. In addition, the flow in the isothermal shock front is thermally unstable because the cooling rate increases with increasing density. The resultant tendency to form clumps behind an isothermal shock front has been demonstrated both by a one-dimensional perturbation theory [10] and by two-dimensional numerical calculations [11]. Both in front and behind, the cold shell is likely to break up into smaller fragments, which because of the magnetic field will tend to be filamentary.

According to both the theory and calculations, the supernova shell becomes radiation dominated when r_s is about 15 pc, if n_H is 1 cm^{-3} in the undisturbed gas; the age is then about 4×10^4 years. Subsequently, the remnant expands slowly, the hot inner gas cooling adiabatically, with r_s reaching 40 pc and the pressure approaching the external gas pressure after about 10^6 years. If there is one such explosion every 25 years, the fraction of the galactic disk (assumed 250 pc thick and 10,000 pc in radius) occupied by such remnants is of order 15 percent. If most of the volume of the disk is occupied by an intercloud medium with $n_H \approx 0.1$ cm^{-3}, the supernova remnants radiate less effectively and reach larger radii. Since the supernova remnants will probably persist for substantially more than 10^6 years, it seems likely that an appreciable fraction of the galactic disk is occupied by the hot coronal gas in these remnants [12].

12.3 EFFECT OF EXPLOSIONS ON CLOUDS

When an expanding H II region or supernova shell passes by an existing interstellar cloud, the cloud will be strongly influenced. There will be some acceleration, some compression, and in some cases at least partial ionization. We discuss here the phenomena that may be expected on the basis of rather simple arguments. It is entirely likely that the end result will be the complete transformation of the cloud either by complete ionization or by turbulent disruption. We consider first the fate of an H I cloud subject to the ionizing radiation in an H II region, and next the interaction between a cloud and a passing shock front.

a. H I Cloud Engulfed by an H II Ionization Front

In regions of star formation where early-type stars are born, many H I clouds are likely to be present. Those that are well within the equilibrium radius r_S will experience a weak R-type ionization front passing quickly around them, after which the denser material in the cloud will be ionized gradually by ultraviolet photons. We let J_0 be the flux of such photons per second from the star at the ionization front when this first reaches the cloud. If the cloud is sufficiently close to the star so that absorption of the ionizing radiation by neutral hydrogen or by dust is negligible, then

$$J_0 = \frac{N_u}{4\pi r^2}, \tag{12-32}$$

where N_u is the stellar output of ionizing photons [S5.1b]. Ultraviolet photons in the diffuse radiation field will also reach the cloud, but their number is relatively less for clouds which are close to the central star and we ignore them. The H I cloud is assumed to be at rest initially with respect to the surrounding less dense medium.

We consider now the rate at which the ionization front moves into the cloud. The velocity of the front in the H I gas will be reduced by the increased density of the cloud, and also by absorption of ultraviolet photons in the recombining gas which streams away from the cloud. The ionized hydrogen which is produced in the cloud, on the side facing the star, will be at a much higher pressure than the gas outside the cloud, because of its greater density. As a result, this gas will tend to expand toward the star, forming a relatively dense cloud of ionized H between the cloud and the star. Electrons will recombine with protons in this region, and the resultant H atoms will be reionized by the ultraviolet photons. As a result of this absorption, the flux J of ultraviolet photons reaching the ionization front inside the cloud will be reduced below J_0.

We now compute the reduction of J/J_0 produced by absorption in this "insulating layer" [13]. Let n_i be the number density of H ions at the base of the insulating layer, just behind the ionization front that is eating its way into the H I cloud. The ionized gas will stream out behind the ionization front with a velocity u_i relative to the front; u_i depends on the characteristics of this front, and is likely to be comparable with the sound speed C_{II} in the H II region. Since the velocity of the ionization front relative to the dense neutral gas ahead is relatively small, u_i is nearly equal to the initial velocity of the ionized gas with respect to the cloud. The decreasing pressure away from the cloud will accelerate the ionized gas, producing a velocity increment of order C_{II}. To simplify the problem drastically we let $u_i = C_{\mathrm{II}}$, an exact result for a D-critical front [S12.1a], and ignore the additional outward acceleration. If we also assume spherical divergence of the expanding gases, equation (12-1) and the equation of continuity give

$$n_i(R) = \frac{J}{C_{\rm II}}\left(\frac{R_i}{R}\right)^2, \tag{12-33}$$

where R is the distance from the cloud center and R_i is the value of R at the ionization front.

The difference between J_0 and J is simply the number of recombinations in a column 1 cm^2 in cross section extending through the insulating layer toward the central star. Hence we have

$$J_0 - J = \int_{R_i}^{\infty} \alpha\left[n_i(R)\right]^2 dR = \frac{\alpha J^2 R_i}{3C_{\rm II}^2}, \tag{12-34}$$

where the integration has been carried out after substituting equation (12-33) for n_i. In equation (12-34) α will be intermediate between $\alpha^{(1)}$ and $\alpha^{(2)}$, since the amount of diffuse ultraviolet radiation which escapes from the cloud will vary with the thickness of the insulating layer. While equation (12-34) is strictly valid only along the line from the cloud center to the star, we shall obtain approximate results by applying this equation to the entire hemisphere illuminated by the central star.

Equation (12-34) is quadratic in J_0/J, yielding the solution

$$\frac{2J_0}{J} = 1 + \left\{1 + \frac{\alpha N_u R_i}{3\pi r^2 C_{\rm II}^2}\right\}^{1/2}, \tag{12-35}$$

where equation (12-32) has been used to eliminate J_0. A typical value of R_i is about 0.2 pc, and if this value is inserted into equation (12-35), with N_u taken from Table 5.3, α equated to 4×10^{-13} cm^3 s^{-1}, and $C_{\rm II}$ set equal to 10 km s^{-1}, J_0/J equals 8 for a cloud 9 pc from an O7 star. The corresponding value of n_i immediately behind the ionization front is about 100 cm^{-3}, and the density of neutral atoms in the cloud immediately ahead of the ionization front must exceed 3×10^4 cm^{-3} [see equation (12-7)]. A preceding shock presumably compresses the gas to this high density. At smaller values of r, n_i increases about as $1/r$.

One may reasonably identify the observed bright rims in H II regions [see Fig. 1.2] with the luminous insulating layers of ionized gas discussed here [14]. The observed rim thickness averages between 15 and 20 percent of the observed radius of curvature, which may be set equal to R_i; this result is consistent with a surface brightness varying as the inverse cube of the projected distance from the cloud center, for $R > R_i$, a result obtained when n_i^2 given by equation (12-33) is integrated along a line of sight. Moreover, a spectroscopic measurement of a bright condensation in M16, presumably an insulating layer seen nearly face on, gives a measured velocity of approach relative to the surrounding H II gases of 13 km s^{-1} [15]; the value of 10 km s^{-1} for $C_{\overline{\text{II}}}$ in Table 11.1 is evidently of this same order. The simplified theory presented here seems to give a reasonable fit to the observed data.

These luminous streams of ionized gas from H I clouds toward neighboring O-type stars will have two effects. First, they will accelerate the clouds by their recoil. Second, they will reduce the mass M of the H I cloud, perhaps even evaporating it completely. These two effects are related, since the momentum gained by the cloud per second from a backward stream of mean vector velocity $\mathbf{V}$ is VdM/dt; hence if the cloud velocity v is initially zero, at which time $M = M_0$, integration of the equation of motion gives the usual result for rocket acceleration,

$$M = M_0 e^{-v/V}. \tag{12-36}$$

The change of M per unit time is equal to the rate at which gas is ionized at the I front, and we have

$$\frac{dM}{dt} = -\pi R_i^2 J \mu_i, \tag{12-37}$$

where μ_i is the mean mass of gas per atom ionized, equal to 1.40 m_{H} if helium remains neutral.

To trace the evolutionary history of a cloud which finds itself within a few parsecs of a newly born O star requires additional assumptions, particularly on the change of projected area with mass. If we assume arbitrarily not only that the cloud holds together but also that its cross section remains constant with time, we find that there is an initial critical mass M_e which will just evaporate as the cloud remnant reaches the boundary of the H II zone. Clouds less massive than M_e initially will evaporate entirely within the H II region, whereas more massive clouds will survive, some of them escaping at appreciable velocities. Simultaneous solution of equations (12-36) and (12-37), with J expressed in terms of r by use of equation (12-35), approximated for $J \ll J_0$, indicates [16] that $M_e/\pi R_i^2$, the initial ratio of mass to projected area for a cloud which just evaporates at $r = r_S$, is proportional to $(N_u/R_i)^{1/2}\ln(r_S/r_0)$, where r_S/r_0 is the ratio of the Strömgren radius to the initial distance of the cloud from the star. For the value of N_u given in Table 5.3 for an O7 star, for

$r_S/r_0 = 10$, and for R_i between 0.2 and 10 pc, the value of $M_e/\pi R_i^2$ varies from 70 to 10 $M_\odot$ pc^{-2}, corresponding to a mean visual extinction, A_V, through the cloud of 3 to 0.4 mag (see Table 7.2). Evidently an average diffuse cloud well within the H II region is likely to evaporate entirely. The tendency for O stars to be surrounded by H II regions of appreciable n_e, in contrast to the low-density H II regions detected around some B stars, may result at least in part from the rapid evaporation of clouds near the much more luminous O stars [17]. The clumpiness of n_e observed in the Orion nebula, together with the increase of n_e toward the center, may both be explained by the evaporation of H I clouds in this relatively young system [18].

b. H I Cloud Engulfed by a Shock Front

When a shock front passes over a cloud, the pressure in the external medium surrounding the cloud is increased by this "external shock," and a "cloud shock" will start propagating inward from the surface of the cloud. In general, the cloud shock will radiate strongly and can be regarded as isothermal. Even if the external shock is a supernova shell in its adiabatic phase, the greater density of the cloud material will tend to decrease the temperature behind the shock into the range at which radiation is dominant.

The pressure which drives the isothermal cloud shock will be greatest on the front or upstream side of the cloud, which is first struck by the oncoming external shock. The gas streaming on behind the external shock, at a velocity v_{e2} and a density ρ_{e2}, will be nearly stopped at the cloud; the pressure at the stagnation point will be increased above the pressure immediately behind the external shock by about half an order of magnitude [19]. Hence the cloud shock will be strongest on the upstream side of the cloud, and the cloud as a whole will receive a net acceleration in the direction traveled by the shock. In addition, the much greater strength of the cloud shock in front will tend to flatten the cloud, distorting a sphere into a pancake perpendicular to the shock direction.

We consider now the velocity V_c of this isothermal cloud shock. Equations (10-6) and (10-24) give $\rho_{c1} V_c^2$ for p_{c2}, the pressure behind the cloud shock which in turn will be of the same order as the pressure behind the external shock, given by equation (10-23). Thus if we neglect factors of order unity, we have

$$\rho_{c1} V_c^2 = \rho_{e1} V_e^2, \tag{12-39}$$

where we have let ρ_{c1} denote the density in front of the cloud shock, whereas ρ_{e1} and V_e denote the density in the medium ahead of the external shock and the velocity of this shock. If ρ_{c1}/ρ_{e1} is of order 100, V_c/V_e will be about 0.1.

This general mechanism may be responsible [19] for the large spread of radial velocities observed in the Cas A supernova remnant, where some gas clouds have velocities about 6000 $km\,s^{-1}$, whereas for others the radial velocities average about 200 $km\,s^{-1}$. The intensity of the Hα and N II lines emitted from these low-velocity clouds indicates collisional excitation with n_H about equal to 10^5 cm^{-3}. The high-velocity clouds could be remnants from the original supernova explosion. The low-velocity emitting regions may be identified with denser H I clouds which have been engulfed in a strong supernova shock and both heated and somewhat accelerated by the resultant cloud shock. A detailed theoretical discussion of this problem would consider the compression of the cloud produced by shocks from the sides and back. Since a typical diffuse cloud has insufficient gravitational binding energy to hold it together [see Table 7.2], such a cloud is likely to be disrupted completely by the resultant field of flow, when this becomes turbulent because of various instabilities.

These theoretical results may be compared with numerical computations of an H I cloud subject to a passing external shock [20]; to make the problem two-dimensional, axial symmetry was assumed. The cloud was assumed initially to be a sphere 15 pc in radius with $n_H = 1.5$ cm^{-3}. The velocity V_e of the external nonradiating shock through the low-density medium surrounding the cloud was taken to be about 29 $km\,s^{-1}$ (corresponding to a Mach number $\mathfrak{M}$ of about 2.6), with v_{e2} equal to 18 $km\,s^{-1}$; the initial ratio of cloud–intercloud densities was 79. The final configuration of the cloud, 5.9×10^6 years after the shock first reached the front surface of the cloud, is shown in Fig. 12.1, which shows clearly the distortions produced by several instabilities. The Helmholtz instability [S10.3a] appears first, with the flow of intercloud gas past the denser material producing ripples. Subsequently, even greater distortions appear, with long tongues of the compressed cold gas penetrating the intercloud medium, perhaps resulting from instabilities of Rayleigh-Taylor type [S10.3a]. The blank region within the cloud represents gas as yet undisturbed. Evidently the surrounding compressed regions are highly irregular, and the cloud as a whole presumably breaks into separate condensations, some of which may be sufficiently dense to undergo gravitational collapse [S13.3b]. The thermal instabilities within an isothermal shock [S12.2d], which do not appear in these numerical calculations, might be expected to produce further irregularities on smaller scales.

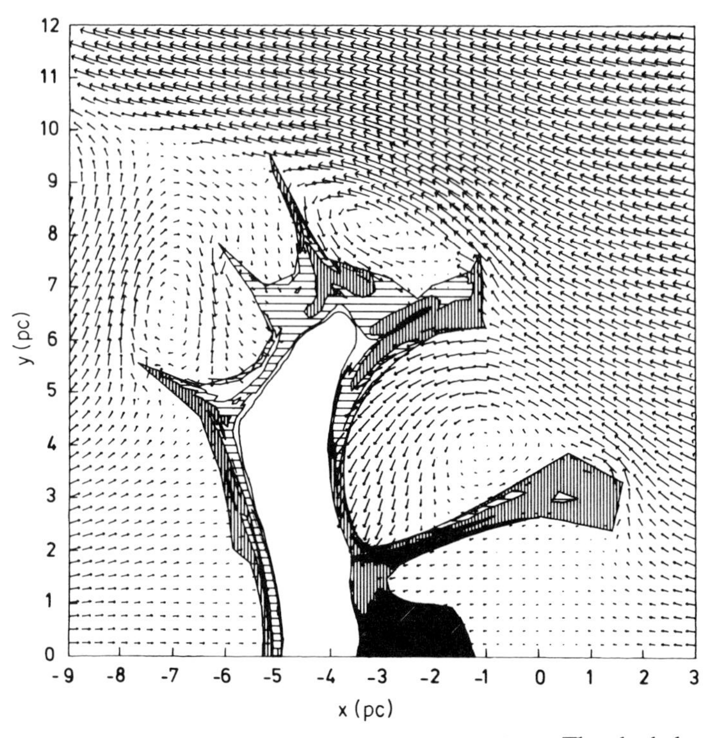

Figure 12.1 H I cloud compressed by a passing shock [20]. The shaded areas of this map, with x and y in parsecs, represent density contours of 20, 40, 100, and 200 H atoms cm^{-3} at a time 5.9×10^6 years after the shock first reaches the cloud, initially a uniform sphere of radius 15 pc, centered at $x = y = 0$. Arrows represent the direction and magnitude of the velocity relative to the gas in front of the shock. Rotational symmetry about the x axis is assumed throughout.

REFERENCES

1. W. G. Mathews and C. R. O'Dell, *Ann. Rev. Astron. Astroph.,* **7**, 67, 1969.
2. P. O. Vandervoort, *Ap. J.,* **139**, 889, 1964.
3. D. C. V. Mallik, *Ap. J.,* **197**, 355, 1975.
4. B. M. Lasker, *Ap. J.,* **149**, 23, 1967.
5. D. G. Wentzel, *Ann. Rev. Astron. Astroph.,* **12**, 71, 1974.
6. L. I. Sedov, *Similarity and Dimensional Methods in Mechanics* (translated from the Russion edition by M. Friedman) Academic Press (New York), 1959.
7. R. A. Chevalier, *Ap. J.,* **188**, 501, 1974.
8. D. P. Cox and E. Daltabuit, *Ap. J.,* **167**, 113, 1971.
9. V. N. Mansfield and E. E. Salpeter, *Ap. J.,* **190**, 305, 1974.
10. R. McCray, R. F. Stein, and M. Kafatos, *Ap. J.,* **196**, 565, 1975.

11. R. A. Chevalier and J. C. Theys, *Ap. J.,* **195**, 53, 1975.
12. D. P. Cox and B. W. Smith, *Ap. J.* (*Lett.*), **189**, L105, 1974.
13. J. H. Oort and L. Spitzer, *Ap. J.,* **121**, 6, 1955.
14. S. R. Pottasch, *B.A.N.,* **14**, 29, 1958; No. 482.
15. G. Courtès, P. Cruvellier, and S. R. Pottasch, *Ann. d'Astroph.,* **25**, 214, 1962.
16. L. Spitzer, *Diffuse Matter in Space,* 1966, Wiley-Interscience (New York), Section 5.5, p. 206.
17. B. Elmergreen, *Ap. J.,* **205**, 405, 1976.
18. J. E. Dyson, *Astroph. Space Sci.,* **1**, 388, 1968.
19. C. F. McKee and L. L. Cowie, *Ap. J.,* **195**, 715, 1975.
20. P. R. Woodward, *Ap. J.,* **207**, 484, 1976.

13. Gravitational Motion

The influence of gravitational fields on the motion of interstellar gas is treated here for three idealized situations. First, the motion in the neighborhood of a mass concentration, either a star or some more massive object, will be considered. Second, motions in the galactic plane under the influence of large-scale density waves, such as are presumably associated with spiral arms, will be discussed. Third, a brief analysis will be given of certain processes occurring when gravitational instability sets in, triggering the complex series of events leading to star formation.

13.1 ACCRETION

A massive object surrounded by an extended gas cloud will tend to accrete material. The resultant rate of mass increase can be of importance for stars in close binary systems, for supermassive objects which have been proposed as energy sources for quasars, and for galaxies moving through intergalactic matter. The accretion rate dM/dt is given by different equations under different physical conditions.

a. Uniform Streaming of a Cold Gas

The simplest situation is that in which the gas is cold and pressure forces are negligible. First we simplify the problem even further and assume that the mean free path is effectively infinite. If the accreting mass M is a sphere of radius R, and if we consider particles approaching with a velocity V at infinity, we may define p_c as the critical value of the impact parameter p [S2.1] such that the particle just grazes the surface of the sphere. If $v_\theta(R)$ denotes the velocity at this grazing collision, conservation of angular momentum and energy gives

$$p_c V = R v_\theta(R), \tag{13-1}$$

$$\tfrac{1}{2}V^2 = \tfrac{1}{2}\left[v_\theta(R)\right]^2 - \frac{GM}{R}. \tag{13-2}$$

The collisional cross section is πp_c^2, and eliminating $v_\theta(R)$ yields

$$\frac{dM}{dt} = \pi R^2 \rho_1 V\left(1 + \frac{2GM}{RV^2}\right), \tag{13-3}$$

where ρ_1 is the density at infinity. If, as is generally the case, $2GM/R$ much exceeds V^2, dM/dt varies as R/V.

A much larger accretion rate is obtained if the mean free path in the cold gas is taken to be very short. Then streams of gas will be deflected around the central mass, and collide inelastically with each other on the far side, on the axis of cylindrical symmetry, or "accretion axis," which is parallel to the approach velocity $\mathbf{V}$. As a result of such collisions the transverse velocity v_θ will be eliminated, with only the radial velocity v_r conserved. The condition that at the distance r_a the radial velocity is just equal to the escape velocity yields

$$\tfrac{1}{2}\left[v_r(r_a)\right]^2 = \frac{GM}{r_a}. \tag{13-4}$$

To obtain a simple result we ignore mutual interaction of fluid elements after they have reached the accretion axis, and conclude that the gas reaching this axis at $r < r_a$ is accreted, with other gas all escaping. Combining equation (13-4) with the equation for the total energy, just before collisions of the intersecting streams, gives the result that $v_\theta(r_a) = V$, from which we obtain, using the constancy of angular momentum [see equation (13-1)],

$$r_a = p_c, \tag{13-5}$$

where p_c is now the critical impact parameter for the fluid elements which are just accreted.

To evaluate p_c we use the familiar equation for the orbit under inverse-square forces

$$r = \frac{p^2 V^2}{GM}\,\frac{1}{1 + \varepsilon\cos\theta}, \tag{13-6}$$

where θ is the orbital angle measured from the direction of pericenter. The incoming streams approach initially from the direction for which r is infinite and $\cos\theta = -1/\varepsilon$. Hence on the accretion axis $\cos\theta = 1/\varepsilon$, and equations (13-5) and (13-6) combine to give

$$p_c = \frac{2GM}{V^2}, \tag{13-7}$$

and the accretion rate becomes

$$\frac{dM}{dt} = \frac{4\pi(GM)^2\rho_1}{V^3}. \tag{13-8}$$

Evidently dM/dt found from equation (13-8) exceeds that obtained from equation (13-3) by the factor $2GM/RV^2$ when this quantity is large.

In an actual gas, the collisions of the intersecting streams will not only modify the radial velocity along the accretion axis but will also increase the temperature and produce a pressure which modifies the fluid flow. We give below the relatively simple analytical solution for the accretion rate resulting when the initial streaming velocity V is zero [1]. Numerical solutions are then presented for the more general case of finite V.

b. Spherical Adiabatic Inflow

We consider a steady state in which $\mathbf{v}$ is radial, and $\mathbf{v}$, p, and ρ are functions of r only. As in equation (10-6) we assume

$$\frac{p}{p_1} = \left(\frac{\rho}{\rho_1}\right)^{\gamma}, \tag{13-9}$$

where p_1 and ρ_1 are evaluated at infinite r. This equation is valid for adiabatic flow, and with use of an effective value of γ [S10.1] may provide a useful approximation for a radiating gas. If we take v to be positive for infall, conservation of mass [equation (10-2)] yields

$$4\pi r^2 \rho v = \frac{dM}{dt}. \tag{13-10}$$

Equation (10-1) together with equation (13-9) may also be integrated, giving Bernoulli's equation in the form

$$\frac{v^2}{2} + \frac{C_1^2}{\gamma-1}\left[\left(\frac{\rho}{\rho_1}\right)^{\gamma-1} - 1\right] = -\phi(r) = \frac{GM}{r}, \tag{13-11}$$

where C_1 is the value at infinity of the sound speed C introduced in equation (10-6). If we use equation (13-9) again, C in general is given by the relation

$$C^2 = \gamma \frac{p_1}{\rho_1}\left(\frac{\rho}{\rho_1}\right)^{\gamma-1} = C_1^2\left(\frac{\rho}{\rho_1}\right)^{\gamma-1}. \tag{13-12}$$

The gravitational potential for a point mass M has been introduced in equation (13-11).

To obtain a single equation relating v and r we eliminate ρ from equation (13-11) with the use of equation (13-10). For simplification of the resulting expression we substitute for v the Mach number $\mathfrak{M}$, defined in equation (10-21) by the ratio of v to the local sound speed C. We define the constant λ by the expression, analogous to equation (13-8),

$$\frac{dM}{dt} = \frac{4\pi\lambda(GM)^2\rho_1}{C_1^3}, \tag{13-13}$$

and introduce a dimensionless radius ξ by the relationship

$$r = \xi \frac{GM}{C_1^2}. \tag{13-14}$$

After some algebra the transformed version of equation (13-11) becomes [1]

$$\begin{aligned} F(\mathfrak{M}) &\equiv \frac{\mathfrak{M}^{2\alpha/(\gamma-1)}}{2} + \frac{1}{(\gamma-1)\mathfrak{M}^{\alpha}} \\ &= \frac{1}{\lambda^{\alpha}}\left[\frac{\xi^{2\alpha}}{\gamma-1} + \frac{1}{\xi^{1-2\alpha}}\right] \equiv \frac{1}{\lambda^{\alpha}} G(\xi), \end{aligned} \tag{13-15}$$

where

$$\alpha \equiv \frac{2(\gamma-1)}{\gamma+1}. \tag{13-16}$$

Equation (13-15) determines the Mach number $\mathfrak{M}$ for each value of ξ from infinity down to zero (or some very small value near a stellar surface, for example). We now show that a realistic solution is possible only for one particular value of λ, which we denote by λ_c. For $1<\gamma<5/3$ [corresponding from equation (13-16) to $0<\alpha<\frac{1}{2}$], the functions $G(\xi)$ and $F(\mathfrak{M})$ are each the sum of one positive and one negative power of ξ or $\mathfrak{M}$, respectively, and must each possess a minimum value. It is readily seen that $F(\mathfrak{M})$ has its minimum value, equal to $1/\alpha$, at $\mathfrak{M}=1$; we denote by ξ_m the value of ξ at which $G(\xi)$ has its minimum. The quantity λ_c^α is defined as the ratio of these two minimum values, giving

$$\lambda_c^\alpha = \frac{G(\xi_m)}{F(1)} = \alpha G(\xi_m). \tag{13-17}$$

We now investigate the variation of $\mathfrak{M}$ with ξ and the resultant nature of the fluid inflow for different values of λ/λ_c. If λ exceeds λ_c, then the minimum value of $G(\xi)/\lambda^\alpha$ on the right-hand side of equation (13-15) is reduced below $F(1)$, the minimum value of $F(\xi)$ on the left-hand side. Hence for ξ in the neighborhood of ξ_m equation (13-15) cannot be satisfied for any $\mathfrak{M}$. Values of $\mathfrak{M}$ for such a case are shown by the dashed lines in Fig. 13.1. Since $\mathfrak{M}$ must be defined for all ξ, it follows that such values of λ are excluded. On the other hand, if λ is less than λ_c, the minimum value of $G(\xi)/\lambda^\alpha$ now exceeds $F(1)$, and for $\mathfrak{M}$ in the neighborhood of unity equation (13-15) cannot be satisfied for any ξ. Solutions are now possible, but $\mathfrak{M}$ near unity is excluded and the velocity must be everywhere subsonic ($\mathfrak{M}<1$) or everywhere supersonic ($\mathfrak{M}>1$), as shown by the dotted lines in Fig. 13.1. In the present problem the fluid velocity v must vanish at infinity, hence solutions with $\lambda<\lambda_c$ must be everywhere subsonic. In such solutions, $\mathfrak{M}$ rises to a maximum value at $\xi=\xi_m$, and then decreases toward zero as the accreting mass is approached. Near a star such configurations correspond to an adiabatic stellar atmosphere in nearly hydrostatic equilibrium, with a small inflow velocity and large pressures and densities. We shall not consider such solutions here, although to exclude them conclusively would require a detailed analysis of

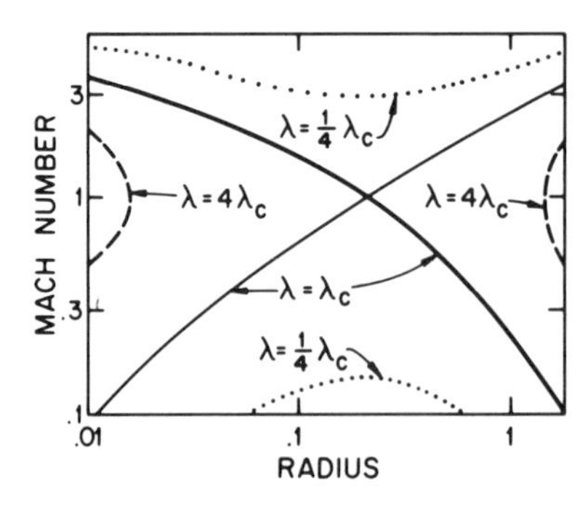

Figure 13.1 Velocity field for spherical accretion [1]. The Mach number $\mathfrak{M}$ is plotted against the radius ξ in units of C^2/GM, for three values of the dimensionless constant λ, determining the accretion rate [equation (13-13)]. The constant γ in equation (13-9) has been set equal to 1.4. The intersecting solid lines represent possible solutions for $\lambda=\lambda_c$; the heavier line satisfies the boundary conditions at $\xi=0$ and ∞.

boundary conditions at small $\mathfrak{M}$ and perhaps also of the stability of any overall solution obtained.

In the special case $\lambda=\lambda_c$, $\mathfrak{M}=1$ at $\xi=\xi_m$, and a solution is possible which is subsonic at large ξ, and supersonic at $\xi<\xi_m$, see Fig. 13.1. In such solutions, $\mathfrak{M}$ approaches infinity as ξ approaches zero, and the infall velocity v approaches the free-fall velocity $(2GM/r)^{1/2}$. The radius at which $\mathfrak{M}=1$ is sometimes called the "sonic point" and plays an important role in these flows (see also Section 13.2). If the accreting mass is a star, the inward supersonic flow will presumably produce a strong shock somewhere above the stellar surface. This type of solution is the one of chief physical interest and we therefore take λ to equal λ_c. Values of λ_c computed [1] from equations (13-15) and (13-17) are given in Table 13.1.

Table 13.1. Values of Accretion Parameter λ_c

γ	1.0	1.2	1.4	1.5	5/3
λ_c	1.120	0.871	0.625	0.500	0.250

The case in which $\gamma=5/3$ differs from the others in that ξ_m is now zero. No supersonic flow is possible under these conditions; we assume that the flow for which $\mathfrak{M}$ approaches unity at small ξ, corresponding to $\lambda=\lambda_c$, is the most likely situation physically.

When an H II region surrounds the star, with an H I zone outside the ionization front, the accretion rate is roughly given by equation (13-13) with λ taken from Table 13.1, but with ρ_1 and C_1 set equal to their values within the H II region rather than at infinity [2]. Accretion by normal stars after their formation is not likely to be very important. However, inflow of matter from one star into a binary component is believed to be important in X-ray sources. Moreover, the above analysis is similar to the corresponding analyses for outgoing stellar or solar winds [3].

c. Uniform Adiabatic Streaming

We turn now to the situation where pressure is considered, but where the gas has an initial streaming velocity V at infinity. In this case, solutions can be obtained by detailed numerical integrations of the equations of motion for specific values of V/C_1, which we denote by $\mathfrak{M}_1$. Such computations have been carried out [4] for $\mathfrak{M}_1$ equal to 0.6, 1.4, and 2.4, with γ set equal to 5/3; axial symmetry was assumed, with v_ϕ, the velocity component about the symmetry axis, set equal to zero. The full time-dependent equations were integrated over time until a steady state was obtained. For the two supersonic cases, bow shocks appeared, as shown in Fig. 13.2 for $\mathfrak{M}_1=2.4$. In all three cases, stagnation points appeared on the

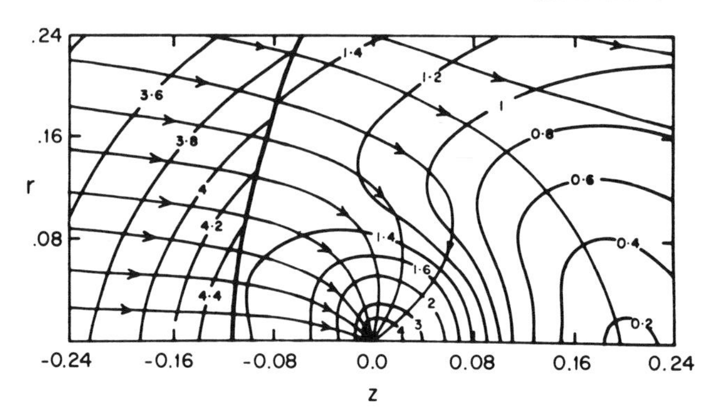

Figure 13.2 Streamlines and velocity magnitude contours for accretion flow [4]. The lines with arrows are streamlines for axially symmetrical flow into a point mass when the Mach number $\mathfrak{M}$ at infinity is 2.4 and $\gamma=5/3$. The heavy solid line shows the position of the shock front. The other curves indicate values of v/C_1, where v is the fluid velocity and C_1, the sound speed at infinity. The values of the axial and radial coordinates z and r are in units of GM/C_1^2.

accretion axis, and all the fluid with smaller impact parameters streamed into the star.

As pointed out above, no supersonic solutions exist for the spherical accretion problem if $\gamma=5/3$, although for $\lambda=\lambda_c$, $\mathfrak{M}$ approaches unity at small ξ; this solution for $\lambda=\lambda_c$ gives the least value of the pressure p for low ξ. In the numerical computations the boundary conditions at small ξ were arbitrarily chosen to dispose of the inflowing fluid in a nonphysical way, and gave a flow pattern which for small ξ is in close agreement with the spherical case for $\lambda=\lambda_c$. Thus the question of which value of λ gives the most realistic solution, for the case $\gamma=5/3$, remains unanswered. Also there is serious question concerning possible instability of these converging flows against three-dimensional irregularities.

For values of $\mathfrak{M}$ equal to 0.6 and 1.4, the accretion rates given by these numerical computations are within about 10 percent of the values obtained from equation (13-13) with $\lambda=\lambda_c=0.25$. For $\mathfrak{M}=2.4$, the accretion rate is less by about 13 percent than is predicted by equation (13-8). Apparently within about this accuracy dM/dt is given by equation (13-8) or (13-13), whichever gives the smaller value. Less accurate computations for $\gamma=1$ [5] are consistent with this result.

13.2 SPIRAL DENSITY WAVES

A plausible explanation for the spiral arms observed in many flattened galaxies is that these are density waves, which have a pattern rotating like a rigid body with an angular "pattern speed" Ω_p [6]. Throughout the regions where these waves exist, Ω_p is assumed less than Ω_c, where $r\Omega_c(r)$ is the "circular velocity" at which centrifugal force balances the mean gravitational force of the galaxy at the distance r from the center. Thus the stars, moving around the galactic center at about the circular velocity, will move through the pattern of density waves. The gravitational potential produced by the concentration of stars in the arms will maintain the excess stellar concentration there, although in time the waves may decay because of various dissipative processes.

While the existence of spiral density waves, as normal modes of a flattened galactic stellar disk, has been demonstrated theoretically, the conditions for excitation and maintenance of such dynamical perturbations are not yet fully explored. Furthermore, in regions where the spiral pattern is confused, as in the outer regions of M31 and perhaps in our own galaxy also, the situation may well be more complicated than has been considered in dynamical theories. Nevertheless, in galaxies such as M51 which show a simple two-arm structure (see Fig. 1.5), the spiral density wave theory may well be applicable.

It is therefore of interest to investigate the motions of the interstellar gas in such a spiral density wave. As we shall see below, it appears that large-scale shock fronts may appear near the inner edges of the spiral arms, accounting, perhaps, for at least some of the excess gas density observed in the arms, and for the concentration there of bright, young early-type stars which have presumably formed from this gas. We derive below the relevant equations for this situation, and present some of the results obtained. We assume that the spiral wave pattern is fixed by stellar motions, and that the resultant gravitational potential ϕ is given. Since the total mass of the interstellar matter in the solar neighborhood is a small fraction of the total mass, this assumption is likely to be realistic, though the dissipation of energy by the interstellar gas, moving through a spiral gravitational field, may have important effects on the overall level of these density waves.

a. Equations for Gas Motion in a Spiral Disk

We consider a thin disk, in which the gravitational potential is the sum of an axisymmetric term, $\phi_0(r)$, and a spiral term, $\phi_1(\mathbf{r})$. Since the spiral gravitational field is fixed in a reference frame rotating with a pattern speed Ω_p, we derive the equations for the gas velocity components, v_r and v_θ, measured in this frame. Equation (10-1) will be valid in this frame if we add $\mathbf{r}\Omega_p^2 - 2\mathbf{\Omega}_p \times \mathbf{v}$ to the right-hand side, where $\mathbf{r}$ is the vector distance from the axis of symmetry. Since we are interested in a steady state, we set the time derivatives equal to zero. If we introduce $v_{\theta 0}$, the circular velocity in the assumed frame of reference, equal to

$$v_{\theta 0}=r\left[\Omega_c-\Omega_p\right]=r\left[\left(\frac{\partial \phi_0}{r\partial r}\right)^{1/2}-\Omega_p\right], \tag{13-18}$$

and express v_θ as $v_{\theta 0}+v_{\theta 1}$, we obtain

$$v_r\frac{\partial v_r}{\partial r}+\frac{v_\theta}{r}\frac{\partial v_r}{\partial \theta}-\frac{v_{\theta 1}^2}{r}=-\frac{C^2}{\rho}\frac{\partial \rho}{\partial r}-\frac{\partial \phi_1}{\partial r}+2\Omega_c v_{\theta 1}, \tag{13-19}$$

$$v_r\left(\frac{\partial v_{\theta 1}}{\partial r}+\frac{v_{\theta 1}}{r}\right)+\frac{v_\theta}{r}\frac{\partial v_{\theta 1}}{\partial \theta}=-\frac{C^2}{\rho r}\frac{\partial \rho}{\partial \theta}-\frac{\partial \phi_1}{r\partial \theta}-\frac{\kappa^2}{2\Omega_c}v_r. \tag{13-20}$$

Equation (10-7) has been used to eliminate p. The various terms in $\Omega_p v_\theta$ and $\Omega_p v_r$ have cancelled out from these equations, and a term in Ω_c^2 has been cancelled by the $\partial\phi_0/r\partial r$ term. The quantity κ is the epicyclic frequency, given by

$$\kappa^2=\frac{2\Omega_c}{r}\frac{d}{dr}\left(r^2\Omega_c\right); \tag{13-21}$$

κ decreases from $2\Omega_c$ for solid-body rotation to Ω_c for Keplerian revolution. The velocity v_z perpendicular to the disk and the variation of all quantities with z have been ignored. These two equations, together with equation (10-2) of continuity, are to be solved for v_r, $v_{\theta 1}$, and ρ.

Since we are interested in the appearance of shock fronts, we cannot assume that $v_{\theta 1}$ and v_r are both small relative to $v_{\theta 0}$ or C. Instead we adopt a different approximation, based on the picture that the inclination i of the spiral arms to circles of constant radius is small, and that the spiral arms are closely spaced in r. We then define a system of orthogonal curvilinear coordinates, η, ξ, at the radius r, aligned with the spiral arms and defined by

$$\begin{aligned} r d\eta &= dr\cos i + r d\theta \sin i,\\ r d\xi &= -dr\sin i + r d\theta\cos i. \end{aligned} \tag{13-22}$$

Along a curve of constant η it is readily seen that r is proportional to $\exp(-\theta\tan i)$, giving a logarithmic spiral. Since i is assumed small, $d\eta$, transverse to the local spiral arm, is nearly in the same direction as dr. Since the arms are assumed closely spaced, the η derivatives are much larger than all other terms, except for the two Coriolis force terms, $2\Omega_c v_{\theta 1}$ in equation (13-19) and $-\kappa^2 v_r/2\Omega_c$ in equation (13-20). Neglecting all the smaller terms, and eliminating $\partial\rho/\partial\eta$ by use of the equation of continuity, we obtain [7]

$$\frac{\partial v_{\eta 1}}{r\partial\eta}=\frac{1}{v_\eta\left(1-C^2/v_\eta^2\right)}\left[-\frac{\partial\phi_1}{r\partial\eta}+2\Omega_c v_{\xi 1}\right], \tag{13-23}$$

$$\frac{\partial v_{\xi 1}}{r\partial\eta}=-\frac{v_{\eta 1}}{v_\eta}\frac{\kappa^2}{2\Omega_c}. \tag{13-24}$$

Here v_η is the sum of $v_{\eta 0}+v_{\eta 1}$, and similarly for v_ξ, where $v_{\eta 0}$ and $v_{\xi 0}$ are the components of $v_{\theta 0}$ in the η and ξ directions, respectively, giving

$$v_{\eta 0}=v_{\theta 0}\sin i, \qquad v_{\xi 0}=v_{\theta 0}\cos i; \tag{13-25}$$

$v_{\theta 0}$ is given in equation (13-18).

Let us consider now the limiting form of these equations when $v_{\eta 1}$, $v_{\xi 1}$, and ϕ_1 are small, and v_η in the denominators may be regarded as constant and equal to $v_{\eta 0}$. Differentiating equation (13-23), with neglect of $\partial\Omega_c/\partial\eta$, and eliminating $\partial v_{\xi 1}/\partial\eta$ with equation (13-24) yields

$$\left(v_{\eta 0}^2-C^2\right)\frac{\partial}{r\partial\eta}\left(\frac{\partial v_{\eta 1}}{r\partial\eta}\right)+\kappa^2 v_{\eta 1}=-v_{\eta 0}\frac{\partial}{r\partial\eta}\left(\frac{\partial\phi_1}{r\partial\eta}\right). \tag{13-26}$$

In a two-arm spiral, η increases by $\pi\sin i$ from one spiral arm to the next [for $dr=0$, $d\theta=\pi$ in equation (13-22)]; hence if sinusoidal variations between arms are assumed, ϕ_1 and $v_{\eta 1}$ will vary as $\sin(2\pi\eta/\pi\sin i)$ or about as $\sin(2\eta/i)$ when i is small. In this same limit $(\partial/\partial\eta\times\partial v_{\eta 1}/r\partial\eta)$ equals $\partial^2 v_{\eta 1}/r\partial\eta^2$ and similarly for the derivative of ϕ_1. If we use equation (13-25), then equation (13-26) gives

$$v_{\eta 1}\left[-\left(\frac{2v_{\theta 0}}{r}\right)^2\left(1-\frac{C^2}{v_{\eta 0}^2}\right)+\kappa^2\right]=-v_{\eta 0}\frac{\partial^2\phi_1}{r^2\partial\eta^2}. \tag{13-27}$$

Since $\Omega_c-\Omega_p$ is believed to be less than $\kappa/2$ in the region where density waves are possible [6], $v_{\theta 0}$ is less than $r\kappa/2$ [see equation (13-18)]. As a result, the second term in brackets in equation (13-27) will exceed the first if $v_{\eta 0}$ exceeds C and the flow is supersonic. In the absence of Coriolis forces, only the first term would be present, and for supersonic flow the acceleration toward the potential minimum in the spiral arms would decrease the gas density there. In fact, the second term, resulting from Coriolis forces, exceeds the first and has opposite sign with the result that for small $v_{\eta 1}$ and supersonic flow the gas density is increased in the potential minimum of the spiral arms.

b. Occurrence of Shock Fronts

The flow pattern for finite $v_{\eta 1}$ which results from equations (13-23) and (13-24) has been investigated numerically, assuming that ϕ_1 varies as $\sin(2\eta/i)$ [8]. Typical results are shown in Fig. 13.3, computed for parameters believed to be representative of the solar neighborhood ($r=10$ kpc, $r\Omega_c=250$ km s^{-1}, $r\kappa=313$ km s^{-1}, $r\Omega_p=135$ km s^{-1}, $i=6.7°$, and $C=8.6$ km s^{-1}). Values of v_η are plotted as functions of θ (for a fixed r). The minimum of the spiral potential ϕ_1 is taken to be at $\theta=90°$. Different curves are plotted for various values of F, the ratio of the spiral field $\partial\phi_1/\partial r$ to the axisymmetric field $\partial\phi_0/\partial r$. For the lowest values of F, v_η changes rather little with θ and equation (13-27) is applicable, but for F exceeding 0.8 percent, shocks appear, moving to lower θ (upstream), and then with increasing F moving back toward the center of the spiral arm. The velocity component v_ξ averages about 115 km s^{-1}, and changes with θ by less than 10 km s^{-1} for F less than 5 percent.

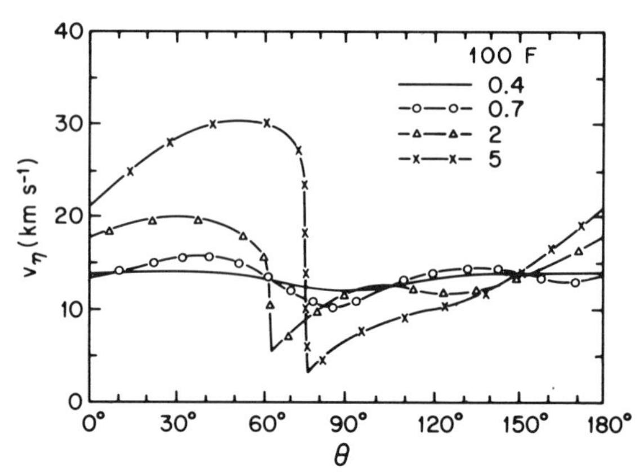

Figure 13.3 Steady gas flows in spiral density waves. The curves [8] show v_η, the component of **v** normal to the spiral wave pattern. The average value $v_{\eta 0}=13$ km s^{-1} results from the difference between the angular rotational velocities of galactic rotation Ω_c and of spiral pattern rotation Ω_p. The additional component $v_{\eta 1}$ is different for the different indicated values of F, the ratio of the spiral gravitational field to the corresponding axisymmetric galactic field.

For this periodic problem, shocks appear whenever "sonic points" are present; that is, when $v_{\eta}^2 = C^2$, and the flow changes from sonic to supersonic or vice versa. In general, transitions from sonic to supersonic flow need not give rise to shocks, as illustrated by the results for spherically symmetric accretion in the previous section. However, the converse process normally produces shocks in practice, presumably as a result of irreversible processes [8]. The results plotted in Fig. 13.3 were obtained from a solution of the time-dependent equations, and shocks occurred naturally whenever sonic points appeared. Other computations [7] have taken into account a number of terms which were ignored in the derivation of equations (13-23) and (13-24) and have shown that these approximations do not greatly affect the results. Solutions of the type shown in Fig. 13.3 have also been obtained [9] from the steady-state equations, using non-linear analytical formulae for low amplitudes and numerical calculations when shocks are present.

The increase of magnetic field strength in the shock produced by a spiral density wave may be expected to enhance synchrotron radiation from the region of the spiral arms, thus explaining, perhaps, the relatively intense radio emission from the spiral arms of M51 (see Fig. 1.5). Since the shock is displaced toward the upstream, or inner, edge of the spiral arm, the concentration of the radio emission toward the inner edge of the M51 arm finds a ready explanation on the basis of this picture.

The theory has been extended [10] to include the possibility that the interstellar medium may possess two phases—a warm phase and a cold phase—which can exist simultaneously for certain pressures [S6.2b]. On this picture, the warm intercloud medium is compressed behind the shock and condenses into cold clouds, triggering star formation on a large scale. Because of the time delay between cloud condensation and the appearance of bright stars, one would expect the early-type stars to lie some distance behind the shock. However, if much of the galactic disk is filled with coronal gas at high T [S11.1], with $C \gg v_{\eta 0}$, spiral density waves in the gas may be unimportant.

13.3 GRAVITATIONAL CONDENSATION AND STAR FORMATION

The high luminosities of early-type stars require that these stars have formed recently, and the concentration of these objects to spiral arms (Fig. 1.4), where dust and gas are also concentrated, indicates that new stars form from interstellar matter [11]. Moreover, the observations suggest that new stars, especially those of relatively high mass, are formed in groups (Fig. 1.2), perhaps as a result of the minimum mass for gravitational collapse found in equations (11-28) and (11-32). However the formation of low-mass stars singly from isolated individual clouds cannot be excluded observationally.

The detailed analysis of star formation is a complex topic, as well as a somewhat uncertain one. Here we discuss only a few of the dominant physical processes, in particular gravitational instability of a system in equilibrium, gravitational collapse of an isolated cloud, and a number of processes important in late stages of star formation, including fragmentation, and reduction of angular momentum and magnetic flux.

a. Gravitational Instability

A gas in pressure equilibrium may be unstable against the growth of pressure disturbances as a result of self-gravitational attraction of the denser regions in the gas. Such an instability may be analyzed [S10.3], with the linearized equations. If we use a subscript 0 for the equilibrium value of p, ρ, and v, and a subscript 1 for the perturbed quantities, the linearized form of equations (10-1) and (10-2) for the perturbed quantities becomes

$$\frac{\partial \mathbf{v}_1}{\partial t} + \mathbf{v}_0 \cdot \nabla \mathbf{v}_1 + \mathbf{v}_1 \cdot \nabla \mathbf{v}_0 = -\nabla \phi_1 - C^2 \nabla \left(\frac{\rho_1}{\rho_0} \right), \tag{13-28}$$

$$\frac{\partial \rho_1}{\partial t} + \mathbf{v}_0 \cdot \nabla \rho_1 + \mathbf{v}_1 \cdot \nabla \rho_0 = -\rho_1 \nabla \cdot \mathbf{v}_0 - \rho_0 \nabla \cdot \mathbf{v}_1. \tag{13-29}$$

We have assumed here that $\partial v_0/\partial t = \partial \rho_0/\partial t = B = 0$ and have used equation (10-7), with $C_0 = C_1 \equiv C$, assumed to be constant. In addition Poisson's law, equation (10-3), yields

$$\nabla^2 \phi_1 = 4\pi G \rho_1. \tag{13-30}$$

In general ρ_0 and $\mathbf{v}_0$ will be functions of position, and equations (13-28) through (13-30) are not easily solved.

The pioneering work by Jeans [12] made a drastic simplification of the problem by assuming that ρ_0 is also a constant and that v_0 vanishes. On this assumption equations (13-28) and (13-29) are simplified; if one takes the divergence of equation (13-28), and uses the resultant equation to eliminate $\nabla \cdot \mathbf{v}_1$ from equation (13-29), eliminating $\nabla^2 \phi_1$ by use of equation (13-30), the resultant equation for ρ_1 may be solved very simply. For an assumed plane wave of the form

$$\rho_1 = K e^{i(\kappa x + \omega t)} \tag{13-31}$$

one gets

$$\omega^2 = \kappa^2 C^2 - 4\pi G\rho_0 \equiv C^2\left(\kappa^2 - \kappa_J^2\right). \tag{13-32}$$

If the wave number κ is less than the critical value κ_J, ω^2 is negative, and the disturbance grows exponentially. Hence any small sinusoidal disturbance with a wavelength exceeding $2\pi/\kappa_J$ will be gravitationally unstable. The mass of material that starts to condense gravitationally must exceed the "Jeans mass" M_J, usually defined as ρ_0 times the cube of this critical wavelength. Evidently for $\gamma = 1$ in equation (10-6) for C,

$$M_J \equiv \left(\frac{2\pi}{\kappa_J}\right)^3 \rho_0 = \left(\frac{\pi kT}{\mu G}\right)^{3/2} \frac{1}{\rho_0^{1/2}}. \tag{13-33}$$

This approach has the great advantage of simplicity, but equations (13-31) and (13-32) do not strictly represent a solution of any physical problem, since the assumed equilibrium conditions do not satisfy the basic equations. Equation (10-1) shows that $\nabla^2\phi_0$ must vanish in a hypothetical uniform homogeneous medium with $v_0 = 0$. However, equation (10-3) indicates that $\nabla^2\phi_0$ cannot vanish if ρ_0 is finite. While the condition that $\kappa < \kappa_J$ is the correct criterion for instability of a collapsing sphere under certain special conditions [S13.3c], in other cases the Jeans result can be misleading qualitatively as well as quantitatively [13].

A simple problem which permits an exact analysis and shows gravitational instability is the perturbed motion of an isothermal gaseous slab or disk of infinite radius, subject to its self-gravitational attraction. In the equilibrium state v_0 is assumed to vanish, whereas ρ_0 and ϕ_0 are functions of z only. Equations (10-1), (10-3), and (10-6) may be combined to give

$$\frac{d}{dz}\left(\frac{1}{\rho_0}\frac{d\rho_0}{dz}\right) = -\frac{4\pi G}{C^2}\rho_0. \tag{13-34}$$

The solution of this equation is [14]

$$\rho_0 = \rho_0(0)\,\mathrm{sech}^2\left(\frac{z}{\mathrm{H}}\right) = \rho_0(0)(1 - w^2), \tag{13-35}$$

where $\rho_0(0)$ is the value of the unperturbed density ρ_0 at the midplane, $z = 0$;

$$\mathrm{H} \equiv \frac{M}{2\rho_0(0)} = \left(\frac{kT}{2\pi G\mu\rho_0(0)}\right)^{1/2}, \tag{13-36}$$

and

$$w \equiv \tanh \frac{z}{\mathrm{H}} = \frac{M(z)}{M}. \tag{13-37}$$

For positive z, the quantity $M(z)$ is the mass of the disk per unit area between $-z$ and $+z$, whereas M denotes $M(\infty)$, the total mass of the disk per unit area.

Instead of solving the linearized equations, we consider now the perturbation which neither oscillates nor grows exponentially, corresponding to ω equal to zero in equation (13-31) and to a velocity field independent of time. Such a state of "marginal stability" generally separates regions of stability, where small perturbations are damped, from instability regions, where small perturbations grow exponentially. Situations in which instability occurs by exponential growth of oscillations (called "overstability") must be treated by other means [15]. In general overstability does not appear in conservative systems (in which viscosity, resistivity, and thermal conductivity vanish), provided that v_0, the velocity in the unperturbed state, is zero or can be transformed away. Since we are considering a conservative system, with v_0 equal to zero, the marginal state in which the perturbed quantities are independent of time will separate regions of stability and instability.

Defining κ_c as the value of κ in the marginal state, we may write for this state

$$\frac{\rho_1(w)}{\rho_0(w)} = e^{i\kappa_c x}\theta(w), \tag{13-38}$$

independent of t and of the coordinate y; the variable w defined in equation (13-37) replaces z. In equation (13-28) the left-hand side vanishes in the marginal state; if the divergence of this equation is taken, with $\nabla^2\phi_1$ eliminated by use of equation (13-30), one obtains [16] after straightforward calculation, using equations (13-35) and (13-37),

$$\frac{d^2\theta}{dw^2} - \frac{2w}{1-w^2}\frac{d\theta}{dw} + \theta\left[\frac{2}{1-w^2} - \frac{\nu^2}{(1-w^2)^2}\right] = 0, \tag{13-39}$$

where

$$\nu = \kappa_c \mathrm{H}. \tag{13-40}$$

The general solution of equation (13-39) is

$$\theta(w) = A_1\left(\frac{1+w}{1-w}\right)^{\nu/2}(\nu - w) + A_2\left(\frac{1-w}{1+w}\right)^{\nu/2}(\nu + w). \qquad (13\text{-}41)$$

Since $\theta(w)$ must remain finite as w approaches ± 1, ν must equal unity. From equations (13-40) and (13-36) we obtain

$$\kappa_c^2 = \frac{1}{\mathrm{H}^2} = \frac{2\pi G\mu\rho_0(0)}{kT}. \qquad (13\text{-}42)$$

Physically one would expect instabilities to appear on the long wavelength side of the marginal state, and we may conclude that gravitational instability appears for κ less than κ_c. The corresponding minimum length of an unstable region in the plane of the disk is about $\pi\mathrm{H}$; since the thickness of the disk is somewhat less than this value, motions in the z direction only cannot produce instability, as may be shown directly from the linearized equations [16].

In the classical Jeans analysis the amplification rate $-i\omega$ increases to a constant value, $(4\pi G\rho_0)^{1/2}$, as κ falls below κ_J [see equation (13-32)]. In the present case, however, $-i\omega$ rises to a maximum value for κ about half κ_c [17] and falls toward zero as κ approaches zero, corresponding to the physical fact that the potential gradient produced by density perturbations of constant amplitude in an infinite disk does not increase as the wavelength in this plane grows indefinitely.

When the disk has a finite radius and rotates around its axis the situation is more complicated. It is well known that the density of a rotating gas must generally exceed some lower limit to permit gravitational contraction. For example, the condition that the centrifugal force at the equator of a gaseous sphere of radius r and uniform density ρ rotating with the angular velocity Ω be less than the self-gravitational force gives immediately

$$\rho > \frac{3\Omega^2}{4\pi G}. \qquad (13\text{-}43)$$

Analyses of a rotating disk [18, 19], taking into account the shearing effect of galactic rotation, show that instabilities appear if ρ is about two times the value obtained from equation (13-43). Thus for conditions in the solar neighborhood, gravitational instability appears only if ρ exceeds about 5×10^{-24} gm/cm^3, about twice the value in Table 11.1.

The condensations produced by such instabilities in a rotating disk have a very large scale, with dimensions comparable to the thickness of the galactic disk; the corresponding masses are roughly 10^6 $M_{\odot}$. Evidently these considerations are of importance primarily in forming large aggregations, or cloud complexes. As we have seen [S11.3b], a magnetic field of order 3×10^{-6} G will not hinder the gravitational contraction of such large masses. However, the concentration of gas in magnetic valleys or troughs [S11.2c], near the galactic midplane, may be an equally effective method for forming such large aggregations. Since we know from observations that such large concentrations of gas exist within spiral arms, the method by which these cloud complexes form is not a necessary part of star formation theory.

b. Gravitational Collapse of a Sphere

When a cloud departs from equilibrium, either because of gravitational instability or because its mass or external pressure exceeds the limits for possible equilibria, contraction will generally follow. If the temperature is assumed constant, the contraction of a spherical cloud of radius R will be an accelerating process. The gravitational force per cm^3 varies as ρ/R^2 or as M/R^5, whereas the force due to the pressure gradient varies as $p(0)/R$ or $MT(0)/R^4$, where $p(0)$ and $T(0)$ are the pressure and temperature at the center. Thus if the contraction is isothermal, pressure forces cannot retard the collapse. Magnetic forces also are unimportant if the mass exceeds a critical value [S11.3b], although as we shall see below angular momentum can stop the contraction. If the temperature rises adiabatically, with $\gamma > 4/3$, pressure forces can also arrest the contraction, but as long as the cloud has a sufficiently low opacity to remain isothermal, a cloud that starts contracting spherically as a result of its self-gravitational attraction will contract with an increasing acceleration if its angular momentum is sufficiently low.

This type of accelerating isothermal collapse is not possible for contraction in one or even two dimensions. For the contraction of a plane parallel sheet, for example, the acceleration g perpendicular to the sheet for any layer is unaffected by changes in the sheet thickness, whereas the corresponding pressure force per gram, $\nabla p/\rho$, tends to increase as the sheet is compressed, reducing the acceleration as the contraction proceeds. The two-dimensional case is intermediate in that the gravitational and pressure forces change in the same way with cylindrical radius r if $T(0)$ remains constant.

We compute first the collapse time for a cold sphere of uniform density $\rho(t)$ on the assumption that initially, when $t=0$, the sphere is at rest. We let

$r(t)$ be the radius of a particular mass shell as a function of time. The value of $r(0)$ will be denoted by a, and we omit the argument t from ρ and r. The equation of motion then becomes

$$\frac{d^2r}{dt^2} = -\frac{GM(a)}{r^2} = -\frac{4\pi G\rho(0)a^3}{3r^2}, \tag{13-44}$$

where $M(a)$ is the mass interior to the initial radius of the shell; evidently the mass inside the shell stays constant during the collapse, if the shells are assumed not to cross each other. Multiplying equation (13-44) by dr/dt and integrating gives the energy integral

$$\frac{dr}{adt} = -\left[\frac{8\pi G\rho(0)}{3}\left(\frac{a}{r}-1\right)\right]^{1/2}. \tag{13-45}$$

If we make the substitution $r/a = \cos^2\beta$, equation (13-45) yields [20]

$$\beta + \tfrac{1}{2}\sin 2\beta = t\left[\frac{8\pi G\rho(0)}{3}\right]^{1/2}, \tag{13-46}$$

where we let t vanish initially, when dr/dt is zero. Evidently β is the same for all mass shells at any one time, and all shells reach the center at the same time, when β equals $\pi/2$, and after a "free-fall time" given by

$$t_f = \left\{\frac{3\pi}{32G\rho(0)}\right\}^{1/2} = \frac{4.3\times 10^7}{\{n_{\mathrm{H}}(0)\}^{1/2}} \text{ years.} \tag{13-47}$$

We consider next the situation where the initial density is a function of initial radius a. In this case we denote by $\rho_m(a,t)$ the mean density at the time t within the mass shell of initial radius a. We assume again that different shells do not cross each other; hence $M(a)$ is again constant with time. Equations (13-44) through (13-47) then are valid as before, provided that $\rho_m(a,0)$ replaces $\rho(0)$ throughout. The free-fall time is now different for different mass shells, and the condition that different shells do not cross will be satisfied provided that $\rho_m(a,0)$ decreases with increasing a; that is, that the initial density decreases outward. The inner regions will then collapse first, with the outer ones falling in at progressively later times. We assume that the shells do not "bounce"; that is, that all mass shells reaching the center stay there. Under these conditions it is readily seen that the density distribution must become very peaked toward the center. An analytical study [21] shows that the density-radius profile for a cold gas tends to approach a variation as $r^{-12/7}$. For isothermal collapse, the density varies asymptotically [22] as r^{-2}.

More detailed results on the spherical collapse of clouds have been obtained [22, 23] with extensive numerical computations. These mostly ignore both rotation and magnetic fields, but the full dynamical and thermal equations can be taken into account. These computations confirm that the central regions always collapse before the outer layers, giving rise to a sharply peaked density profile. In part, this effect results from a moderate density concentration of the cloud towards its center before collapse, as discussed above. In part, this results from the effects of finite pressure, which were ignored in the analysis of a freely falling cold sphere; the finite pressure at the outer spherical boundary of the cloud remains less than the steadily increasing pressure within, and inhibits the contraction of the outer layers. The thermal computations show [22] that for a typical cloud the temperature decreases as the cloud contracts. Absorption in the outer shell of ultraviolet photons and of low-energy cosmic ray particles, if present, reduces the kinetic temperature of the inner regions down to 10°K or less for densities between roughly 10^4 and 10^8 gm cm^{-3}. Only at values of n_H exceeding 10^{10} cm^{-3} does the gas become sufficiently opaque in the far infrared, as a result of absorption by dust grains and molecules, so that the gas temperature at the center starts to rise by compressional heating. Evidently, spherical free fall of the central regions of the clouds will continue until the gas begins to approach densities and temperatures typical of stellar atmospheres.

These model collapsing clouds have been used to provide one scenario for star formation. According to this picture, a stellar core develops at the center of a cloud and grows in mass by spherical accretion, with the infalling material stopped at a shock front. If the cloud mass is relatively small, much of the material will collect in the star, which will slowly evolve by gravitational contraction toward the main sequence. If the cloud has a large mass, the stellar core will grow to a mass appreciably greater than $M_\odot$ and soon develop so large a luminosity that the incoming gas will be ionized and heated, terminating the inward flow. Radiation pressure on the grains [S9.3] may also help to terminate the inflow. On the basis of this picture formation of stars with masses much greater than roughly 50 $M_\odot$ appears unlikely [23].

c. Fragmentation

The previous discussion has neglected the instability of the contracting cloud against gravitational condensations of subunits within the cloud. In an idealized cold cloud, condensations of all scales are unstable [S13.3a]. In an actual cloud, at finite T, the value of the Jeans mass M_J in equation (13-33) is likely to be comparable with the total mass of the cloud when collapse begins. However, as the collapse continues the local density at each point will increase; since also T tends to decrease with increasing ρ, M_J will decrease steadily. Thus within a collapsing cloud, smaller masses can be expected to start condensing, a process known as fragmentation [24]. As long as T stays small, this process can be expected to continue, with fragments condensing into yet smaller fragments. This process provides an attractive mechanism by which many stars with masses in the range observed can form from a cloud of many solar masses. The concentration of the observed young stars in groups and the requirement that the cloud mass exceed about 700 $M_\odot$ for gravitational collapse to occur in a cloud with $n_{H0}=20$ cm^{-3} and $B=2.5$ μG [equation (11-32)] both indicate that the fragmentation process probably plays an important part in star formation.

Fragmentation will continue until the gas becomes so opaque that the radiative cooling time will be longer than the free-fall time. The collapse then becomes adiabatic, and if γ, the ratio of specific heats, exceeds 4/3, M_J in equation (13-33) will decrease no further. The opacity per gram produced by grains and molecules is nearly independent of density; hence the optical thickness τ through a fragment varies as ρ times the thickness of the fragment, which varies about as $1/\kappa_J$ or $\rho^{-1/2}$. Hence τ increases as $\rho^{1/2}$, and for sufficiently great density the opacity will bring fragmentation to a halt. We denote by M_F the mass of the smallest fragment when fragmentation stops.

A simple if approximate lower limit [25] on M_F may be obtained from the fact that the rate of energy radiated from a fragment of radius R cannot exceed that from a blackbody of surface area $4\pi R^2$ and temperature T, where T is a mean temperature of the cloud. Fragmentation can continue only if the rate of release of gravitational energy from a fragment of mass M, which is about equal to GM^2/Rt_f, is about equal to the rate at which energy is radiated. Hence we have as a condition on continuing fragmentation

$$4\pi R^2\sigma T^4 > \frac{GM^2}{Rt_f}, \tag{13-48}$$

where σ is the Stefan-Boltzmann constant and t_f is again the free-fall time. Equation (13-48) is somewhat approximate since some of the gravitational energy released goes into kinetic energy instead of heat.

We may use equation (13-47) to express t_f in terms of M and R, eliminating ρ by setting $M \approx 4\pi R^3\rho/3$. Similarly the condition that $M = M_J$ can be used to express R in terms of M and T, giving

$$\frac{GM}{R} \approx \left(\frac{4\pi}{3}\right)^{1/3} \frac{\pi kT}{\mu} = 5.1\frac{kT}{\mu}. \tag{13-49}$$

Again, μ is defined as the mean mass per atom. Equation (13-48) now yields as a condition for continuing fragmentation

$$M > \frac{(5.1)^{9/4}}{\pi 2^{1/4}}\left(\frac{k}{\mu}\right)^{9/4} \frac{T^{1/4}}{G^{3/2}\sigma^{1/2}} = \frac{0.027\,T^{1/4}}{(\mu/m_{\mathrm{H}})^{9/4}} M_{\odot}. \tag{13-50}$$

The lowest mass M_F of the fragments formed is not likely to be much less than the right-hand side of this equation. If $\mu/m_{\mathrm{H}} \leqslant 2$, and $T \geqslant 10°\mathrm{K}$, we have [25]

$$M_F \geqslant 0.01\ M_{\odot} \tag{13-51}$$

More detailed computations [26], taking into account the dominant contribution of dust grains to the infrared opacity, show that the radiative cooling time equals the free-fall time and fragmentation ends for $M_F = 0.007\ M_{\odot}$. In these theoretical models the grains and gas within fragments for which $M > M_F$ are heated to temperatures between 4 and 10°K by compressional heating. In view of the uncertainties in the analysis leading to equation (13-51), the agreement between the value of M_F obtained from this equation and that obtained from the more detailed analysis of radiative cooling is about as good as might be expected. The value of n_{H} found when the fragments become opaque and further fragmentation ceases is about $10^{10}\ \mathrm{cm}^{-3}$; the corresponding value of N_{H} along a spherical radius is $4 \times 10^{24}\ \mathrm{cm}^{-2}$, one-fifth the column density of molecules in the Earth's atmosphere.

While the fragmentation concept provides a plausible method for producing small stars from massive clouds, more detailed study indicates a number of problems. The initial cloud must have both angular momentum and magnetic flux, and these may prevent the approach to free fall which is assumed in the fragmentation picture. A perturbation analysis of the collapsing homogeneous sphere [20] shows that the condensation of smaller units does not grow exponentially; instead the linearized theory shows that the ratio of ρ_1, the density in a small scale condensation, to $\rho_0(t)$, the density of the uniformly collapsing sphere, varies asymptotically as $[\rho_0(t)]^{1/2}$. Finite pressure may readily be included in this analysis, if the external pressure is assumed to equal $C^2\rho_0(t)$, the zero-order internal pressure. The criterion for instability is then exactly the Jeans result, $\kappa < \kappa_J$ [28], and the rate of growth of the instability is reduced below its value for the cold sphere [20]. It is not evident that the smaller condensations will contract fast enough to separate out and avoid coalescence with each other as the entire cloud contracts toward the center [27].

d. Transfer of Angular Momentum

In an isolated cloud the total angular momentum J will remain constant. The transfer of angular momentum from one part of the cloud to another must have major effects on gravitational condensation and star formation.

To illustrate the importance of angular momentum, we consider the simple case of a cloud of radius $R(t)$, rotating with a uniform angular velocity $\Omega(t)$. If the cloud contracts homologously, then the conservation of J requires that

$$\Omega(t)R^2(t) = \Omega(0)R^2(0). \tag{13-52}$$

The centrifugal force per unit mass at the equator, equal to $\Omega^2 R$, varies as $1/R^3$, whereas the gravitational force varies as $1/R^2$. Evidently at some equilibrium cloud radius R_c centrifugal and gravitational forces will balance at the equator. If for each element of fluid its initial angular momentum per unit mass remains constant in time, the outer layers of the cloud on the equator can contract no further transverse to **J**. Continuing contraction parallel to **J** remains possible, and the cloud may tend to evolve toward a flattened disk of radius R_c. It is clear that the fragmentation process described above becomes completely altered when a cloud has appreciable angular momentum. Fragmentation can apparently still occur [29], with condensations forming in flattened disks and then becoming themselves flattened.

It can readily be seen that if a star is formed from material in an interstellar cloud, the angular momentum is almost inevitably very much greater than is consistent with a single star of normal radius. To consider an extreme case, with minimum angular momentum, let us suppose that a condensation forms by contraction parallel to **J** of gas in a typical diffuse cloud (see Table 7.2). A cylinder 10 pc in length and 0.2 pc in radius, with $n_H = 20$ cm^{-3}, contains about one solar mass. To form a star of solar density the radius must decrease by a factor about 10^{-7}, increasing Ω by about 10^{14}. If we assume that initially Ω had the normal value resulting from galactic rotation, about 10^{-15} s^{-1}, the resulting rotation period is roughly a minute, giving a centrifugal force exceeding gravity by some four orders of magnitude at the equator.

There are at least two ways in which the internal rotational momentum of a gaseous condensation (the angular momentum about axes passing through the center of gravity of the condensation) can be decreased. The first is to transfer the angular momentum from rotation to orbital motion. Thus fission of a rotating protostar into two protostars may make it possible for each of the two protostars to continue contracting, especially if either tidal forces or magnetic fields synchronize the rotation with the revolution, reducing the rotational angular momentum as each protostar contracts. The second way is to transfer the angular momentum from the rotating condensation to the surrounding interstellar gas, either by turbulent convection or by magnetic stresses. We consider here one process by which a magnetic field may produce this effect.

We compute the time scale for retardation of rotation for a highly idealized model in which the magnetic field is initially uniform and constant [30]. A spherical cloud with a relatively high density and a radius R is surrounded by a medium of lower density ρ. The magnetic field **B** extends through both the cloud and the surrounding gas; the angular velocity of the cloud is $\mathbf{\Omega}$, assumed to be parallel to **B**, whereas in the lower density gas the angular velocity is initially zero. Thus initially there is a discontinuity in $\mathbf{\Omega}$ along the lines of force at the condensation radius. Evidently the field becomes twisted, and one may show that the kink in the lines of force travel outward along the lines of force at the Alfvén velocity V_A. This outgoing wave will accelerate the surrounding material up to the angular velocity Ω. The amount of material accelerated per second in the cylindrical shell of radius r and thickness dr is $2\pi\rho r V_A dr$; multiplication by Ωr^2 gives the angular momentum gained by the shell. If the moment of inertia of the condensation equals $2MR^2/5$, its value for a sphere of uniform density, we obtain, integrating over dr,

$$\frac{2MR^2}{5}\frac{d\Omega}{dt} + \pi\rho V_A R^4 t\,\frac{d\Omega}{dt} = -\,\pi\rho V_A R^4 \Omega, \tag{13-53}$$

including the loss of angular momentum by the two corotating cylinders of interstellar gas*, each of radius R and length $V_A t$, on each side of the cloud; solid-body rotation is assumed, together with constant R and ρ. If we define

$$t_B \equiv \left(\frac{-\Omega}{d\Omega/dt}\right)_{t\,=\,0} = \frac{2M}{5\pi R^2 \rho V_A}, \tag{13-54}$$

equation (13-53) gives Ω varying as $(1 + t/t_B)^{-1}$.

*This effect was pointed out by B. Draine (1983, informal communication).

The ratio M/BR^2 is essentially $(M/M_c)G^{-1/2}$ [see equations (11-26) and (11-27)] and is unaffected by the contraction of a cloud across the field, or by fragmentation, if the initial mass and the initial flux are both divided equally among a number of fragments. From equation (13-47) it follows that if M/M_c does not differ much from unity, t_B is roughly equal to the free-fall time t_f at the density ρ. However, this density outside the condensation is assumed to be substantially less than the density inside, giving a value for t_B appreciably greater than t_f for the condensation itself. For this retardation mechanism to be effective we must apparently assume that the condensations remain as extended interstellar clouds for a period substantially greater than their free-fall time.

In actual situations the magnetic topography is undoubtedly more complex than the simple situation envisaged here. Studies of magnetic braking with **B** perpendicular to **J** indicate a relatively more rapid retardation than when **B** and **J** are parallel [29, 31].

e. Decrease of Magnetic Flux

A uniform magnetic field within a cloud has a certain similarity to rotation in that both B and Ω vary as $1/R^2$ for contraction of the cloud in a direction perpendicular to $\mathbf{B}$ or to $\mathbf{\Omega}$. The energy densities per cm^3 are entirely different in the two cases, however, varying as $1/R^4$ in the former, and $1/R^5$ in the latter, assuming isotropic contraction. We have already seen that as a result of this difference these two fields have somewhat dissimilar effects on at least the initial phases of gravitational collapse; the magnetic field leads to a lower limit on the mass which can collapse, whereas the rotation gives rise to a lower limit on the equatorial radius R_e of an isolated contracting cloud, rotating like a solid body. In the magnetic case as in the angular momentum case, fragmentation can apparently still occur [29], through a hierarchy of flattened disks, provided that initially M exceeds M_c [S11.3b].

To decrease the magnetic flux through the gas is not quite so vital for star formation as to decrease the rotational angular momentum; if contraction parallel to **B** brings together all the gas lying within some tube of force extending through the cloud, the total flux is rather large for a star but not impossibly so. To demonstrate this, we consider a cylinder 10 pc long and 0.2 pc in radius in a cloud of density 5×10^{-23} g/cm^3; as we have seen before, such a cylinder contains a total mass of about 1 $M_\odot$. The magnetic field in the star formed from this material will exceed that in the original cloud by 10^{14}, the same factor found for Ω. If B were 3×10^{-6} G in the initial cloud, in a star of solar type the magnetic field would equal 3×10^8 G, giving a magnetic energy somewhat exceeding the negative gravitational energy. The discrepancy between objects formed in this way and the observed stars is not nearly so great for a magnetized cloud as was found above for a rotating cloud.

While in theory star formation can occur even if the lines of force are permanently frozen into the gas [S10.1], there are several processes which can reduce the magnetic flux through the gas and which must be considered in a detailed theory of star formation. One of these is the change of field topology by reconnection of lines of force [29]. This process can occur where two lines of force intersect at a point of zero magnetic field strength, or where two oppositely directed lines of force are pushed close together by gravitational or hydromagnetic forces. Another process, which we describe here, is that in which the magnetic field remains frozen to the ionized particles or plasma, consisting of electrons and positively charged atomic and molecular ions, but the plasma and field drift through the gas of neutral atoms. This process is sometimes called "ambipolar diffusion," the name given to a related process observed in the laboratory, where electrons and positive ions, held together by electrostatic forces, drift through the neutral atoms. In the interstellar plasma it is magnetic rather than electrostatic forces that keep ions and electrons together.

We compute here the magnitude of w_D, the drift velocity of the ions and the magnetic field relative to the gas of neutral atoms. The magnetic field provides the primary driving force for w_D, since these forces act only on the ions, and not directly on the neutral atoms. If the lines of force are assumed straight and parallel, this force equals $-\nabla B^2/8\pi$ per cm^3, and in a quasisteady state must be balanced by the exchange of momentum in collisions, chiefly those between positive ions and neutral hydrogen atoms. This mean momentum exchange may be computed from the slowing-down time t_s, given in equation (2-14). Since these two forces on the positive ions must be equal and opposite, we obtain

$$-\frac{1}{8\pi}\nabla B^2 = \frac{n_i m_i w_D}{t_s} = n_i n_{\rm H} m_{\rm H} \langle u\sigma_s \rangle w_D, \qquad (13\text{-}55)$$

where u is the random velocity of hydrogen atoms relative to positive ions, and n_i and m_i are the particle density and mass of the positive ions, which we assume to have about the mass of carbon, although under some conditions hydrogen ions may predominate; the appropriate value of $\langle u\sigma_s \rangle$ is given in Table 2.1. Collisions of helium atoms with ions also contribute, but since their random velocity is only half that of hydrogen and their cross section is also about half [S2.2], they increase the right-hand side of equation (13-55) by about 10 percent and we shall neglect such collisions.

While not explicitly stated in equation (13-55), the force which the magnetic field exerts on the charged particles, and which is transmitted to the neutral atoms by ion-neutral collisions, must be balanced by an equal and opposite force on the neutrals if equation (13-55) is to be applicable. This force, which may result from gravity or from pressure gradients, for example, is needed to produce a quasiequilibrium state in which the drift velocity w_D is a small perturbation, producing a steady separation. As an example of such a quasiequilibrium we consider the idealized case in which an infinite cylinder of gas, mostly of neutral atoms, is supported against its self-gravitational force by the magnetic force on the ions. The ions and the magnetic field will then expand relative to the neutral gas at the velocity w_D, with the neutral gas slowly contracting. Instead of analyzing these motions in detail we may compute an approximate "diffusion time" t_D, defined as the ratio of the cylinder radius r to the relative velocity w_D between neutrals and ions. If we assume that the density ρ is uniform throughout the cylinder, then we obtain

$$t_D \equiv \frac{r}{w_D} = \frac{\langle u\sigma_s \rangle}{2\pi G m_{\mathrm{H}}} \times \frac{n_i}{n_{\mathrm{H}}} \times \frac{1}{(1+4n_{\mathrm{He}}/n_{\mathrm{H}})^2}, \tag{13-56}$$

where we have assumed that n_i/n_{H}, the ratio of the positive ion density to the overall hydrogen density, is much less than unity. The contribution of helium atoms to the gas density has been taken into account; as before, we let $n_{\mathrm{He}}/n_{\mathrm{H}}$ equal 0.1. With $\langle u\sigma_s \rangle$ set equal to 2.2×10^{-9} cm^{3}s^{-1}, we find

$$t_D = 5.0 \times 10^{13} \frac{n_i}{n_{\mathrm{H}}} \text{ years.} \tag{13-57}$$

While the assumptions here are somewhat idealized, equation (13-57) should give at least the order of magnitude of the time required for magnetic flux to diffuse out of an interstellar cloud. If n_i/n_{H} is 5×10^{-4}, a minimum value for a normal H I cloud, the diffusion time t_D is 2.5×10^{10} years, too long to be of interest for interstellar conditions.

Within a relatively dense cloud, n_i/n_{H} may fall to a very low value, since the ultraviolet radiation responsible for most of the ionization will be completely absorbed in the outer layers when the cloud density has increased by a few orders of magnitude above its initial level. Detailed computations [31] show that cosmic-ray ionization, with $\zeta_{\mathrm{H}} \approx 10^{-17}$ s^{-1}, maintains n_i/n_{H} high enough so that t_D somewhat exceeds the free-fall time t_f [S5.3b]. Thus at $n_{\mathrm{H}} \approx 10^6$ cm^{-3}, $n_i/n_{\mathrm{H}} \approx 10^{-7}$, giving $t_D = 5 \times 10^6$ years as compared with 4×10^4 years found for t_f from equation (13-47). A similar disparity is found at $n_{\mathrm{H}} \approx 10^{10}$ cm^{-3}, at the point where a protostar of mass M_F [S13.3c] becomes opaque to infrared radiation.

We conclude that the reduction of magnetic field made possible by plasma drift (ambipolar diffusion) is unimportant during the free-fall time. If magnetic forces, combined perhaps with centrifugal forces, maintain a cloud or fragment in hydrostatic equilibrium, this process can significantly reduce the magnetic flux, permitting gradual contraction and possibly the resumption of free fall when the flux has fallen to a sufficiently low value [32].

REFERENCES

1. H. Bondi, *M.N.R.A.S.,* **112**, 195, 1952.
2. L. Mestel, *M.N.R.A.S.,* **114**, 437, 1954.
3. T. E. Holzer and W. I. Axford, *Ann. Rev. Astron. Astroph.,* **8**, 31, 1970.
4. R. Hunt, *M.N.R.A.S.,* **154**, 141, 1971.
5. K. N. Dodd, *Proc. Camb. Phil. Soc.,* **49**, 486, 1953.
6. C. C. Lin, *Spiral Structure of Our Galaxy,* IAU Symp. No. 38, W. Becker and G. Contopoulos, Editors, D. Reidel Publ. Co. (Dordrecht, Holland), 1970, p. 377.
7. W. W. Roberts, *Ap. J.,* **158**, 123, 1969.
8. P. R. Woodward, *Ap. J.,* **195**, 61, 1975.
9. F. H. Shu, V. Milione, and W. W. Roberts, *Ap. J.,* **183**, 819, 1973.
10. F. H. Shu, V. Milione, W. Gebel, C. Yuan, D. W. Goldsmith, and W. W. Roberts, *Ap. J.,* **173**, 557, 1972.
11. L. Spitzer, *Stars and Stellar Systems,* Vol. 7, University of Chicago Press (Chicago), 1968, p. 1.
12. J. H. Jeans, *Astronomy and Cosmogony,* Dover reprint, 1961.
13. L. Mestel, *Quart. J. Roy. Astron. Soc., London,* **6**, 161, 265, 1965.
14. L. Spitzer, *Ap. J.,* **95**, 329, 1942.
15. S. Chandrasekhar, *Hydrodynamics and Hydromagnetic Stability,* Oxford University Press (London), 1961, Chapt. 1.
16. P. Ledoux, *Ann. d'Ap.,* **14**, 438, 1951.
17. R. Simon, *Ann. d'Ap.,* **28**, 40, 1965.
18. V. V. Safranov, *Dokl. Akad. Nauk USSR,* **130**, 53, 1960; *Ann. d'Ap.,* **23**, 979, 1960.
19. P. Goldreich and D. Lynden-Bell, *M.N.R.A.S.,* **130**, 125, 1965.
20. C. Hunter, *Ap. J.,* **136**, 594, 1962.
21. M. V. Penston, *M.N.R.A.S.,* **144**, 425, 1969.
22. R. B. Larson, *Star Formation,* IAU Symp. No. 75, T. de Jong and A Maeder, Editors, D. Reidel Publ. Co. (Dordrecht, Holland), 1977, p. 249.
23. R. B. Larson, *Ann. Rev. Astron. Astroph.,* **11**, 219, 1973.
24. F. Hoyle, *Ap. J.,* **118**, 513, 1953.
25. M. J. Rees, *M.N.R.A.S.,* **176**, 483, 1976.
26. C. Low and D. Lynden-Bell, *M.N.R.A.S.,* **176**, 367, 1976.
27. D. Layzer, *Ap. J.,* **137**, 351, 1963.
28. D. Lynden-Bell, *Dynamical Structure and Evolution of Stellar Systems,* Geneva Observatory, 1973, p. 129.
29. L. Mestel, *Quart. J. Roy. Astron. Soc., London,* **6**, 161, 1965.

30. R. Ebert, S. vonHoerner, and S. Temesvary, *Die Entstehung von Sternen durch Kondensation Diffuser Materie,* J. Springer (Heidelberg), 1960, p. 311.
31. L. Mestel, *Star Formation,* IAU Symp. No. 75, T. de Jong and A. Maeder, Editors, D. Reidel Publ. Co. (Dordrecht, Holland), 1977, p. 213.
32. T. Nakano, *Publ. Astr. Soc. Japan,* **28**, 355, 1976.

Symbols

a		Radius of grain.
		Radiation density constant, 7.56×10^{-15} erg cm^{-3} deg^{-4}.
A	A_{kj}	Einstein probability coefficient for downward spontaneous transitions [S3.2].
	A_λ	Interstellar extinction in magnitudes at wavelength λ, equation (7-1).
	A_V	Interstellar extinction in the visible [S7.3a].
	A_r	Recapture constant for electron–proton collisions, equation (5-10).
b		Galactic latitude (b^{II}).
		Velocity spread parameter for a Maxwellian distribution, which equals $2^{1/2}$ times dispersion of radial velocities, equations (3-20) and (3-21).
	b_j	Ratio of particle density to that in "equivalent thermodynamic equilibrium, (ETE)," equation (2-25).
B	$\mathbf{B}$	Magnetic field in gauss; $B_\perp$ and $B_\parallel$, components of $\mathbf{B}$ perpendicular and parallel, respectively, to line of sight.
	$B_\nu(T)$	Planck function for intensity of radiation in thermodynamic equilibrium, equation (3-4).
	B_{jk}	Einstein probability coefficients for induced radiative transitions.
c		Velocity of light, 2.998×10^{10} cm s^{-1}.
C		Speed of sound, equation (10-7); C_{I} and C_{II}, the values of C in H I and H II regions [S10.1].

Symbol		Meaning
d		Subscript referring to interstellar dust grains; for example, n_d, ρ_d, and Z_d.
D	D_m	Dispersion measure, equation (3-63).
e		Charge of proton, 4.803×10^{-10} esu.
		Base of Napierian logarithms.
		Subscript referring to electrons; n_e and T_e.
E		Energy in ergs.
	$\bar{E}_2$	Mean kinetic energy per photoelectron.
	E_{rj}, E_j	Energy of atom in level j in stage of ionization r.
	E_{jk}	Positive energy difference between levels j and k.
	E_m	Emission measure in pc cm^{-6}, equation (3-35).
	E_{B-V}	Color excess on the $B-V$ system.
f		Subscript denoting field particles [S2.1].
	$f(w)$	Velocity distribution function; $f^{(0)}(w)$, Maxwellian function, equation (2-17).
	f_r	Partition function for atom in stage of ionization r, equation (2-29).
	f_e	Partition function for a free electron, equation (2-31).
	$f_{jk}, f_{j\nu}$	Oscillator strength, equations (3-25) and (5-4).
F		Flux of particles in or out of dust grains per unit projected area per second.
	F_e	Photoelectric flux from grains, equation (6-19).
	$F_{L\alpha}$	Flux of Lα photons absorbed by grain, equation (9-10).
	F_r	Force of radiation pressure on grains, equation (9-19).
	$F(\phi)$	Phase function for scattering of light by grains through an angle ϕ, equations (7-3) and (7-8).
	$\mathcal{F}$	Flux of radiation; $\mathcal{F}_\nu$, flux per unit frequency interval.
g		Acceleration of gravity; g_z, acceleration perpendicular to galactic plane.
	g_{rj}, g_j	Statistical weight of level j for atom in stage of ionization r.
	g_{ff}, g_{nf}	Gaunt correction factors for free–free and bound–free transitions, equations (3-55) and (5-7).
G		Gravitational constant, 6.67×10^{-8} $cm^3 s^{-2} g^{-1}$.
		Subscript referring to interstellar gas; for example, p_G.
		Rate of thermal energy input to dust grains per unit projected area.

	G_r	Energy input G from photon absorption, equation (9-1).
	G_c	Energy input G from particle collisions, equation (9-2).
	$G_{L\alpha}$	Energy input G from Lα absorption, equation (9-12).
h		Planck constant, 6.626×10^{-27} ergs.
H		Effective half-thickness of galactic disk.
	H	Subscript referring to hydrogen atoms; for example, n_H and m_H.
i		Subscript referring to positive ions; for example, n_i and m_i.
I	I_ν	Specific intensity of radiation at frequency ν, [S3.1].
j		Subscript usually referring to lower level in an atomic transition.
		Current density in emu.
	j_ν	Emissivity of matter per cm^3 at frequency ν.
J		Particle flux of ionizing photons per cm^2 per second.
		Quantum number for total angular momentum in atoms [S4.1a] and for rotational momentum in molecules [S4.3b].
k		Boltzmann constant, 1.381×10^{-16} erg deg^{-1}.
		Number of clouds in the line of sight per kpc.
		Rate of coefficient for chemical reaction [S5.2c and S5.3b].
		Subscript usually referring to upper level in an atomic transition.
	$\langle k \rangle$	Mean value of dissociation probability of H_2 molecule following absorption in Lyman or Werner bands.
K		Arbitrary constant.
	K_{Jm}	Absorption correction factor for upward transition probability β_{Jm} in H_2, equation (5-44).
l		Galactic longitude (l^{II}).
		Parameter characterizing Maxwellian velocity distribution, equation (2-18).
L		Path length.
		Luminosity; $L_\nu d\nu$, stellar luminosity within the frequency interval $d\nu$.

L_r — Energy radiated from dust grain per unit projected area per second.

m — Particle mass; m_e and m_p, mass of electron and proton.

Index of refraction, Chapter 7.

M — Total mass.

Magnetic moment [S8.3].

M_c — Critical mass below which gravitational collapse and fragmentation are impossible in a magnetic field, equation (11-28).

$\mathfrak{M}$ — Total magnetic energy of system, equation (10-11).

Mach number, v/C, equation (10-21).

n — Density of particles per unit volume; n_e, n_i, n_a, and n_d, values of n for electrons, ions, neutral atoms, and dust grains.

Principal quantum number.

$n_j(X^{(r)})$ — Particle density of atoms of element X (or of molecules) ionized r times and in level j. Alternatively, $r+1$ may be expressed as a Roman numeral, with H I and Ca II replacing $H^{(0)}$ and $Ca^{(1)}$.

n_j — Particle density of atoms (or molecules) in level j.

n_j^* — Value of n_j in "equivalent thermodynamic equilibrium" (ETE).

n_X — Particle density of element X in all stages of ionization, including atoms in molecules.

n_f — Value of n for field particles.

N — Number of particles in a column along the line of sight 1 cm^2 in cross section, or "column density."

$N_j(X^{(r)})$ — Column density of particles of type X ionized r times and in level j, equation (3-17); see $n(X^{(r)})$.

N_u — Number of ultraviolet photons shortward of Lyman limit radiated by star per second, equation (5-21).

p — Pressure; p_G, p_R, and p_B, pressure due to gas, cosmic rays, and magnetic field, equations (11-7) to (11-9); p_m, maximum external pressure at which an isothermal sphere is in equilibrium, equation (11-29).

Collision parameter; that is, distance of closest approach in the absence of mutual forces.

Subscript referring to protons; that is, n_p and T_p.

P		Power per cm^3; P_c, power lost by collisions among clouds; P_u and P_s, power from absorption of ultraviolet radiation and from supernova shells.
		Polarization of radiation, equation (8-1).
	P_{jk}	Probability per second of transition from level j to k induced by photon pumping [S4.3a].
	$P(v)dv$	Fraction of particles whose radial velocity lies within the range dv.
Q		Stokes parameter, equation (8-7).
	Q_e	Efficiency factor for extinction by solid particles; Q_{eE} and Q_{eH}, values of Q_e for axis of spheroid parallel to **E** and **H**, respectively, in polarized radiation.
	Q_a	Efficiency factor for pure absorption by solid particles.
	Q_s	Efficiency factor for scattering.
r		Radius; distance from central star; r_i and r_s, radius of an ionized region or shock front.
	r_S	Radius of H II region in radiative equilibrium, in the absence of dust grains.
R		Radius of a cloud.
		Rate coefficient for formation of H_2 by collisions of H atoms with grains.
	R_m	Rotation measure, equation (3-72).
	$(R_{jk})_Y$	Probability per second of transition j to k induced by process Y, equation (4-1).
	R_V	Ratio of visual extinction to color excess E_{B-V}, equation (7-20).
s		Path length, especially along ray of electromagnetic radiation.
		Integrated value of s_ν, equation (3-23).
		Subscript referring to solid material in interstellar grains; that is, ρ_s and T_s.
	s_ν	Atomic cross section for absorption of radiation of frequency ν.
	s_u	Integrated value of s_ν, uncorrected for stimulated emission, equation (3-24).
S		Surface; dS element of surface area.
	$S_u(r)$	Net ultraviolet flux in photons per second flowing out through a shell of radius r; $S_u(0)$, ultraviolet luminosity of star in photons per second, $\equiv N_u$.

t	Time in seconds.
	Subscript denoting test particles [S2.1].
t_s	Slowing-down time, equation (2-1) [S2.1 and S2.2].
t_T	Cooling time, equation (6-2).
t_m	Time in which a grain collides with its own mass of gas, equation (8-19).
t_f	Free-fall time for a cold uniform sphere, equation (13-47).
T	Temperature in degrees Kelvin; kinetic temperature of gas.
T_b	Observed brightness temperature.
T_c	Color temperature of stars for ultraviolet radiation.
T_E	Temperature in equilibrium.
T_R	Temperature of universal blackbody radiation $= 2.7°$K.
T_s	Temperature of solid material in grains.
u	Relative velocity in collisions between particles.
	Velocity of fluid relative to shock front or ionization front.
u_R, u_D	Critical values of u for an ionization front, equations (12-3) and (12-4).
U	Electric potential in esu; U(V), in Volts
	Stokes parameter, equation (8-8).
	Energy density; U_R energy density of cosmic rays.
$U_\nu, U(\nu)$	Energy density for photons per unit frequency interval, equation (4-3).
U_λ	Energy density for photons per unit wavelength interval.
v	Fluid or macroscopic velocity.
V	Volume; dV, volume element.
	Phase velocity of a wave.
V_A	Velocity of Alfvén wave, equation (10-9).
V_i, V_s	Velocity of ionization or shock front.
w	Random velocity.
W	Dilution factor for radiation, equation (4-13).
W_λ	Equivalent width of an absorption line in wavelength units, equation (3-47).
x	Ratio of grain circumference to wavelength [S7.1].
	Fraction of H atoms ionized [S5.1].

y	y_e	Photoelectric efficiency, or quantum yield [S6.2b and S9.2b].
Y		Relative abundance of helium by mass.
z		Distance from galactic plane.
Z		Ion charge in units of proton charge.
α		Recombination coefficient; $\alpha^{(m)}$, recombination coefficients for all levels with $n \geqslant m$ [S5.1a].
	α_{mn}	Production coefficient for optical recombination lines [S3.3a and 4.2a].
β		Ratio of H ionization energy to kT, equation (5-13).
	β_{jf}	Probability per second of photoelectric ionization of an atom in state j.
	β_{Jm}	Probability per second that H_2 molecule in rotational level J absorbs a photon and jumps to some upper level m, equation (4-36).
	$\beta(J)$	Sum of β_{Jm} over all upward transitions induced by photon absorption, equation (4-40).
	β_0	Value of $\beta(J)$ outside a cloud.
γ		Ratio of specific heats [S10.1].
	γ_{jk}	Rate coefficient for excitation from level j to k by collisions.
Γ		Total gain of kinetic energy of interstellar gas per cm^3 per second.
	$\Gamma_{\zeta\eta}$	Component of Γ due to collisions between particles of types ζ and η; subscripts e, i, p, H, d, and R denote electrons, ions, protons, neutral H atoms, dust grains, and cosmic rays, respectively.
δ	δ_k	Radiation damping constant divided by 4π, equation (3-44).
ΔX		Increment of X.
ε		Efficiency; ε_u and ε_s efficiency of accelerating interstellar clouds by ultraviolet radiation and by supernova shells.
	ε_{ff}	Energy radiated per cm^3 per second by free-free transitions.
ζ	ζ_H	Probability per second of collisional ionization of H atom by cosmic rays [S5.1d].
η		Electrical resistivity in emu, equal to 10^9 times the resistivity in Ohm-cm.

θ — Angle; θ_p, position angle of plane of vibration of polarized light, in galactic coordinates.

κ — Wave number, equal to $2\pi/\lambda$.

- $\boldsymbol{\kappa}$ — Unit vector in the direction of wave propagation.
- κ_J — Critical wave number for gravitational instability on Jeans theory, equation (13-32).
- κ_ν — Absorption coefficient per cm^3 for radiation of frequency ν.

λ — Wavelength in cm.

Accretion constant, equation (13-13).

Λ — Quantity whose natural logarithm appears in formula for ion–ion collisions, equation (2-5).

Total loss of kinetic energy of interstellar gas per cm^3 per second.

- $\Lambda_{\zeta n}$ — Component of Λ due to collisions between particles of types ζ and η; subscripts e, i, H, and H_2 denote electrons, ions, neutral H atoms, and H_2 molecules, respectively.

μ — Mean mass of the gas per particle.

- μ_i — Mean mass of the gas per positive ion.
- μ_{jk} — Dipole matrix element for transition between levels j and k.

ν — Frequency in cycles per second (Hz).

- ν_1 — Frequency at Lyman limit in neutral H, $3.29\times10^{15}\,s^{-1}$.

ξ — Sticking probability in a collision of an atom with a grain; ξ_a, value of ξ for neutral atoms.

ρ — Mass density in $g\,cm^{-3}$.

- ρ_s — Internal density within grain.
- ρ_{I}, ρ_{II} — Density in H I and H II regions.

σ — Cross section for collisions between particles.

- σ_{cj} — Cross section for capture of an electron in level j.
- σ_d — Geometrical cross section of a dust grain.

Σ Σ_d — Mean projected area of dust grains per H atom, equation (7-23).

τ τ_ν — Optical depth at frequency ν, equation (3-2).

- $\tau_{\nu r}$ — Optical thickness of a region.
- τ_0 — Optical thickness along the line of sight at the center of an absorption line.

ϕ $\phi(r)$ — Gravitational potential as a function of position r.

Symbol	Meaning
$\phi(\Delta\nu)$	Profile of absorption coefficient, equation (3-14).
$\phi_a(\Delta\nu)$	Average of $\phi(\Delta\nu)$ along line of sight, equation (3-16).
ϕ_1, ϕ_2	Functions of β occurring in equation (5-14) for α and $\alpha^{(2)}$, respectively.
χ	Correction factor for absorption coefficient s in the absence of thermodynamic equilibrium, equation (3-30).
χ_1, χ_2	Functions of β occurring in Λ_{ep}, equation (6-8).
ψ	Function characterizing the mean photoelectron energy $\bar{E}_2$, equation (6-7).
ω	Solid angle.
	Angular frequency $2\pi\nu$; angular velocity of grain.
ω_B	Gyration frequency around magnetic lines of force, equation (9-20).
Ω	Angular velocity of gas or stellar system.
Ω_p	Angular velocity of spiral pattern in galaxy.
$\Omega(j,k)$	Collision strength for transitions between levels j and k, equation (4-10).
∇X	Gradient of X.
$\nabla^2 X$	Laplacian of X.
$\langle X \rangle$	Mean value of X, especially the value averaged over a Maxwellian velocity distribution or over space.
*	Superscript denoting values in "equivalent thermodynamic equilibrium"; for example, n_j^* [S2.4].

Index